MATHEMATICS RESEARCH DEVELOPMENTS

ANALYTICAL AND NUMERICAL METHODS FOR PRICING FINANCIAL DERIVATIVES

MATHEMATICS RESEARCH DEVELOPMENTS

Additional books in this series can be found on Nova's website under the Series tab.

Additional E-books in this series can be found on Nova's website under the E-books tab.

FINANCIAL INSTITUTIONS AND SERVICES

Additional books in this series can be found on Nova's website under the Series tab.

Additional E-books in this series can be found on Nova's website under the E-books tab.

MATHEMATICS RESEARCH DEVELOPMENTS

ANALYTICAL AND NUMERICAL METHODS FOR PRICING FINANCIAL DERIVATIVES

DANIEL SEVCOVIC
BEÁTA STEHLÍKOVÁ
AND
KAROL MIKULA

Nova Science Publishers, Inc.
New York

Copyright © 2011 by Nova Science Publishers, Inc.

All rights reserved. No part of this book may be reproduced, stored in a retrieval system or transmitted in any form or by any means: electronic, electrostatic, magnetic, tape, mechanical photocopying, recording or otherwise without the written permission of the Publisher.

For permission to use material from this book please contact us:
Telephone 631-231-7269; Fax 631-231-8175
Web Site: http://www.novapublishers.com

NOTICE TO THE READER

The Publisher has taken reasonable care in the preparation of this book, but makes no expressed or implied warranty of any kind and assumes no responsibility for any errors or omissions. No liability is assumed for incidental or consequential damages in connection with or arising out of information contained in this book. The Publisher shall not be liable for any special, consequential, or exemplary damages resulting, in whole or in part, from the readers' use of, or reliance upon, this material. Any parts of this book based on government reports are so indicated and copyright is claimed for those parts to the extent applicable to compilations of such works.

Independent verification should be sought for any data, advice or recommendations contained in this book. In addition, no responsibility is assumed by the publisher for any injury and/or damage to persons or property arising from any methods, products, instructions, ideas or otherwise contained in this publication.

This publication is designed to provide accurate and authoritative information with regard to the subject matter covered herein. It is sold with the clear understanding that the Publisher is not engaged in rendering legal or any other professional services. If legal or any other expert assistance is required, the services of a competent person should be sought. FROM A DECLARATION OF PARTICIPANTS JOINTLY ADOPTED BY A COMMITTEE OF THE AMERICAN BAR ASSOCIATION AND A COMMITTEE OF PUBLISHERS.

Additional color graphics may be available in the e-book version of this book.

Library of Congress Cataloging-in-Publication Data

Sevcovic, Daniel.
  Analytical and numerical methods for pricing financial derivatives / Daniel Sevcovic.
     p. cm.
  Includes index.
  ISBN 978-1-61728-780-0 (hardcover)
  1. Derivative securities--Prices--Mathematical models. 2. Options (Finance)--Prices--Mathematical models. I. Title.
  HG6024.A3S46 2010
  332.64'57--dc22
                          2010026267

Published by Nova Science Publishers, Inc. † New York

Contents

Preface   xi

Introduction   xiii

1   The Role of Protecting Financial Portfolios   1
    1.1.   Stochastic Character of Financial Assets   2
    1.2.   Using Derivative Securities as a Tool for Protecting Volatile Stock Portfolios   4
        1.2.1.   Forwards and Futures   4
        1.2.2.   Plain Vanilla Call and Put Options   5

2   Black–Scholes and Merton Model   9
    2.1.   Stochastic Processes and Stochastic Differential Calculus   9
        2.1.1.   Wiener Process and Geometric Brownian Motion   10
        2.1.2.   Itō's Integral and Isometry   13
        2.1.3.   Itō's Lemma for Scalar Random Processes   16
        2.1.4.   Itō's Lemma for Vector Random Processes   18
    2.2.   The Black–Scholes Equation   19
        2.2.1.   A Stochastic Differential Equation for the Option Price   20
        2.2.2.   Self-financing Portfolio Management with Zero Growth of Investment   21
    2.3.   Terminal Conditions   23
        2.3.1.   Pay–off Diagrams for Call and Put Options   24
        2.3.2.   Pay–off Diagrams for Combined Option Strategies   24
    2.4.   Boundary Conditions for Derivative Prices   28
        2.4.1.   Boundary Conditions for Call and Put Options   28
        2.4.2.   Boundary Conditions for Combined Option Strategies   29

3   European Style of Options   33
    3.1.   Pricing Plain Vanilla Call and Put Options   33
    3.2.   Pricing Put Options Using Call Option Prices and Forwards, Put-Call Parity   38
    3.3.   Pricing Combined Options Strategies: Spreads, Straddles, Condors, Butterflies and Digital Options   40
    3.4.   Comparison of Theoretical Pricing Results to Real Market Data   41
    3.5.   Black–Scholes Equation for Pricing Index Options   42

4 Analysis of Dependence of Option Prices on Model Parameters — 49
- 4.1. Historical Volatility of Stocks — 49
  - 4.1.1. A Useful Identity for Black–Scholes Option Prices — 51
- 4.2. Implied Volatility — 52
- 4.3. Volatility Smile — 55
- 4.4. Delta of an Option — 56
- 4.5. Gamma of an Option — 58
- 4.6. Other Sensitivity Factors: Theta, Vega, Rho — 60
  - 4.6.1. Sensitivity with Respect to a Change in the Interest Rate – Factor Rho — 60
  - 4.6.2. Sensitivity to the Time to Expiration – Factor Theta — 61
  - 4.6.3. Sensitivity to a Change in Volatility – Factor Vega — 62

5 Option Pricing under Transaction Costs — 65
- 5.1. Leland Model, Hoggard, Wilmott and Whalley Model — 65
- 5.2. Modeling Option Bid–Ask Spreads by Using Leland's Model — 70

6 Modeling and Pricing Exotic Financial Derivatives — 75
- 6.1. Asian Options — 75
  - 6.1.1. A Partial Differential Equation for Pricing Asian Options — 76
  - 6.1.2. Dimension Reduction Method and Numerical Approximation of a Solution — 78
- 6.2. Barrier Options — 79
  - 6.2.1. Numerical and Analytical Solutions to the Partial Differential Equation for Pricing Barrier Options — 83
- 6.3. Binary Options — 84
- 6.4. Compound Options — 85
- 6.5. Lookback Options — 87

7 Short Interest Rate Modeling — 93
- 7.1. One-Factor Interest Rate Models — 93
- 7.2. The Density of Itō's Stochastic Process and the Fokker–Plank Equation — 97
  - 7.2.1. Multidimensional Version of the Fokker-Planck Equation — 98
- 7.3. Two- and Multi-factor Interest Rate Models — 103
- 7.4. Calibration of Short Rate Models — 105
  - 7.4.1. Maximum Likelihood Method for Estimation of the Parameters in the Vasicek and CIR Models — 105

8 Pricing of Interest Rate Derivatives — 109
- 8.1. Bonds and Term Structures of Interest Rates — 109
  - 8.1.1. One–Factor Equilibrium Models, Vasicek and CIR Models — 110
  - 8.1.2. Two-factor Equilibrium Models — 117
  - 8.1.3. Non-arbitrage Models, the Ho-Lee and Hull–White Models — 122
- 8.2. Other Interest Rate Derivatives — 124

9 American Style of Derivative Securities — 129
9.1. Pricing of American Options by Solutions to Free Boundary Problems . . . 131
9.2. Pricing American Style of Options by Solutions to a Linear Complementarity Problem 136
9.3. Pricing Perpetual Options 138

10 Numerical Methods for Pricing of Simple Derivatives — 141
10.1. Explicit Numerical Finite Difference Method for Solving the Black–Scholes Equation 141
    10.1.1. Discrete Methods Based on Binomial and Trinomial Trees 145
10.2. Implicit Numerical Method for Solving the Black–Scholes Equation 149
10.3. Compendium of Numerical Methods for Solving Systems of Linear Equations 152
    10.3.1. LU Decomposition Method 152
    10.3.2. Gauss–Seidel Successive Over-relaxation Method 153
10.4. Methods for Solving Linear Complementarity Problems 156
    10.4.1. Projected Successive Over-relaxation Method 156
    10.4.2. Numerical Solutions of the Obstacle Problem 158
10.5. Numerical Methods for Pricing of American Style Options 159
    10.5.1. Identification of the Early Exercise Boundary for American Options 161
    10.5.2. Implied Volatility for American Options 164
    10.5.3. Source codes for Numerical Algorithms 164

11 Nonlinear Extensions of the Black–Scholes Pricing Model — 167
11.1. Overview of Nonlinear Extensions to the Black–Scholes Option Pricing Model 167
11.2. Risk Adjusted Pricing methodology Model 170
    11.2.1. Nonlinear Black–Scholes Equation for the Risk Adjusted Pricing Model 172
    11.2.2. Derivation of the Gamma Equation 174
    11.2.3. Pricing of European Style of Options by the RAPM Model 175
    11.2.4. Explanation of the Volatility Smile by the RAPM Model 177
11.3. Modeling Feedback Effects 178
    11.3.1. The Case of the Standard Black–Scholes Delta Hedging Strategy . . 180
    11.3.2. Zero Tracking Error Strategy and the Nonlinear Black–Scholes . . 181
    11.3.3. The Gamma Equation for Frey's Nonlinear Model 181
11.4. Modeling Investor's Preferences 182
    11.4.1. The Gamma Equation and a Numerical Approximation Scheme . . 188
11.5. Jumping Volatility Model and Leland's Model 189
11.6. Finite Difference Scheme for Solving the Γ Equation 191

12 Transformation Methods for Pricing American Options — 195
12.1. Transformation Methods for Plain Vanilla Options 195
    12.1.1. Fixed Domain Transformation for the American Call Option 196
    12.1.2. Reduction to a Nonlinear Integral Equation 198
    12.1.3. Numerical Experiments 201

12.1.4. Early Exercise Boundary for the American Put Option 204
12.1.5. Analytical Approximation Valuation Formula by Zhu 207
12.2. Transformation Method for a Class of Nonlinear Equations of the Black–Scholes Type 208
12.2.1. Alternative Representation of the Early Exercise Boundary 211
12.2.2. Numerical Iterative Algorithm for Approximation of the Early Exercise Boundary 211
12.2.3. Numerical Approximations of the Early Exercise Boundary 214
12.3. Early Exercise Boundary for American Style of Asian Options 218
12.3.1. American-Style of Asian Call Options 221
12.3.2. Fixed Domain Transformation 221
12.3.3. A Numerical Approximation Operator Splitting Scheme 224
12.3.4. Computational Examples of the Free Boundary Approximation . . 226

13 Calibration of Interest Rate and Term Structure Models   231
13.1. Generalized Method of Moments 231
13.2. Nowman's Parameter Estimates 233
13.3. Method Based on Comparison with Entire Market Term Structures 236
13.3.1. Parameters Reduction Principle 237
13.3.2. The Loss Functional 239
13.3.3. Non-linear Regression Problem for the Loss Functional 240
13.3.4. Evolution Strategies 241
13.3.5. Calibration Based on Maximization of the Restricted Likelihood Function 242
13.3.6. Qualitative Measure of Goodness of Fit and Non-linear $\mathcal{R}^2$ Ratio . 243
13.3.7. Results of Calibration 244
13.4. Overview of Other Methods 246

14 Advanced Topics in the Term Structure Modeling   247
14.1. Approximate Analytical Solution for a Class of One-factor Models 247
14.1.1. Uniqueness of a Solution to the PDE for Bond Prices 248
14.1.2. Error Estimates for the Approximate Analytical Solution 251
14.1.3. Improved Higher Order Approximation Formula 253
14.1.4. Comparison of Approximations to the Exact Solution for the CIR Model 255
14.2. Mathematical Analysis of the Two-Factor Vasicek Model 255
14.2.1. Statistical Properties of Bond Prices and Interest Rates 257
14.2.2. Averaged Bond Price, Term Structures and Their Confidence Intervals 261
14.2.3. Relation of Averaged Bond Prices to Solutions of One-Factor Models 262
14.3. The Two-Factor Cox Ingersoll Ross Model 264
14.3.1. Distribution of Bond Prices and Interest Rates 265
14.3.2. Averaged Bond Prices, Term Structures and Their Confidence Intervals 268
14.3.3. Relation of Averaged Bond Prices to Solutions of One-Factor Interest Rate Models 270

14.4. The Fong-Vasicek Model with a Stochastic Volatility 276
    14.4.1. Qualitative Properties of Bond Prices and Term Structures 276
    14.4.2. Distribution of Stochastic Bond Prices and Term Structures for the Fong–Vasicek Model 278
    14.4.3. Averaged Bond Prices, Term Structures and Their Confidence Intervals 280
    14.4.4. Relation of Averaged Bond Prices to Solutions to One-Factor Interest Rate Models 282
14.5. Stochastic CIR Model Describing Volatility Clustering 284
    14.5.1. Empirical Evidence of Existence of Volatility Clusters and Their Modeling 284
    14.5.2. Generalized CIR Model with Rapidly Oscillating Stochastic Volatility and Its Asymptotic Analysis 286

References   293

List of Symbols   305

Index   307

Preface

The aim of this book is to acquaint the reader with basic facts and knowledge of pricing financial derivatives. It is focused on qualitative and quantitative analysis of various derivative securities. It can be used as a comprehensive textbook for graduate and undergraduate students having some background in the theory of stochastic processes, theory of partial and ordinary differential equations and their numerical approximation. It may also attract the attention of university students and their teachers with specialization on mathematical modeling in economy and finance. The book contains special topics reflecting state of the art in the pricing of derivative securities like qualitative analysis of the early exercise boundary, multi-factor interest rate models and transformation methods for pricing American style of options. These topics might be of interest to specialists in the mathematical theory of pricing financial derivatives.

Introduction

The aim of this book is to acquaint the reader with basic facts and knowledge of pricing financial derivatives. We focus our attention on qualitative analysis and practical methods of their pricing. The extensive expansion of various financial derivatives dates back to the beginning of seventies. The analysis of derivative securities was motivated by pioneering works [14, 84] due to economists Myron Scholes and Robert Merton and theoretical physicist Fisher Black. They derived and analyzed a pricing model nowadays referred to as the Black–Scholes model. The approach was indeed revolutionary as it brought the method of pricing derivative securities by means of solutions to partial differential equations. The Black–Scholes methodology enables us to price various derivatives of underlying assets as functions depending on time remaining to expiry and the underlying asset price.

The book is thematically divided into several chapters. The first ten of them can be considered as a standard introduction to pricing derivative securities by means of solutions to partial differential equations. We were deeply inspired by a comprehensive book [122] by Wilmott, Dewynne and Howison. In these introductory chapters we made an attempt to give the reader a balanced presentation of modeling issues, analytical parts as well as practical numerical realizations of derivative pricing models. We furthermore put a special attention to the comparison of theoretically computed results to financial market data. The remaining four chapters mostly represent our own research contributions to the subject of financial derivatives pricing.

In the first chapter we present a descriptive analysis of stochastic evolution processes of underlying assets and their derivative securities. We discuss basic types of derivative securities like plain vanilla options and forward contracts. Although this chapter does not have a strict mathematical character, its aim is to verbally highlight the importance of studying and analyzing financial derivatives as useful and necessary financial instruments for hedging and protecting volatile portfolios. The second chapter focuses on the standard Black–Scholes model for pricing derivative securities. Its mathematical formulation is a partial differential equation of the parabolic type, whose solution represents the price of a derivative contract. The chapter is also devoted to the presentation of the basic knowledge and facts of stochastic differential calculus, which is needed throughout the rest of the book. It is a basis for derivation of a broad class of financial derivatives pricing models. The third chapter is devoted to the classical theory of pricing European call and put options. We derive an explicit pricing formula refereed to as the Black–Scholes or Feynman-Kac formula. The content of the fourth chapter is focused on qualitative analysis of dependence of the option prices on various model parameters. We present basic concepts of analyzing financial markets including a notion of historical and implied volatilities, in particular. Then we concentrate on various sensitivity factors like Delta, Gamma, Theta, Vega and Rho of a financial derivative. These factors can be used in managing portfolios consisting of options, underlying assets and riskless money market instruments. We analyze the dependence of the sensitivity factors on underlying asset price and other model parameters. Modeling transaction costs is a main topic of the fifth chapter. In the sixth chapter we introduce basic classes of exotic derivatives and we discuss qualitative and practical aspects of their pricing. In more detail, we analyze Asian derivatives, barrier options and look-back options. In the seventh chapter, we are interested in modeling of the short interest rate. It can be considered as a nontradable underlying asset for a wide range of the so-called interest rate derivatives. In the subsequent chapter we deal with practical issues of pricing interest rate derivatives. We present the methodology of pricing bonds and other interest rate derivatives for single and multi-factor models. A special attention is put on non-arbitrage models of interest rates, such as the Vasicek or Cox–Ingersoll-Ross model and their multi-factor generalizations. The American style of financial derivatives is studied in the ninth chapter. These derivatives are characterized by a possibility of an early exercising of the financial derivative. We show that the problem of pricing American derivatives can be transformed to a mathematical problem of finding the free boundary for the Black–Scholes parabolic partial differential equations defined on a time dependent domain. In financial terminology, this free boundary is referred to as the early exercise boundary. In the tenth chapter, we present stable and robust numerical approximation methods for solving the Black–Scholes partial differential equation by means of explicit and implicit finite difference methods. We show how to solve the problem of pricing the American style of derivatives numerically by the so-called projected successive over relaxation algorithm. The eleventh chapter contains an overview of recent topics on pricing derivative securities. We present various nonlinear generalizations of the classical Black–Scholes theory. We show that, in the presence of transaction costs and risk from unprotected portfolio, the resulting pricing model is a nonlinear extension of the Black–Scholes equation in which the diffusion coefficient is no longer constant and it may depend on the option price itself. A similar nonlinear generalization of the Black–Scholes equation often arises when modeling illiquid and incomplete markets, in the presence of a dominant investor in the market, etc. We also show how to solve these nonlinear Black–Scholes models numerically. The twelfth chapter is devoted to modern transformation methods for pricing American style of derivative securities. These methods are capable of reducing the problem to construction of the early exercise position as a solution to a nonlinear integral equation. The last two chapters deal with advanced topics in modeling of interest rates. In the thirteenth chapter we focus on calibrating issues of standard one-factor interest rate models. We concentrate on estimation of the model parameters for Cox–Ingersoll–Ross model. The fourteenth chapter consists of two main parts. First, we consider an analytic approximation of bond prices in one-factor models, in which a closed explicit formula is not available yet. Then we study two-factor models, distributions of bond prices and interest rates with respect to the unobserved parameters of the model and their averaged values.

The book is designed to provide a bridge between theoretical and practical aspects of derivative securities pricing. We hope that the methodology of pricing derivative securities by means of analytical and numerical solutions to partial differential equations may attract the attention of students as well as mathematicians, engineers and practitioners having some experience with analysis and numerical solving of partial differential equations. It contains study materials which can be taught in basic and advanced courses on financial derivatives for undergraduate as well as graduate university students. The organization and presentation of the material reflects our experience with teaching the subject of financial derivatives at Comenius University and Slovak University of Technology in Bratislava, Slovakia. We thank our colleagues M. Takáč, T. Bokes and S. Kilianová for their valuable comments that helped us to improve presentation of the material contained in this book.

March 2010

Daniel Ševčovič, Beáta Stehlíková and Karol Mikula
authors

Chapter 1

The Role of Protecting Financial Portfolios

In the last decades, we have witnessed rapid expansion and development of various types of companies - starting from classical enterprises and ending with modern technological dot-com companies. One of the basic indicators of successful management and future expectations of a company is represented by the value of stock assets of the company. At the same time, it brings a defined profit in the form of dividends that are being paid to holders of stocks. From the point of view of a company development, stocks are often sources of further capitalization of the company. Although the stock price need not necessarily reflect the real value of the company, it is one of the best indicators of its present state, perspective and future development.

One of the most important problems in managing asset portfolios is the problem of effective portfolio allocation of the investment between stocks and bonds. Stocks usually bring higher returns. On the other hand, they represent a risky type of assets. Secure bonds (e.g., treasure bills) usually have lower yields, but they are less risky and volatile when compared to stocks. Investors are therefore looking for an optimal risk profile structure of their portfolios. A basic tool for protecting (hedging) an investor against risk is the so-called financial derivative. The origin of plain financial derivatives can be dated back to the 19th century. Historically first derivative security contracts were closely related to agricultural contracts for purchasing a crop. These types of derivative contracts were made during the winter season and gave the farmers a possibility of further investments and estimation of necessary amount of parcels of arable land. In modern language, these contracts can be interpreted as one of the basic type of a financial derivative called a forward. The last three decades constitute a turning point in trading financial derivatives. Derivative securities are mostly written on stocks, exchange rates or commodities. Among basic types of financial derivatives belong options and interest rate derivatives.

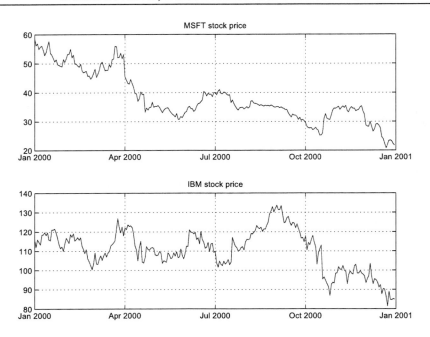

Figure 1.1. Time evolution of Microsoft (top) and IBM (bottom) stock prices in 2000.

1.1. Stochastic Character of Financial Assets

By examining financial data streams, we can get an idea of the stochastic evolution of the underlying asset prices, such as stocks, indices, interest rates and other assets. Their time development is often unstable and volatile, having fluctuations of larger or smaller sizes. These random changes are often caused by the influence of extensive trading in stock exchange markets. The stock prices are formed by supply and demand for those assets. When analyzing the time series, we often observe a certain trend in the stock prices and, at the same time, a fluctuating component of the price evolution. The trend part usually corresponds to a long term trend in the stock price, mostly influenced by a position and future expectations of the company. On the other hand, the fluctuating part can be due to balancing of demand and supply in the market.

In Fig.1.1 and Fig.1.2 we can see the time evolution of stock prices of Microsoft and IBM companies in the years 2000 and 2007-2008. The total trading volume of transactions is shown in bottom parts of both plots. The next Fig. 1.3 depicts evolution of the industrial Dow–Jones index.

The purpose of previous financial market data examples was to persuade ourselves about the stochastic evolution of various stock prices and indices on the market. We will deal with modeling of the stochastic behavior of stock prices in the following chapter. From the practical point of view, it should be emphasized that one of principal goals of investors is to minimize their possible losses from sudden decrease of stock prices. One of the most effective tools how to achieve this goal consists in usage of modern hedging instruments such as various derivative securities.

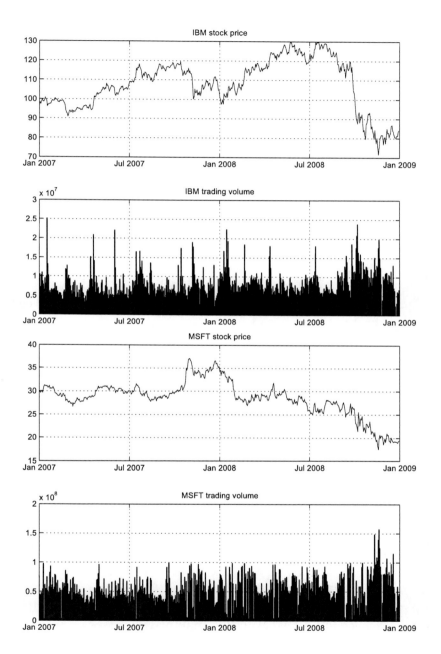

Figure 1.2. Time evolution of Microsoft (top) and IBM (bottom) stock prices in 2007 and 2008 and their trading volumes.

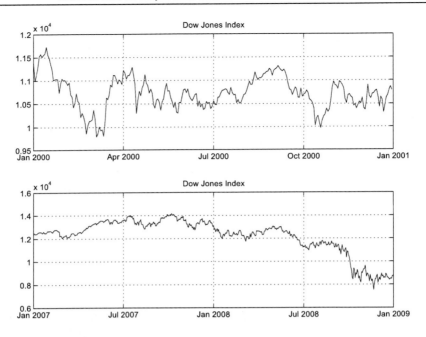

Figure 1.3. Time evolution of the Dow–Jones DJIA index in years 2000 (top) and 2007/08 (bottom).

1.2. Using Derivative Securities as a Tool for Protecting Volatile Stock Portfolios

In this section we discuss the importance of derivative securities in achieving the goal of stability of a portfolio book with respect to volatile fluctuations of the underlying asset prices. First, we consider the so-called forward and option types of financial derivatives.

1.2.1. Forwards and Futures

Historically first hedging tools for protecting investors against the risk arising from volatile underlying assets were the so-called forward contracts. A forward is an agreement between two parties — a holder and writer — representing both the right and, at the same time, obligation to accomplish the forward contract between the writer of the contract and its future holder. It consists in purchasing the underlying asset (usually a stock) at the precisely specified expiration time for a precisely specified expiration price. As it will be obvious from Chapter 3, pricing a forward is relatively simple and it is based on the idea of pricing the forward as the price of the underlying asset minus the expiration price discounted by a continuous interest rate. This simple pricing formula is, in fact, a consequence of the definition of a forward contract as a right and, at the same time, obligation to realize the contract.

1.2.2. Plain Vanilla Call and Put Options

In contrary to forward contracts, a call (put) option represents a kind of a derivative contract in which a holder of the contract has the right but not obligation to purchase (to sell) the underlying asset for the predetermined expiration price at the specified expiration time. Hence call or put options do not have an obligatory character, i.e., they give the holder certain freedom in their use. For a better illustration, let us consider a simple example of a call option. Suppose that we own a call option on purchasing the IBM stock for the exercise price USD 60 which can be exercised in three months. Let the present spot price be USD 55. Suppose that after three months of the lifetime of an option the underlying stock price rises up to USD 70. Then exercising such an option gives us a net profit of USD 10, which is just the difference between the spot underlying stock price USD 70 and the expiration price USD 60 of the option. In such a case we say that the option is *in-the-money* since exercising yields a nontrivial profit. On the other hand, if after three months the underlying stock price falls and achieves only the level of, say, USD 58, then for option holders, such an option becomes useless and it makes no sense to exercise it. In this case we say that the option is *out-of-money*, since it does not gain any profit. Note that the right but not obligation to exercise the option gives the option holder a certain advantage when compared to those not having this right. Hence, the right to purchase the underlying stock has a certain value. This value has to be paid by the future holder at the time of writing the contract. The writer asks for the so-called option premium for holder's right for future purchase of the underlying stock. One of the basic problems in the theory of financial derivatives is the question what is the fair price of such a derivative contract.

The basic types of derivative contracts are represented by the so-called plain vanilla options. The European style of a call option is a derivative contract in which the holder of such an option has the right but not obligation to purchase the underlying stock at the specified expiration time $t = T$ of maturity, for the predetermined expiration price E. On the other hand, the European style put option is a derivative contract in which the option holder has the right but not obligation to sell the underlying stock at the exercise time $t = T$ for the expiration price E. For both types of options the key problem is to find a value V of the call (put) option at the time $t = 0$ of signing the derivative contract.

In Table 1.1 we depict real market prices of call and put option on the underlying Microsoft stock from October 21, 2008. For example, a call option with expiration on December 19, 2008, and the expiration price $E = 15$ (USD) had the bid price 5.20 (bid price is an offer to purchase the option) and ask price 5.30 (ask price is an offer to sell the option). The underlying stock price was $S = 20.12$. The difference between the spot and expiration prices, i.e., $S - E$, was 5.12. It means that the option price was slightly higher than the price at the time of expiration. This difference can be explained since there are remaining four weeks until expiration. The underlying stock price is subject to stochastic fluctuations. Therefore, there is a certain risk of the rise of its price. Similarly, for the put option with exercise price $E = 25$, having its spot price in the interval from 4.85 to 4.95, we can observe that its price is slightly higher than just the difference $E - S = 4.78$.

Table 1.1. Prices of call and put options with various expiration prices on the Microsoft stock from October 26, 2008, 9:41am. The stock price $S = 20.12$ USD and options expired on Friday, December 19, 2008

Call options							Strike Price	Put options						
Symbol	Last	Change	Bid	Ask	Volume	Open Int		Symbol	Last	Change	Bid	Ask	Volume	Open Int
MQFLE.X	15.20	0.00	15.10	15.20	42	34	5.00	MQFXE.X	N/A	0.00	N/A	N/A	0	0
MQFLB.X	10.15	0.00	10.10	10.20	74	2541	10.00	MQFXB.X	0.03	0.00	0.02	0.04	97	3473
MQFLM.X	7.20	0.00	7.15	7.25	95	187	13.00	MQFXM.X	0.07	0.00	0.05	0.07	459	2994
MQFLN.X	6.15	0.00	6.15	6.25	55	211	14.00	MQFXN.X	0.10	0.00	0.07	0.10	204	2147
MQFLC.X	5.06	0.11	5.20	5.30	11	1348	15.00	MQFXC.X	0.14	0.00	0.13	0.14	5	8183
MQFLO.X	4.35	0.00	4.25	4.35	263	368	16.00	MQFXO.X	0.20	0.02	0.19	0.21	2	337
MQFLQ.X	3.40	0.00	3.30	3.40	122	4157	17.00	MQFXQ.X	0.32	0.02	0.33	0.34	11	8395
MQFLS.X	1.83	0.05	1.89	1.92	36	7567	19.00	MQFXS.X	0.83	0.06	0.77	0.80	169	31116
MQFLU.X	1.28	0.02	1.27	1.29	56	8886	20.00	MQFXD.X	1.14	0.06	1.13	1.16	109	23562
MQFLU.X	0.78	0.09	0.75	0.78	105	72937	21.00	MQFXU.X	1.83	0.23	1.65	1.68	1	72472
MSQLN.X	0.40	0.04	0.41	0.43	350	16913	22.00	MSQXN.X	2.58	0.23	2.30	2.36	3	4495
MSQLQ.X	0.21	0.01	0.20	0.22	125	20801	23.00	MSQXQ.X	3.10	0.00	3.05	3.15	30	3840
MSQLD.X	0.09	0.02	0.09	0.11	92	12207	24.00	MSQXD.X	3.80	0.00	3.95	4.05	167	3871
MSQLE.X	0.04	0.02	0.04	0.05	165	14193	25.00	MSQXE.X	4.90	0.00	4.85	4.95	157	2075
MSQLR.X	0.02	0.00	0.02	0.03	161	9359	26.00	MSQXR.X	6.15	0.00	5.85	5.95	210	1795
MSQLS.X	0.02	0.00	N/A	0.03	224	3643	27.00	MSQXS.X	7.00	0.00	6.85	6.95	45	1156
MSQLT.X	0.02	0.00	N/A	0.02	59	2938	28.00	MSQXT.X	7.55	0.00	7.80	7.95	24	874
MSQLF.X	0.01	0.00	N/A	0.02	10	1330	30.00	MSQXF.X	10.54	0.00	9.85	10.00	26	124

When analyzing prices of options with exercise prices higher and lower than the present spot price of the underlying stock, we can see that the option price is composed of two parts. One of them is the so-called intrinsic value of an option given by the value $\max(S-E,0)$ in the case of a call option and $\max(E-S,0)$ in the case of a put option. The remaining part of the option price represents the risk premium valuing the risk arising from the stochastic (volatile, fluctuating) character of evolution of the underlying asset price during the entire time interval remaining to expiration.

Chapter 2

Black–Scholes and Merton Model

In this chapter we derive and analyze a partial differential equation describing evolution of an option price depending on the underlying asset price and time remaining to maturity. The mathematical model is called the Black–Scholes and Merton model and the resulting equation is refereed to as the Black–Scholes equation. A key role in the derivation of the Black–Scholes equation is modeling the stochastic behavior of underlying assets. A basic tool for describing such a random evolution of the asset price is a concept of the so–called random Markov processes. Although there is a wide range of various types of Markov processes, the Wiener process and its generalization Brownian motion play a crucial role in modeling stochastic evolution of asset prices. We present a stochastic differential equation describing the Wiener process and Brownian motion. We furthermore derive a basic tool of the stochastic analysis – Itō's lemma. It turned out to be very useful in financial modeling. In particular, application of Itō's lemma is a key step in forthcoming derivation of the Black–Scholes partial differential equation. In derivation of the model we have to adopt several economical principles such as market completeness and liquidity, risk aversion of an investor, nonexistence of arbitrage opportunities. At the end of the chapter, we discuss various option strategies: from simple plain vanilla call and put options to more complex combined option strategies.

2.1. Stochastic Processes and Stochastic Differential Calculus

A stochastic process is a t–parametric system of random variables $\{X(t), t \in I\}$, where I is an interval of reals or a discrete set of indices. The Markov process is a stochastic process having the Markov property of the so-called memorylessness. By this we mean that its future random values $X(t)$ for $t > s$ conditioned to the present state $X(s)$ are independent of the history of previous random values $X(u)$ for $u < s$. From practical point of view, if the process $\{X(t), t \in I\}$ is a Markov process, then, for any time s we can restart generation of the process $\{X(t), t \in I, t > s\}$ *ab initio*, i.e., starting from the given initial value s without knowing the past history of the process. Assumption of a Markov character of asset prices is in agreement with the so–called weak form of market efficiency, since only present values of assets can be used in order to generate the future values. If I is a discrete set a Markov process defined on I is also refereed to as the Markov chain.

2.1.1. Wiener Process and Geometric Brownian Motion

Definition 2.1. *A Brownian motion $\{X(t), t \geq 0\}$ is a t–parametric system of random variables, for which*

i) *all increments $X(t+\Delta) - X(t)$ have a normal probability distribution with the expected value $\mu\Delta$ and dispersion (or variance) $\sigma^2\Delta$,*

ii) *for any partition $t_0 = 0 < t_1 < t_2 < t_3 < \cdots < t_n$ of the interval $(0, t_n)$, all increments $X(t_1) - X(t_0), X(t_2) - X(t_1), \ldots, X(t_n) - X(t_{n-1})$ are independent random variables with parameters according to the point i),*

iii) *$X(0) = 0$ and trajectories $\{X(t), t \geq 0\}$ are continuous almost surely.*

A Brownian motion with parameters $\mu = 0, \sigma^2 = 1$ is called the Wiener process. Clearly, the Wiener process as well as the Brownian motion are Markov processes.[1]

Figure 2.1. Norbert Wiener (1884-1964) and Robert Brown (1773-1858).

When analyzing the definition of a Brownian motion, the following question naturally arises: what is the reason for the expected value and dispersion of increments $X(t+\Delta) - X(t)$ to be proportional to Δ and not to some other function of Δ? At this point we make attempt to provide a convincing argument giving an answer to this question. Let us consider arbitrary partition of a given interval $[0, t]$, i.e., $0 = t_0 < t_1 < \cdots < t_n = t$. Then

$$X(t) - X(0) = \sum_{i=1}^{n} X(t_i) - X(t_{i-1}),$$

and hence the expected values and dispersions on the left hand and right hand sides must be equal. For the expected values of the term $X(t) - X(0)$, from the definition we obtain that

$$E(X(t) - X(0)) = \mu(t - 0) = \mu t.$$

On the other hand, the expected value of the random variable $\sum_{i=1}^{n} X(t_i) - X(t_{i-1})$ is given by

$$E\left(\sum_{i=1}^{n} X(t_i) - X(t_{i-1})\right) = \sum_{i=1}^{n} E(X(t_i) - X(t_{i-1})) = \sum_{i=1}^{n} \mu(t_i - t_{i-1}) = \mu t$$

---

[1] Norbert Wiener, 1884-1964, mathematician and statistician. He worked in the fields of mathematical analysis and probability theory. He is also considered to be a founder of modern cybernetics.

and hence the expected values of $X(t) - X(0)$ and $\sum_{i=1}^{n}(X(t_i) - X(t_{i-1}))$ are equal. Notice that without the assumption that every increment $X(t_i) - X(t_{i-1})$ has the expected value exactly equal to $\mu(t_i - t_{i-1})$, we would not be able to derive this equality. Now we concentrate on analysis of the dispersion of random variables $X(t) - X(0)$ and $\sum_{i=1}^{n}(X(t_i) - X(t_{i-1}))$. By the definition, we have

$$Var(X(t) - X(0)) = \sigma^2(t - 0) = \sigma^2 t.$$

Recall that for independent random variables A, B it holds: $Var(A+B) = Var(A) + Var(B)$. Since we assume independence of increments $X(t_i) - X(t_{i-1})$ for $i = 1, 2, \ldots, n$, it holds that

$$Var\left(\sum_{i=1}^{n} X(t_i) - X(t_{i-1})\right) = \sum_{i=1}^{n} Var(X(t_i) - X(t_{i-1})) = \sum_{i=1}^{n} \sigma^2(t_i - t_{i-1}) = \sigma^2 t.$$

Again, it should be noted that without the assumption that each increment $X(t_i) - X(t_{i-1})$ has dispersion exactly equal to $\sigma^2(t_i - t_{i-1})$ and the Markov property, the equality between dispersions of $X(t) - X(0)$ and $\sum_{i=1}^{n}(X(t_i) - X(t_{i-1}))$ could be violated.

From the preceding definition, it immediately follows that, if $\{w(t), t \geq 0\}$ is a Wiener process then for its first two statistical moments (the expected value and dispersion), we have:

$$E(w(t)) = 0, \qquad Var(w(t)) = t. \tag{2.1}$$

Moreover, the cumulative distribution function of a Wiener process is given by

$$\text{Prob}(w(t) < x) = \frac{1}{\sqrt{2\pi t}} \int_{-\infty}^{x} e^{-\xi^2/2t} d\xi. \tag{2.2}$$

A sample of five numerical realizations of a Wiener process is shown in Fig. 2.3. An experimental evidence of linear dependence (2.2) between the variance $Var(w(t))$ and time t are presented in Fig. 2.4.

A Brownian motion $\{X(t), t \geq 0\}$ with parameters μ and σ can be also analyzed by means of its increments $dX(t) = X(t + dt) - X(t)$ where dt is an infinitesimal small quantity. According to definition i), for their expected value and dispersion, it holds that $E(dX(t)) = \mu dt$ and $Var(dX(t)) = \sigma^2 dt = \sigma^2 Var(dw(t))$. It means that a Brownian motion can be characterized by its deterministic and fluctuating components. Its increments $dX(t)$ can be expressed in the following form of a total differential

$$dX(t) = \mu dt + \sigma dw(t), \tag{2.3}$$

where $\{w(t), t \geq 0\}$ is a Wiener process. Equation (2.3) is called stochastic differential equation.

Definition 2.2. *If $\{X(t), t \geq 0\}$ is a Brownian motion with parameters μ, σ and $y_0 \in \mathbb{R}^+$, then the system of random variables $\{Y(t), t \geq 0\}$,*

$$Y(t) = y_0 e^{X(t)}, \quad t \geq 0,$$

is called a geometric Brownian motion.

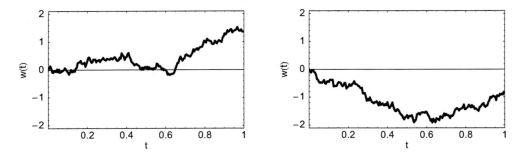

Figure 2.2. Two different random realizations of a Wiener process.

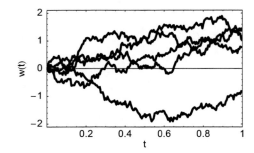

Figure 2.3. Five trajectories of a Wiener process displayed together.

Any geometric Brownian motion is again a Markov process. Based on the explicit form of the probability distribution function of a Wiener process (2.2) we are yet able to compute its first two statistical moments:

$$E(Y(t)) = y_0 e^{\mu t + \frac{\sigma^2 t}{2}}, \qquad Var(Y(t)) = y_0^2 e^{2\mu t + \sigma^2 t}(e^{\sigma^2 t} - 1). \qquad (2.4)$$

To simplify computation of moments (2.4) it suffices to consider only the case when $y_0 = 1$. Then for the cumulative distribution function $G(y,t) = \text{Prob}(Y(t) < y)$ of the geometric Brownian motion $Y(t)$ we have that $G(y,t) = 0$ for $y \leq 0$. It is a consequence of positivity of the variable $Y(t)$. For $y > 0$ it holds that

$$G(y,t) = \text{Prob}(Y(t) < y) = \text{Prob}\left(Z(t) < \frac{-\mu t + \ln y}{\sigma}\right),$$

where $Z(t)$ is a random variable, $Z(t) = (-\mu t + \ln Y(t))/\sigma$. Clearly, $dZ(t) = dw(t)$ and hence $Z(t) = Z(0) + w(t) = w(t)$ because $Z(0) = 0$. This way we have shown that $Z(t)$ is indeed a Wiener process. Using the knowledge of the cumulative distribution function of a Wiener process (2.2), for the distribution function $G(y,t)$ of random variable $Y(t)$ we obtain $G(y,t) = 0$ for $y \leq 0$ and

$$G(y,t) = \frac{1}{\sqrt{2\pi t}} \int_{-\infty}^{\frac{-\mu t + \ln y}{\sigma}} e^{-\xi^2/2t} d\xi \qquad \text{for } y > 0.$$

Since $E(Y(t)) = \int_{-\infty}^{\infty} y g(y,t) \, dy$ and $E(Y(t)^2) = \int_{-\infty}^{\infty} y^2 g(y,t) \, dy$, where $g(y,t) = \frac{\partial}{\partial y} G(y,t)$, by computation of these integrals we obtain that

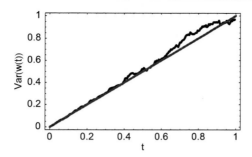

Figure 2.4. Time evolution of the dispersion of a Wiener process computed from simulation of 1000 individual trajectories of the Wiener process.

$$
\begin{aligned}
E(Y(t)) &= \int_{-\infty}^{\infty} y g(y,t)\,dy = \int_{0}^{\infty} y g(y,t)\,dy \\
&= \frac{1}{\sqrt{2\pi t}} \int_{0}^{\infty} y e^{-\frac{(-\mu t + \ln y)^2}{2\sigma^2 t}} \frac{1}{\sigma y}\,dy \\
&\quad (\xi = (-\mu t + \ln y)/(\sigma\sqrt{t})) \\
&= \frac{e^{\mu t}}{\sqrt{2\pi}} \int_{-\infty}^{\infty} e^{-\frac{\xi^2}{2} + \sigma\sqrt{t}\xi}\,d\xi = \frac{e^{\mu t + \frac{\sigma^2}{2} t}}{\sqrt{2\pi}} \int_{-\infty}^{\infty} e^{-\frac{(\xi - \sigma\sqrt{t})^2}{2}}\,d\xi \\
&= e^{\mu t + \frac{\sigma^2}{2} t}.
\end{aligned}
$$

In the same way we obtain the dispersion (2.4).

In what follows, we will say that a random variable $\{Y(t), t \geq 0\}$ has a lognormal distribution with the expected value and dispersion given by (2.4). Throughout the rest of this book, a Wiener process will be denoted by either $\{w(t), t \geq 0\}$ or $\{W(t), t \geq 0\}$. Its increments over a short time interval dt will be denoted by dw, i.e., $dw(t) = w(t+dt) - w(t)$. According to the definition of a Wiener process the increments $dw(t)$ are independent in time t. Their expected value is zero, i.e., $E(dw(t)) = 0$ and their dispersion $Var(dw(t)) = dt$. The increment dw can be therefore written as

$$dw = \Phi\sqrt{dt}, \qquad \text{where } \Phi \sim N(0,1),$$

i.e., Φ is a random variable with the standardized normal distribution.

2.1.2. Itō's Integral and Isometry

Important technical tools in analysis of stochastic processes are the so-called Itō's integral and Itō's isometry. Construction of Itō's integral is very similar to the definition of the Riemann–Stieltjes integral of functions of a real variable.

First, we notice that it follows from the definition of a Wiener process $\{w(t), t \geq 0\}$ that the random variable $w(t)$ has a normal distribution with a zero mean and dispersion t, i.e., $w(t) \sim N(0,t)$. This equality can be rewritten as:

$$\int_{0}^{t} dw(\tau) = w(t) - w(0) = w(t) \sim N(0,t).$$

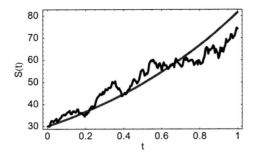 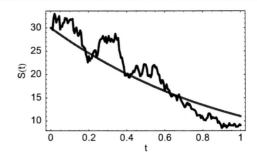

Figure 2.5. Two trajectories of a geometric Brownian motion $dS = \mu S dt + \sigma S dw$ with the positive drift $\mu = 1 > 0$ (left) and negative drift $\mu = -1 < 0$ (right) together with the drift function $t \mapsto S_0 e^{\mu t}$ (exponential curves). We set $S_0 = 30$ and $\sigma = 0.49$ in both examples.

It means that for a constant function $f(\tau) \equiv c$ we have

$$\int_0^t f(\tau)dw(\tau) = c \int_0^t dw(\tau) = cw(t) - cw(0)$$
$$= cw(t) \sim N(0, c^2 t) = N(0, \int_0^t f^2(\tau)d\tau).$$

This simple identity gives us an idea, how to define the so–called Itō's integral of a measurable function $f : (0,t) \to \mathbb{R}$ such that $\int_0^t f^2(\tau)d\tau < \infty$. We let

$$\int_0^t f(\tau)dw(\tau) := \lim_{\nu \to 0} \sum_{i=0}^{n-1} f(\tau_i)(w(\tau_{i+1}) - w(\tau_i)),$$

where $\nu = \max(\tau_{i+1} - \tau_i)$ is the norm of a partition $0 = \tau_0 < \tau_1 < \cdots < \tau_n = t$ of the interval $(0,t)$. Convergence is meant in probability. Let the function f be constant on each subinterval $[\tau_i, \tau_{i+1})$. Then, for the expected value of the finite sum $\sum_{i=1}^{n} f(\tau_i)(w(t_{i+1}) - w(t_i))$, it holds:

$$E\left(\sum_{i=0}^{n-1} f(\tau_i)(w(\tau_{i+1}) - w(\tau_i))\right) = \sum_{i=0}^{n-1} f(\tau_i) E(w(\tau_{i+1}) - w(\tau_i)) = 0,$$

because all increments $w(\tau_{i+1}) - w(\tau_i)$ are normally distributed random variables $w(\tau_{i+1}) - w(\tau_i) \sim N(0, \tau_{i+1} - \tau_i)$. Since these increments are also independent and $w(\tau_{i+1}) - w(\tau_i) = \Phi_i \sqrt{\tau_{i+1} - \tau_i}$, where $\Phi_i \sim N(0,1)$, we may conclude for the sum of the independent normally distributed random variables the following identity:

$$E\left(\left[\sum_{i=0}^{n-1} f(\tau_i)(w(\tau_{i+1}) - w(\tau_i))\right]^2\right) = \sum_{i=0}^{n-1} f^2(\tau_i) E(\Phi_i^2)(\tau_{i+1} - \tau_i)$$
$$= \sum_{i=1}^{n} f^2(\tau_i)(\tau_{i+1} - \tau_i).$$

Similarly as in construction of the Riemann–Stieltjes integral, we can pass to the limit as the norm ν of a partition tends to zero and the sequence of simple step functions pointwise

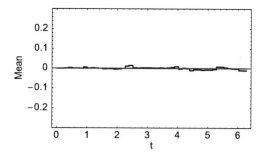
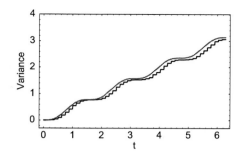

Figure 2.6. A sample of the mean value of Itō's integral $\int_0^t f(\tau)dw(\tau)$ (left) and its dispersion together with a graph of $\int_0^t f(\tau)^2 d\tau$ (right) for the function $f(\tau) = \sin(2\tau)$. A number of partitions of the interval was chosen: $n = 100$.

almost everywhere converges to a measurable function f with a finite integral $\int_0^t f^2(\tau)d\tau < \infty$. This way we obtained the so-called Itō's isometry:

Lemma 2.1. *Let a measurable function $f : (0,t) \to \mathbb{R}$ be such that $\int_0^t f^2(\tau)d\tau < \infty$. Then there exists Itō's integral $\int_0^t f(\tau)dw(\tau)$. It is a normally distributed random variable having $N(0, \sigma^2(t))$ distribution, where $\sigma^2(t) = \int_0^t f(\tau)^2 d\tau$. It means that the following identities hold:*

$$E\left(\int_0^t f(\tau)dw(\tau)\right) = 0,$$

$$E\left(\left[\int_0^t f(\tau)dw(\tau)\right]^2\right) = \int_0^t f(\tau)^2 d\tau.$$

The last identity is called Itō's isometry.

Notice that Itō's isometry holds not only for the measurable functions f, but also for general stochastic processes, which are continuous from the left and locally finite. For such a stochastic process $\{H(\tau), \tau \geq 0\}$ Itō's integral can be again defined as a limit (in probability) of finite sums

$$\int_0^t H(\tau)dw(\tau) := \lim_{\nu \to 0} \sum_{i=0}^{n-1} H(\tau_i)(w(\tau_{i+1}) - w(\tau_i)).$$

Then Itō's isometry has the form:

$$E\left(\left[\int_0^t H(\tau)dw(\tau)\right]^2\right) = E\left(\int_0^t H(\tau)^2 d\tau\right), \qquad (2.5)$$

(see e.g., Øksendal [89]). Moreover, for the mean value of the integral we obtain

$$E\left(\int_0^t H(\tau)dw(\tau)\right) = 0. \qquad (2.6)$$

Further details concerning qualitative and quantitative properties of stochastic processes can be found in the books and survey papers by e.g., Karatzas and Shreeve [71], Papanicolaou [90], Hull [65], Wilmott, Dewynne and Howison [122], Melicherčík et al. [83], Baxter and Rennie [13].

Figure 2.7. Kiyoshi Itō (1915–2008).

2.1.3. Itō's Lemma for Scalar Random Processes

Analysis of functions, representing prices of financial derivatives, whose one or more variables are stochastic random variables satisfying prescribed stochastic differential equations plays a key role in the theory of pricing financial derivatives. In this section, we focus our attention on the question whether there exists a stochastic differential equation describing evolution of a smooth function $f(x,t)$ of two variables, where the variable x itself is a solution to a prescribed stochastic differential equation. The positive answer to this question is given by Itō's lemma. This is a key stone of analysis of stochastic differential equations. According to Wikipedia, Itō's lemma is "the all time most famous lemma".[1]

Lemma 2.2 (Itō's lemma). *Let $f(x,t)$ be a smooth function of two variables. Assume the variable x is a solution to the stochastic differential equation*

$$dx = \mu(x,t)dt + \sigma(x,t)dw,$$

where w is a Wiener process. Then the first differential of the function f is given by

$$df = \frac{\partial f}{\partial x}dx + \left(\frac{\partial f}{\partial t} + \frac{1}{2}\sigma^2(x,t)\frac{\partial^2 f}{\partial x^2}\right)dt,$$

and so the function f satisfies the stochastic differential equation

$$df = \left(\frac{\partial f}{\partial t} + \mu(x,t)\frac{\partial f}{\partial x} + \frac{1}{2}\sigma^2(x,t)\frac{\partial^2 f}{\partial x^2}\right)dt + \sigma(x,t)\frac{\partial f}{\partial x}dw. \qquad (2.7)$$

An intuitive (and from mathematical point of view fairly incomplete) proof of Itō's lemma can be done by expanding the function $f = f(x,t)$ into a Taylor series of the second order. Indeed,

$$\begin{aligned} f(x+dx,t+dt) - f(x,t) &= \frac{\partial f}{\partial t}dt + \frac{\partial f}{\partial x}dx \\ &+ \frac{1}{2}\left(\frac{\partial^2 f}{\partial x^2}(dx)^2 + 2\frac{\partial^2 f}{\partial x \partial t}dx\,dt + \frac{\partial^2 f}{\partial t^2}(dt)^2\right) + \text{h.o.t.} \end{aligned}$$

---

[1] Kiyoshi Itō, 1915-2008, mathematician and statistician, the Gauss prize winner in 2008. He worked in the field of the probability theory and stochastic processes. He proved one of the most important and useful propositions of stochastic differential calculus – Itō's lemma.

Now, it follows from the property $dw = \Phi\sqrt{dt}$, where $\Phi \sim N(0,1)$, that

$$E((dw)^2 - dt) = 0, \quad \text{and} \quad Var((dw)^2 - dt) = [E(\Phi^4) - E(\Phi^2)^2](dt)^2 = 2(dt)^2.$$

By neglecting higher order terms in dt we can approximate the term $(dw)^2$ by dt. We obtain[2]

$$(dx)^2 = \sigma^2(dw)^2 + 2\mu\sigma dw\,dt + \mu^2(dt)^2 \approx \sigma^2 dt + O((dt)^{3/2}) + O((dt)^2).$$

Similarly, the term $dx\,dt = O((dt)^{3/2}) + O((dt)^2)$, and consequently the first order expansion of the differential df with respect to infinitesimal increments dt and dx can be written in the form

$$df = \frac{\partial f}{\partial x}dx + \left(\frac{\partial f}{\partial t} + \frac{1}{2}\sigma^2(x,t)\frac{\partial^2 f}{\partial x^2}\right)dt.$$

The relation (2.7) follows from the above relation for df by substituting the expression $dx = \mu(x,t)dt + \sigma(x,t)dw$ for the differential term dx.

The stochastic differential equation

$$dx = \mu(x,t)dt + \sigma(x,t)dw,$$

describing a general Itō's process, should be understood in the integral sense, i.e.

$$x(t) - x(0) = \int_0^t \mu(x(\tau),\tau)d\tau + \int_0^t \sigma(x(\tau),\tau)dw(\tau).$$

The term $\int_0^t \mu(x(\tau),\tau)d\tau$ is a usual Riemann integral whereas the term $\int_0^t \sigma(x(\tau),\tau)dw(\tau)$ is Itō's integral $\int_0^t H(\tau)dw(\tau)$ (see the previous section) with a random process $H(t)$ defined as $H(t) = \sigma(x(t),t)$. It is important to emphasize that, with regard to the construction of Itō's integral, the increment $w(t_{i+1}) - w(t_i)$ of a Wiener process w and the random variable $x(t_i)$ at any time t_i are independent. Since $E(w(t_{i+1}) - w(t_i)) = 0$ we therefore conclude

$$E(\sigma(x,t)dw) = 0,$$

as well (see also (2.6)). This is one of important features of Itō's process that will be used many times throughout this book.

As an example of application of Itō's lemma, let us consider a Brownian motion $dX = \mu dt + \sigma dw$ and its function $Y(t) = f(X(t),t)$, where $f(X,t) = e^X$. By applying Itō's lemma we obtain

$$dY = \left(\frac{\partial f}{\partial t} + \mu\frac{\partial f}{\partial X} + \frac{1}{2}\sigma^2\frac{\partial^2 f}{\partial X^2}\right)dt + \sigma\frac{\partial f}{\partial X}dw = \left(\mu + \frac{\sigma^2}{2}\right)Y dt + \sigma Y dw.$$

As a consequence, we obtain for the expected value $E(Y(t))$ the following ordinary differential equation

$$dE(Y(t)) = \left(\mu + \frac{\sigma^2}{2}\right)E(Y(t))dt,$$

from which we easily deduce that $E(Y(t)) = E(Y(0))e^{(\mu+\sigma^2/2)t}$, as it was already claimed in (2.4).

---

[2] By the term $f(\xi) = O(g(\xi))$ we have denoted the Landau "big O" symbol representing any function f with the property $|f(\xi)| \leq Cg(\xi)$ for any sufficiently small ξ where $C > 0$ is a constant.

2.1.4. Itō's Lemma for Vector Random Processes

The previous procedure of deriving Itō's lemma for a function of a scalar random variable x can be successfully extended also for the case of C^2 smooth function $f = f(\vec{x},t) : \mathbb{R}^n \times \mathbb{R} \to \mathbb{R}$ of a vector argument $\vec{x} = (x_1, x_2, \ldots, x_n)^T$. Concerning the variables $x_i, i = 1, \ldots, n$ we will assume they satisfy a system of stochastic differential equations

$$dx_i = \mu_i(\vec{x},t)dt + \sum_{k=1}^n \sigma_{ik}(\vec{x},t)dw_k,$$

where $\vec{w} = (w_1, w_2, \ldots, w_n)^T$ is a vector of Wiener processes which have independent increments, i.e.

$$E(dw_i dw_j) = 0 \text{ for } i \neq j, \quad E((dw_i)^2) = dt.$$

The equations for stochastic processes x_i can be written in a vector form as follows:

$$d\vec{x} = \vec{\mu}(\vec{x},t)dt + K(\vec{x},t)d\vec{w},$$

where K is an $n \times n$ matrix

$$K(\vec{x},t) = (\sigma_{ij}(\vec{x},t))_{i,j=1,\ldots,n}.$$

Then, for the increment df of a smooth function $f = f(\vec{x},t)$, we can write its expansion to a Taylor series of the second order. We obtain

$$\begin{aligned}df &= \frac{\partial f}{\partial t}dt + \nabla_x f . d\vec{x} \\ &+ \frac{1}{2}\left((d\vec{x})^T \nabla_x^2 f \, d\vec{x} + 2\nabla_x f \frac{\partial f}{\partial t}d\vec{x}\,dt + \frac{\partial^2 f}{\partial t^2}(dt)^2\right) + \text{h.o.t.},\end{aligned}$$

where $\nabla_x f$, respectively $\nabla_x^2 f$ denote the gradient and the Hess matrix of a function f with respect to variables $x_1, \ldots, x_n$. Similarly as in the derivation of one-dimensional variant of Itō's lemma, the terms $d\vec{x}\,dt$ and $(dt)^2$ are negligible when compared to the term dt. Therefore a crucial part will be again analysis of the term $(d\vec{x})^T \nabla_x^2 f \, d\vec{x} = \sum_{i,j=1}^n \frac{\partial^2 f}{\partial x_i \partial x_j}dx_i dx_j$. From the assumption on the independence of increments dw_i and dw_j for $i \neq j$ we obtain that

$$\begin{aligned}dx_i dx_j &= \sum_{k,l=1}^n \sigma_{ik}\sigma_{jl}dw_k dw_l + O((dt)^{3/2}) + O((dt)^2) \\ &= \left(\sum_{k=1}^n \sigma_{ik}\sigma_{jk}\right)dt + O((dt)^{3/2}) + O((dt)^2),\end{aligned}$$

But it means that the expansion of the differential df with respect to the increments $dt, d\vec{x}$ can be written in the form

$$df = \left(\frac{\partial f}{\partial t} + \frac{1}{2}K : \nabla_x^2 f K\right)dt + \nabla_x f \, d\vec{x}, \qquad (2.8)$$

where the term $K : \nabla_x^2 f K$ is defined as follows:

$$K : \nabla_x^2 f K = \sum_{i,j=1}^{n} \frac{\partial^2 f}{\partial x_i \partial x_j} \sum_{k=1}^{n} \sigma_{ik}\sigma_{jk}. \tag{2.9}$$

The relation (2.8) for the first differential of a smooth function depending on a vector of stochastic processes constitutes the statement of Itō's lemma for functions of a vector argument. This result plays an important role in analysis of multifactor models for pricing interest rate derivatives, analysis of basket and index options.

2.2. The Black–Scholes Equation

In this section we derive a mathematical model for pricing financial derivatives, such as options. Mathematical formulation of this model is represented by the so–called Black–Scholes partial differential equation. It describes the time evolution of a derivative price as a function of the underlying asset price and time remaining to maturity of a derivative.

The derivation of the Black–Scholes differential equation will be shown on the example of a European call option. Recall that the call option is a contract, in which the holder of an option has the right but not obligation to purchase the underlying asset from the writer of an option in the predetermined expiration time $t = T$ at the prescribed expiration (strike) price E. We emphasize that the holder has the right but not obligation to buy the stock. Hence this right has a certain value and at the time of signing the contract $t = 0$. The buyer (holder) of such an option has to pay the so–called option premium V to the writer of a call option. For both sides of the contract, i.e., for the writer or the option as well as for the holder, it is of interest to know, what is the fair price of this premium.

Let us denote:

- S - value of the underlying asset price,

- V - value of a derivative (option) on the underlying asset,

- T - expiration time, i.e., date of expiry (maturity) of the option,

- E - expiration (strike) price of the option,

- t - time, $t \in [0, T]$.

Our goal is to find a mathematical model describing the price of an option $V = V(S,t)$ as a function of the underlying asset price S and the time t. An option premium is then given by the value $V(S,0)$ at the time $t = 0$ of signing the contract.

Concerning a stochastic behavior of the underlying asset process we will suppose that it satisfies a stochastic differential equation for the geometric Brownian motion. The foundation for modeling evolution of asset prices by means of a Brownian motion has been proposed by Louis Bachelier in his famous thesis "La Théorie de la Spéculation" from 1900. He proposed that stock prices are moving similarly as particles in the Brownian motion.

His work has been rediscovered by Paul Samuelson. In [98], Samuelson modified Bachelier's idea of a Brownian motion and he postulated that the stock prices follow a geometric Brownian motion rather than a simple Brownian motion.

Next, derivation of a governing equation for $V = V(S,t)$ consists of two steps. In the first step, we derive a stochastic equation fulfilled by an arbitrary smooth function $V = V(S,t)$ of a stochastic underlying price S and time t. The function V is, in general, called a financial derivate. In the second step, we will construct the so–called self-financing portfolio having zero net investment and consisting of underlying assets, options on these assets and riskless bonds.

2.2.1. A Stochastic Differential Equation for the Option Price

As we have already mentioned in the previous section, in order to model random evolution of the underlying asset price as a function of time $S = S(t)$ we will use the stochastic differential equation representing the geometric Brownian motion.

$$dS = \mu S dt + \sigma S dw, \qquad (2.10)$$

where dS is the change of asset value over the time interval of a length dt, μ represents a trend of underlying asset price evolution and σ is its volatility. By dw we have denoted the differential of a Wiener process. The deterministic process $dS = \mu S dt$ (i.e., $\sigma = 0$) has solution $S(t) = S(0)e^{\mu t}$ representing thus exponential growth (decrease if $\mu < 0$) of asset values observed in financial markets. Furthermore, notice that the stochastic equation (2.10) can be also written in the form

$$\frac{dS}{S} = \mu dt + \sigma dw.$$

The term σdw can be therefore understood as random fluctuation over the trend part of the asset price. Hence the essential information is contained in the relative change dS/S and not in the absolute change in the asset price dS. Moreover, the relativized differential dS/S represents a return on asset. Another reason is that the resulting model has to be invariant with respect to choice of units, i.e., the pricing formula should be currency unit invariant.

In the next step we derive a stochastic differential equation describing the evolution of an arbitrary smooth function (derivative) of asset price and time. Suppose that a function $V = V(S,t)$ is a smooth function of two variables, where S satisfies the stochastic differential equation (2.10). A stochastic differential equation for the function $V = V(S,t)$ can be derived by using the fundamental tool in the theory of random processes – Itō's lemma 2.2). In our case, the variable S satisfies the stochastic differential equation (2.10), i.e., $dS = \mu S dt + \sigma S dw$, and hence $\mu(S,t) = \mu S, \sigma(S,t) = \sigma S$, where μ, σ are constants. Then a function $V(S,t)$ of the stochastic process S satisfies the following stochastic differential equation

$$dV = \left(\frac{\partial V}{\partial t} + \mu S \frac{\partial V}{\partial S} + \frac{1}{2}\sigma^2 S^2 \frac{\partial^2 V}{\partial S^2} \right) dt + \sigma S \frac{\partial V}{\partial S} dw. \qquad (2.11)$$

2.2.2. Self-financing Portfolio Management with Zero Growth of Investment

In this step we focus on construction of a portfolio consisting of underlying assets of the same type, options on this assets and riskless bonds. The idea of self-financing strategy consists in dynamic selling or buying components of the portfolio in such a way, that it is risk-neutral, no further investments are needed (assumption on zero net investments), and selling or buying of one type of assets (stocks, options, bonds) is balanced by buying or selling another assets in the portfolio (self-financing principle). This methodology of deriving the Black-Scholes model is due to Merton. Its difference from the derivation done by Black and Scholes is precisely in considering the self-financed portfolio with zero net investments. Notice that the assumption on effort to create risk–neutral portfolio is a basic pillar in deriving the Black–Scholes equation. This assumption arises from investors' efforts to achieve risk neutral hedging strategies. It is based on the assumption of completeness and liquidity of a market yielding thus possibility of perfect replication of a risk-neutral portfolio. Although basic assumptions of derivation of the Black–Scholes model need not be satisfied in reality, we postulate them in order to derive the standard model. In Chapters 5 and 12 we will present various generalizations of the standard Black-Scholes model by taking into account more realistic concepts like pricing options in incomplete or illiquid markets, presence of a dominant traders, pricing under transaction costs etc.

Let us construct a portfolio consisting of underlying assets, options on these assets and riskless bonds. We will consider the so–called self-financing portfolio, i.e., a portfolio in which the purchase or sale of one of the three components has to be compensated by selling or purchasing another component of the portfolio. More precisely, at time t, the portfolio consists of amount of Q_S stocks with the unit price S, amount of Q_V option with the unit price V and the riskless zero-coupon bonds having the total money value B. If we denote $M_S = SQ_S$, $M_V = VQ_V$, then the assumption of zero net investments means, that the balance equation

$$M_S + M_V + B = 0,$$

has to be satisfied for all times $t \in [0, T]$, i.e.

$$SQ_S + VQ_V + B = 0, \tag{2.12}$$

for $t \in [0, T]$. Now Merton's condition on self-financing of the portfolio can be stated in the following form

$$S dQ_S + V dQ_V + \delta B = 0 \tag{2.13}$$

where dQ_S, dQ_V denote changes in the amount of underlying assets and options. By δB we have denoted a change in the money volume of riskless bonds in the portfolio that have been used in order to finance purchases of assets or options ($\delta B < 0$), or have been gained from selling assets or options ($\delta B > 0$). Recall that for a standalone portfolio of riskless zero-coupon bonds there is a simple pricing formula $B(t) = B(0)e^{rt}$, where $r > 0$ is continuous interest rate. This equation can be written in the differential form as $dB = rB dt$. In the case bonds are dynamically used/gained in self-financing the portfolio we have the total change in the money volume of the bonds dB expressed as:

$$dB = rB dt + \delta B. \tag{2.14}$$

Differentiating relation (2.12), inserting (2.14) into (2.13) and expressing the price B from equation (2.12) we finally obtain the following identity:

$$\begin{aligned} 0 &= d(SQ_S + VQ_V + B) \\ &= SdQ_S + VdQ_V + \delta B + Q_S dS + Q_V dV + rB\,dt \\ &= Q_S dS + Q_V dV - r(SQ_S + VQ_V)\,dt. \end{aligned}$$

After dividing by a nonzero value Q_V of the amount of options in the portfolio, we conclude that:

$$dV - rV\,dt - \Delta(dS - rS\,dt) = 0, \quad \text{where } \Delta = -\frac{Q_S}{Q_V}. \tag{2.15}$$

Recall that both random processes, i.e., the asset price S, as well as the option price V satisfy stochastic differential equations

$$\begin{aligned} dS &= \mu S\,dt + \sigma S\,dw, \\ dV &= \left(\frac{\partial V}{\partial t} + \mu S \frac{\partial V}{\partial S} + \frac{1}{2}\sigma^2 S^2 \frac{\partial^2 V}{\partial S^2}\right) dt + \sigma S \frac{\partial V}{\partial S}\,dw. \end{aligned}$$

Substituting the above expressions for the differentials dS and dV, we obtain, after some manipulations,

$$\left(\frac{\partial V}{\partial t} + \mu S \frac{\partial V}{\partial S} + \frac{1}{2}\sigma^2 S^2 \frac{\partial^2 V}{\partial S^2} - rV - \Delta \mu S + \Delta r S\right) dt + \sigma S \left(\frac{\partial V}{\partial S} - \Delta\right) dw = 0.$$

The purpose of a risk-neutral investor is to combine the portfolio of assets, options and bonds in such a way that the risk of the portfolio is neutralized. Such a behavior of an investor is called risk aversion. Clearly, the only stochastic term in the equation above is represented by the differential dw of the Wiener process. This term vanishes provided that we choose the ratio Δ as follows:

$$\Delta = \frac{\partial V}{\partial S}. \tag{2.16}$$

After substituting this choice of Δ to the remaining determininstic part we obtain the resulting equation

$$\left(\frac{\partial V}{\partial t} + \frac{1}{2}\sigma^2 S^2 \frac{\partial^2 V}{\partial S^2} + rS \frac{\partial V}{\partial S} - rV\right) dt = 0,$$

and hence

$$\frac{\partial V}{\partial t} + \frac{1}{2}\sigma^2 S^2 \frac{\partial^2 V}{\partial S^2} + rS \frac{\partial V}{\partial S} - rV = 0, \tag{2.17}$$

which is refereed to as the Black–Scholes partial differential equation for pricing derivative securities. It was first published in the seminal paper by Black and Scholes [14].

Let us consider a useful generalization of the Black–Scholes equation for the case when the underlying asset is paying nontrivial continuous dividends with an annualized dividend yield $q \geq 0$. In this case, holding the underlying asset with a price S we receive a dividend yield $qSdt$ over any time interval with a length dt. By paying dividends the asset price itself falls. It can be expressed by modifying the drift part of the stochastic differential equation for the asset price. Hence the asset price satisfies the stochastic differential equation

$$dS = (\mu - q)S\,dt + \sigma S\,dw.$$

On the other hand, by receiving dividends, we have new resources for our self-financing portfolio. Their total money volume being $qSQ_S dt$ over the time interval dt. This amount of money can be therefore added as an extra income to the right hand side of the equation (2.14) describing change of the money volume of secure bonds, i.e. $dB = rB dt + \delta B + qSQ_S dt$. This way we have modified equation (2.15) to the following form:

$$dV - rV dt - \Delta(dS - (r-q)S dt) = 0.$$

Repeating the remaining step of derivation of the Black–Scholes equation we end up with the modifies equation (2.17) which includes a continuous dividend yield $q \geq 0$:

$$\frac{\partial V}{\partial t} + \frac{1}{2}\sigma^2 S^2 \frac{\partial^2 V}{\partial S^2} + (r-q)S\frac{\partial V}{\partial S} - rV = 0. \qquad (2.18)$$

We end this section by another useful generalization of the Black–Scholes equation for the case when the underlying asset price is a solution to Itō's stochastic differential equation of the form

$$dS = \mu(S,t)S dt + \sigma(S,t)S dw. \qquad (2.19)$$

As an example of such a generalized process for the underlying asset one can consider the so-called constant elasticity of volatility (CEV) model proposed by Cox and Ross. In this model, the underlying asset price is assumed to follow a stochastic differential equation of the form

$$dS = \mu S dt + \sigma S^\alpha dw, \qquad (2.20)$$

where μ and $\sigma > 0$ are constants and the exponent $\alpha > 0$. Clearly, if $\alpha = 1$ then the process (2.20) is just a geometric Brownian motion.

Following the lines of derivation of the Black–Scholes equation we can see that the assumption that μ and σ are constants plays no role. Hence, for an underlying asset following (2.19) we can derive the corresponding partial differential equation for pricing its derivative security in the form:

$$\frac{\partial V}{\partial t} + \frac{1}{2}\sigma(S,t)^2 S^2 \frac{\partial^2 V}{\partial S^2} + (r-q)S\frac{\partial V}{\partial S} - rV = 0. \qquad (2.21)$$

For example, if S is a solution to the CEV process (2.20) then the corresponding Black–Scholes equation has the form

$$\frac{\partial V}{\partial t} + \frac{1}{2}\sigma^2 S^{2\alpha} \frac{\partial^2 V}{\partial S^2} + (r-q)S\frac{\partial V}{\partial S} - rV = 0. \qquad (2.22)$$

2.3. Terminal Conditions

In the derivation of the Black–Scholes equation so far we have not used our assumption that the derivative security is a call option. In this section, we show that we have to add to the Black–Scholes equation (2.17) an additional terminal condition at expiration time T, determining the type of a derivative contract. Such conditions are called terminal pay–off conditions or pay–off diagrams.

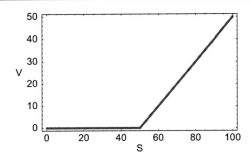 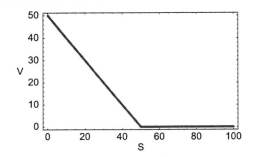

Figure 2.8. Terminal pay–off diagrams for call (left) and put options (right) with the exercise (strike) price $E = 50$.

2.3.1. Pay–off Diagrams for Call and Put Options

In the case of a European call option we have to add the terminal condition at expiry time T to the Black–Scholes equation (2.17), respectively to its modified version (2.18). The terminal pay–off diagram of a call option is easy to understand. Indeed, if the present spot price S of the underlying asset at time T exceeds the exercise price E then the value of the (in-the-money) option premium (if payed at time T) is given just as a difference between the present price S and strike price E, i.e., $S - E$. This follows from our assumption on impossibility of arbitrage opportunities, i.e., possibilities for gaining riskless profit. On the other hand, if the present asset price does not exceed the strike price E then the (out-of-the-money) option has no value, since we will not exercise it. It means that at the time $t = T$ it is easy to price the call option. Its value is given by the terminal pay–off condition

$$V(S,T) = (S-E)^+, \qquad (2.23)$$

which is depicted on Fig. 2.8 (left). Here we have denoted by x^+ the positive part of a real number x, i.e.

$$V(S,T) = (S-E)^+ := \max(S-E,0).$$

If we consider a European put option then derivation of the terminal condition for its price at expiry time T follows from a similar argument as in the case of the call option. If the present asset price S at time T exceeds the exercise (strike) price value E then the value of the (out-of-the-money) option is zero as it would not make no sense to exercise this option and to sell the asset for a lower price E when compared to its market spot value S. On the other hand, if the present price of the asset is lower then the expiration price E, then the (in-the-money) option has a value and it is equal to the difference $E - S$. It means that the terminal pay–off diagram of a put option is the function:

$$V(S,T) = (E-S)^+, \qquad (2.24)$$

which is depicted on Fig. 2.8 (right).

2.3.2. Pay–off Diagrams for Combined Option Strategies

A *bullish spread* option strategy is a combination of purchasing and selling two call options written on the same underlying asset, one with a lower and one with a higher strike price,

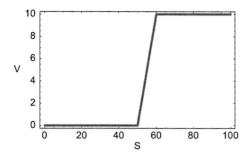 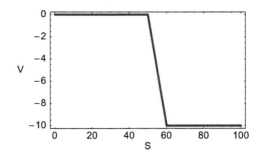

Figure 2.9. Terminal pay–off diagrams of bullish (left) and bearish (right) spreads with the exercise (strike) prices $E_1 = 50, E_2 = 60$.

$E_1 < E_2$. The pay–off diagram is given by

$$V(S,T) = (S-E_1)^+ - (S-E_2)^+. \qquad (2.25)$$

On the other hand, a *bearish spread* strategy is a combination of selling and purchasing two call options written on the same stock, one with a lower and one with a higher strike price, $E_1 < E_2$. The pay–off diagram is given by

$$V(S,T) = -(S-E_1)^+ + (S-E_2)^+. \qquad (2.26)$$

Bullish spread is an option strategy often used in a situation when we expect the growth of the underlying asset price. In the case of a growth of S, it is possible to achieve a defined profit. On the other hand, bearish spread is used when a fall of the asset price is expected. From the pay–off diagrams of both spread strategies it follows that the possible profit and loss are bounded. This is an important property of spreads preferred by investors.

The option strategy *bought straddle* consists in purchasing one call option and one put option written on the same underlying asset, with the same strike prices E and the same maturity T. By employing this strategy the investor is able to limit the loss if the asset price is the strike price. In the opposite case a possible profit is high. A mathematical description of the terminal pay–off condition at the expiration time $t = T$ is:

$$V(S,T) = (S-E)^+ + (E-S)^+. \qquad (2.27)$$

A *sold straddle* option strategy is a combination of selling one call and one put option written on the same underlying asset, with the same strike prices E and the same maturity. Its pay–off is given by

$$V(S,T) = -(S-E)^+ - (E-S)^+. \qquad (2.28)$$

A *butterfly* option strategy is a combined strategy that consists of purchasing two call options, one with a lower strike E_1 and the other one with a higher E_4 strike price and selling two call options with the same strike prices $E_2 = E_3$, where $E_1 < E_2 = E_3 < E_4$. Moreover, we assume $E_1 + E_4 = E_2 + E_3 = 2E_2$. All the options are written on the same

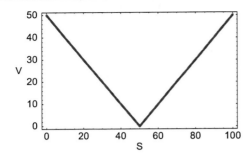 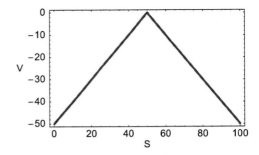

Figure 2.10. Terminal pay–off diagrams of bought (left) and sold (right) straddle strategies for the strike price $E = 50$.

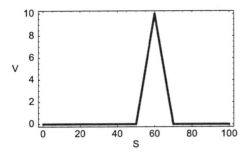 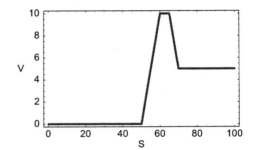

Figure 2.11. Terminal pay–off diagrams of butterfly (left) and condor (right) strategies for strike prices $E_1 = 50, E_2 = E_3 = 60, E_4 = 70$ (butterfly), respectively $E_1 = 50, E_2 = 60, E_3 = 65, E_4 = 70$ (condor).

underlying asset and have the same maturities. The butterfly strategy is based on the investor's expectation about the price stability, it gains a maximal profit if the asset price is in the neighborhood of the value $E_2 = E_3$.

$$V(S,T) = (S - E_1^+) - (S - E_2)^+ - (S - E_3)^+ + (S - E_4)^+ . \qquad (2.29)$$

A *condor* option strategy is a strategy similar to butterfly, but the difference is that the strike prices of sold call options need not be equal, $E_2 \neq E_3$, i.e., $E_1 < E_2 < E_3 < E_4$. The mathematical expression of the terminal pay–off condition is given by the formula (2.29). The condor strategy gains a maximal profit if the asset price is in the interval (E_2, E_3).

A *bought strangle* strategy is a combination of purchasing one call and one put option, where the call option is written on a higher strike price E_2 and put option on a lower strike price E_1, $E_1 < E_2$. Its terminal pay–off diagram is therefore:

$$V(S,T) = (S - E_2)^+ + (E_1 - S)^+ . \qquad (2.30)$$

A *sold strangle* strategy is a combination of selling one call and one put option, where the call option is written on a higher strike price E_2 and put option on a lower strike price E_1, $E_1 < E_2$. The maximal profit of this strategy is achieved in the case the underlying asset

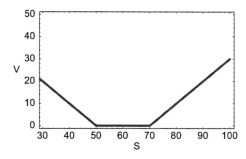 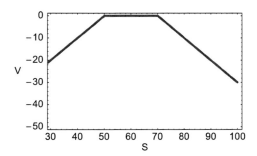

Figure 2.12. Terminal pay–off diagrams for the bought (left) and sold (right) strangle strategies for the strike prices $E_1 = 50, E_2 = 70$.

price is in the interval (E_1, E_2). Its terminal pay–off diagram is given by:

$$V(S,T) = -\max(S - E_2, 0) - \max(E_1 - S, 0). \tag{2.31}$$

At the end of this brief overview of the most common option strategies, we also recall three other simple derivatives that belong to a category of the so–called binary options.

The *cash-or-nothing* strategy is a kind of an option "bet", in which the holder of such an option receives a fixed predetermined amount of money (e.g., one unit amount of money) in the case the underlying asset price S exceeds the given strike price E. It means that the terminal pay–off diagram of such a strategy is:

$$V(S,T) = 1 \quad \text{if } S \geq E, \qquad V(S,T) = 0 \quad \text{if } S < E, \tag{2.32}$$

(see Fig. 2.13).

An *asset-or-nothing* strategy is again a kind of an option "bet", in which the holder of such an option receives the asset value S, provided it exceeds the predetermined strike price E. It means that the terminal pay–off diagram is the following function:

$$V(S,T) = S, \quad \text{if } S \geq E, \qquad V(S,T) = 0, \quad \text{if } S < E, \tag{2.33}$$

(see Fig. 2.13).

A *digital* option is a kind of an option "bet", in which the holder of the option receives a fixed amount (one unit of money) in the case the underlying asset price S belongs to the interval (E_1, E_2). It means that the terminal pay–off diagram is given by:

$$V(S,T) = 1, \quad \text{if } S \in (E_1, E_2), \qquad V(S,T) = 0 \quad \text{otherwise}. \tag{2.34}$$

We refer the reader for other important aspects of option pricing and practical usage of various option strategies to books and lecture notes of Hull [65], Wilmott, Dewynne and Howison [122], Melichercík *et al.* [83], Baxter and Rennie [13].

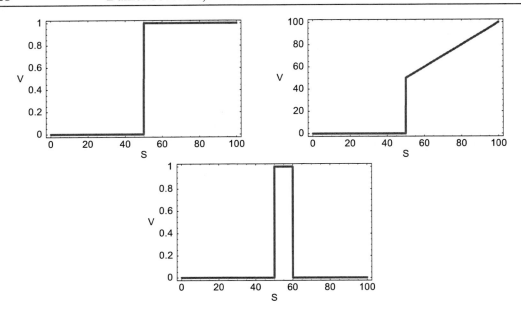

Figure 2.13. Terminal pay–off diagrams of Cash-or-nothing (left) and Asset-or-nothing (right) option strategies for the strike price $E = 50$ and a digital option (bottom) with $E_1 = 50$.

2.4. Boundary Conditions for Derivative Prices

When analyzing partial differential equations of the Black–Scholes type, it is often necessary to prescribe initial (terminal) conditions and boundary conditions. Terminal conditions were already discussed in the previous section and they depend on a chosen option strategy. The aim of this section is to show how to prescribe boundary conditions for basic types of options, such as call and put options.

2.4.1. Boundary Conditions for Call and Put Options

First, we discuss the boundary conditions for a European call options. The domain of possible values of the underlying asset variable S is the interval $[0, \infty)$. A boundary condition for the call option price at the left end $S = 0$ follows from a simple reasoning: the option on the asset that reached its default value $S = 0$ is zero as well, i.e., $V_{ec}(0,t) = 0$. On the other hand, for large values of the assets $S \to \infty$ the option price should approach the value S. More precisely, the option price V for large values of S has to approach the asset price. More precisely, the boundary conditions for European call option on the asset paying no dividends can be stated as follows:

$$V_{ec}(0,t) = 0, \qquad \lim_{S \to \infty} \frac{V_{ec}(S,t)}{S} = 1, \qquad (2.35)$$

for all $t \in (0,T)$. If the underlying asset pays continuous dividends with a dividend yield rate $q \geq 0$ then the boundary conditions have to have the form:

$$V_{ec}(0,t) = 0, \qquad \lim_{S \to \infty} \frac{V_{ec}(S,t)}{Se^{-q(T-t)}} = 1, \qquad (2.36)$$

for all $t \in (0,T)$. It means that the option price for large values of the underlying asset price S approaches the discounted asset price $Se^{-q(T-t)}$.

The boundary conditions for a European put option follow from similar reasoning as in the case of a call option. For large asset prices, the put option loses its value and hence $V_{ep}(\infty,t) = 0$. On the other hand, for the asset in the default state $S = 0$, the value of the put option equals the strike price discounted by the interest rate $r > 0$, i.e.

$$V_{ep}(0,t) = Ee^{-r(T-t)}. \qquad (2.37)$$

In summary, the boundary conditions for a put option can be written as follows:

$$V_{ep}(0,t) = Ee^{-r(T-t)}, \qquad \lim_{S \to \infty} V_{ep}(S,t) = 0, \qquad (2.38)$$

for all $t \in (0,T)$, regardless of the fact whether the asset pays or does not pay dividends.

2.4.2. Boundary Conditions for Combined Option Strategies

If our task is to determine boundary conditions for other option strategies different from plain vanilla call or put options, then we have to decompose the terminal pay–off diagram (for exercise time $t = T$) of the given strategy into a linear combination of several call respectively put options. Then the boundary conditions for remaining times $t \in (0,T)$ will be exactly the same linear combination of boundary conditions of call and put options forming the given combined option strategy.

Problem Section and Exercises

1. In a more detail, derive the formula for the dispersion of a geometric Brownian motion (2.4), i.e.
$$Var(Y(t)) = y_0^2 e^{2\mu t + \sigma^2 t}(e^{\sigma^2 t} - 1).$$

2. Similarly as in the case of a Brownian motion, the geometric Brownian motion can be also decomposed into its deterministic part and fluctuations part. Formally, by differentiating the expression $Y(t) = y_0 e^{X(t)}$, where $dX(t) = \mu dt + \sigma dw(t)$ we obtain that for the increments $dY(t)$ it holds: $dY = y_0 e^X dX = Y dX$. Hence we formally obtain: $dY(t) = \mu Y(t)dt + \sigma Y(t)dw(t)$ should be satisfied. This derivation is however in contradiction with (2.4). In which point of formal derivation we made a mistake?

3. Let X be a Brownian motion with parameters μ and σ, i.e., $dX(t) = \mu dt + \sigma dw(t)$. Derive a stochastic differential equation for the following functions of the variable X and time t: $f(x,t) = x^2$; $f(x,t) = e^{x+t}$; $f(x,t) = \ln(1+x^2)$.

4. Suppose that the underlying asset price follows a geometric Brownian motion with parameters $\mu = 0.3750$ and $\sigma^2 = 0.0669$. These are estimates of parameters of the geometric Brownian motion from daily Google stock prices, from February 7, 2007 to February 7, 2008. On February 14, 2007 the stock price was USD 459.1.

   - Find the expected value and draw a graph of the probability density function of the asset price in May 7, 2007
   - What is the probability that the asset price after one year will be lower than the present price?
   - What is the probability that after a half year the asset price will be higher than USD 600?

5. The constract between the writer of an option and the buyer (holder) is like follows: The option gives the holder to buy a stock for the strike price E_1 at the expiration time T. In the moment of exercising the option, the holder has to sell to the writer the given stock for the strike price E_2 where $E_1 < E_2$. What is the pay-off diagram of this strategy at the expiration time T? To which known strategy it is similar to? Can this strategy be realized using plain vanilla call or put options?

6. From Itō's lemma we know that the differential of the process $x(t) = \exp(w(t))$ is not just the term $x(t)dw(t)$. Find a function $x(t) = f(w(t),t)$, for which $dx(t) = x(t)dw(t)$. Can f depend on w only?

7. Let the asset price S satisfy the stochastic differential equation $dS = \mu S dt + \sigma S dw$. Find a stochastic differential equation that is satisfied by its present discounted value $\tilde{S}(t) = e^{-rt}S(t)$.

8. What is the dependence of the price of a European call or put option on the exercise (strike) price? Is it an increasing or a decreasing function?

9. Denote by $c(S,E,\tau)$, $p(S,E,\tau)$ the prices of European call and put options with the strike price E where the underlying asset price is S and the time remaining to maturity is $\tau = T - t$. The riskless interest rate is $r > 0$. By constructing a suitable portfolio and eliminating the arbitrage opportunity prove the following properties:

   (a) $c(S,E_1,\tau) \leq c(S,E_2,\tau)$ for $E_2 \geq E_1$,
   (b) $p(S,E_1,\tau) \leq p(S,E_2,\tau)$ for $E_1 \geq E_2$,
   (c) $S - Ee^{-r\tau} \leq c(S,E,\tau) \leq S$,
   (d) $(E_2 - E_1)e^{-r\tau} \leq c(S,E_1,\tau) - c(S,E_2,\tau)$ for $E_1 \geq E_2$,
   (e) $c(S,E,\tau)$ and $p(S,E,\tau)$ are convex functions of strike price E.

10. Assign the following processes to their trajectories shown by the graphs in Fig. 2.14:

    (a) $x_1(t) = 5 + 2t + 3w(t)$,

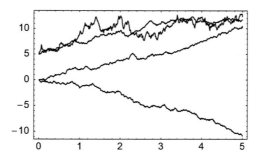

Figure 2.14. Trajectories of random processes.

(b) $x_2(t) = -2t + w(t)$,

(c) $x_3(t) = 5 + 2t + w(t)$,

(d) $x_4(t) = 2t + w(t)$.

11. Find the probability distribution of the following random variables:

   (a) $w(2)$,

   (b) $w(5) - w(3)$,

   (c) $x(t) = 1 + 2t + 3w(t)$, where t is an arbitrary positive number, i.e., the distribution function for the value of the process $x(t) = 2t + 3w(t)$ at time t),

   (d) $w(1) + w(2)$ (note that $w(1)$ and $w(2)$ are not independent),

   (e) $\exp(t + 0.3w(t))$ where t is an arbitrary positive number,

   (f) $\exp(0.05w(t))$ where t is an arbitrary positive number.

12. Suppose that the underlying asset price S is given by the formula $S(t) = S(0)e^{\mu t + \sigma w(t)}$ where $\mu > 0$. Consider the asset price after t years and the probability that it will be higher than the present price. How does this probability depend on t?

Chapter 3

European Style of Options

The aim of this chapter is to derive explicit formulae for pricing European style of options. The basic characteristic of the European style of option contracts is the fact that they can be exercised at the predetermined exercise (maturity) time $t = T$ only. We show that, for this type of option constracts, it is possible to derive an explicit formula for a solution to the Black–Scholes partial differential equation for pricing options. First, we concentrate on plain vanilla call and put options. Then we show how to extend the call/put option pricing formulae to combined option strategies studied in the previous chapter.

3.1. Pricing Plain Vanilla Call and Put Options

With regard to the Black–Scholes model for pricing derivative securities (see previous chapter), the partial differential equation describing the evolution of the price of an option written on a stock paying continuous dividends has the following form

$$\frac{\partial V}{\partial t} + \frac{1}{2}\sigma^2 S^2 \frac{\partial^2 V}{\partial S^2} + (r-q)S\frac{\partial V}{\partial S} - rV = 0, \tag{3.1}$$

$$V(S,T) = \bar{V}(S), \quad S > 0, \ t \in [0,T].$$

The meaning of respective variables is the following: $V = V(S,t)$ is the price of a European option written on the underlying asset having $S > 0$ its present spot price at a time $t \in [0,T]$ and $T > 0$ is the expiration time of the option. The remaining model parameters are: $\sigma > 0$ is a volatility of the stock, i.e., the standard deviation of the stochastic time evolution of the underlying stock prices, $r > 0$ is an interest rate a continuously compounded riskless zero-coupon bond and q is a continuous annualized dividend yield paid by the stock.

Finally, we recall that, in the case of European call option, the terminal condition $\bar{V}(S) = V(S,T)$ at the expiration time is given by the function

$$\bar{V}(S) = (S-E)^+ = \begin{cases} S-E, & \text{for } S \geq E \\ 0, & \text{for } 0 < S < E, \end{cases}$$

where E is the expiration (strike) price at which the option contract is signed. In the case of a put option, the terminal condition reads as follows:

$$\bar{V}(S) = (E-S)^+ = \begin{cases} E-S, & \text{for } 0 < S \le E \\ 0, & \text{for } E < S. \end{cases}$$

The main idea of construction of an explicit solution to equation (3.1) with a given terminal condition consists in a sequence of transformations of this equation into a basic form of a parabolic equation

$$\frac{\partial u}{\partial t} - a^2 \frac{\partial^2 u}{\partial x^2} = 0, \quad (x,t) \in (-\infty, \infty) \times [0,T],$$

with the prescribed initial condition.

1. step - Transformation of time. We transform the time $t \in [0,T]$ such that it flows just in the opposite direction, i.e., from the expiration time T to the initial time $t = 0$. To this end, we introduce a new variable $\tau = T - t$ and set

$$W(S, \tau) = V(S, T - \tau), \quad \text{and so } V(S,t) = W(S, T-t).$$

Using the relation $dt = -d\tau$ equation (3.1) is transformed into:

$$\frac{\partial W}{\partial \tau} - \frac{1}{2}\sigma^2 S^2 \frac{\partial^2 W}{\partial S^2} - (r-q)S \frac{\partial W}{\partial S} + rW = 0, \qquad (3.2)$$

$$W(S,0) = \bar{V}(S), \quad S > 0, \ \tau \in [0,T].$$

2. step - The logarithmic transformation of the underling stock price. It consists in the substitution $S = e^x$, $x = \ln S$ and introducing a new function

$$Z(x,\tau) = W(e^x, \tau), \quad \text{and so } W(S,\tau) = Z(\ln S, \tau).$$

Notice that $S \in (0, \infty)$ if and only if $x \in (-\infty, \infty)$. Using the chain rule for differentiation we obtain

$$\frac{\partial Z}{\partial x} = S \frac{\partial W}{\partial S}, \quad \frac{\partial^2 Z}{\partial x^2} = S^2 \frac{\partial^2 W}{\partial S^2} + S \frac{\partial W}{\partial S} = S^2 \frac{\partial^2 W}{\partial S^2} + \frac{\partial Z}{\partial x}.$$

Equation (3.2) can be then rewritten in the form:

$$\frac{\partial Z}{\partial \tau} - \frac{1}{2}\sigma^2 \frac{\partial^2 Z}{\partial x^2} + \left(\frac{\sigma^2}{2} - r + q\right)\frac{\partial Z}{\partial x} + rZ = 0,$$

$$Z(x,0) = \bar{V}(e^x), \quad -\infty < x < \infty, \ \tau \in [0,T].$$

3. step - Transformation into the basic parabolic partial differential equation $\frac{\partial u}{\partial t} - a^2 \frac{\partial^2 u}{\partial x^2} = 0$. Terms containing the lower order derivatives Z and $\frac{\partial Z}{\partial x}$ can be eliminated by an exponential transformation

$$u(x,\tau) = e^{\alpha x + \beta \tau} Z(x,\tau), \quad \text{i.e. } Z(x,\tau) = e^{-\alpha x - \beta \tau} u(x,\tau),$$

where constants α, β will be specified later. We obtain

$$\frac{\partial Z}{\partial x} = e^{-\alpha x - \beta \tau} \left(\frac{\partial u}{\partial x} - \alpha u \right),$$

$$\frac{\partial^2 Z}{\partial x^2} = e^{-\alpha x - \beta \tau} \left(\frac{\partial^2 u}{\partial x^2} - 2\alpha \frac{\partial u}{\partial x} + \alpha^2 u \right),$$

$$\frac{\partial Z}{\partial \tau} = e^{-\alpha x - \beta \tau} \left(\frac{\partial u}{\partial \tau} - \beta u \right).$$

For the new transformed function u we may therefore conclude that it is a solution to the partial differential equation

$$\frac{\partial u}{\partial \tau} - \frac{\sigma^2}{2} \frac{\partial^2 u}{\partial x^2} + A \frac{\partial u}{\partial x} + Bu = 0,$$

$$u(x, 0) = e^{\alpha x} \bar{V}(e^x),$$

where the coefficients A, B satisfy

$$A = \alpha \sigma^2 + \frac{\sigma^2}{2} - r + q, \quad a \ B = (1 + \alpha) r - \beta - \alpha q - \frac{\alpha^2 \sigma^2 + \alpha \sigma^2}{2}.$$

By a simple algebraic computation, we find that the constants α, β can be chosen in such a way that the terms A, B are vanishing. Indeed,

$$\alpha = \frac{r - q}{\sigma^2} - \frac{1}{2}, \quad \beta = \frac{r + q}{2} + \frac{\sigma^2}{8} + \frac{(r - q)^2}{2\sigma^2}. \tag{3.3}$$

With this choice of coefficients α, β, the resulting equation for the function u has the form

$$\frac{\partial u}{\partial \tau} - \frac{\sigma^2}{2} \frac{\partial^2 u}{\partial x^2} = 0, \tag{3.4}$$

$$u(x, 0) = e^{\alpha x} \bar{V}(e^x), \quad -\infty < x < \infty, \tau \in [0, T].$$

4. step - Applying the Green formula for a solution to the heat equation. It is well-known fact (see e.g., [105, Theorem 4.1.1]), that equation (3.4) has an explicit solution $u(x, \tau)$, which can be written as a convolution of the initial condition with the Green function. The explicit formula has the form of an integral:

$$u(x, \tau) = \frac{1}{\sqrt{2\sigma^2 \pi \tau}} \int_{-\infty}^{\infty} e^{-\frac{(x-s)^2}{2\sigma^2 \tau}} u(s, 0) \, ds.$$

Now, by a sequence of backward substitutions $u \mapsto Z \mapsto W \mapsto V$, we finally obtain:

$$V(S, T - \tau) = e^{-\beta \tau} e^{-\alpha \ln S} u(\ln S, \tau),$$

and hence

$$V(S, T - \tau) = \frac{e^{-\beta \tau}}{\sqrt{2\sigma^2 \pi \tau}} S^{-\alpha} \int_{-\infty}^{\infty} e^{-\frac{(\ln S - s)^2}{2\sigma^2 \tau}} e^{\alpha s} \bar{V}(e^s) \, ds. \tag{3.5}$$

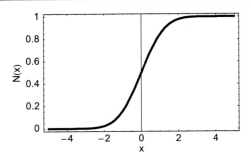
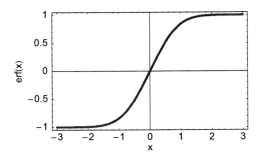

Figure 3.1. A graph of the cumulative distribution function $N(x)$ and the error function erf(x) of a normal distribution.

For the European call option we have $\bar{V}(S) = (S-E)^+$ and so the relation (3.5) can be further simplified as follows:

$$V(S, T-\tau) = \frac{e^{-\beta\tau}}{\sqrt{2\sigma^2\tau}} S^{-\alpha} \frac{1}{\sqrt{\pi}} \int_{\ln E}^{\infty} e^{-\frac{(\ln S - s)^2}{2\sigma^2\tau}} e^{\alpha s}(e^s - E)\,ds.$$

The substitution $y = s - \ln S$ leads to:

$$V(S, T-\tau) = \frac{e^{-\beta\tau}}{\sqrt{2\sigma^2\tau}} \frac{1}{\sqrt{\pi}} \int_{-\ln\frac{S}{E}}^{\infty} e^{-\frac{y^2}{2\sigma^2\tau}} \left(S e^{(1+\alpha)y} - E e^{\alpha y} \right) dy. \quad (3.6)$$

A practical computation using the above formula requires rewriting the price V into a form containing elementary or special functions.

Recall that the cumulative distribution function $N(x)$ and the error function erf(x) of the normal distribution are defined by means of the Euler integral as follows:

$$N(x) = \frac{1}{\sqrt{2\pi}} \int_{-\infty}^{x} e^{-\frac{\xi^2}{2}} d\xi, \quad \frac{1-\text{erf}(x)}{2} = \frac{1}{\sqrt{\pi}} \int_{x}^{\infty} e^{-\xi^2} d\xi. \quad (3.7)$$

The following identities are useful:

$$\text{erf}(-x) = -\text{erf}(x), \quad \frac{1}{2}\left(1 + \text{erf}\left(\frac{x}{\sqrt{2}}\right)\right) = \frac{1}{\sqrt{2\pi}} \int_{-\infty}^{x} e^{-\xi^2/2} d\xi = N(x).$$

for each $x \in \mathbb{R}$.

Let us consider the integral

$$I_1 = \frac{e^{-\beta\tau}}{\sqrt{2\sigma^2\tau}} \frac{1}{\sqrt{\pi}} \int_{-\ln\frac{S}{E}}^{\infty} e^{-\frac{y^2}{2\sigma^2\tau} + (1+\alpha)y} dy.$$

By introducing a substitution $\xi = \frac{y}{\sqrt{2\sigma^2\tau}} - \frac{1+\alpha}{2}\sqrt{2\sigma^2\tau}$ and using relations (3.3) we obtain

$$-\frac{y^2}{2\sigma^2\tau} + (1+\alpha)y = -\xi^2 + (1+\alpha)^2 \frac{\sigma^2\tau}{2} = -\xi^2 + (\beta - q)\tau.$$

Hence

$$I_1 = e^{-q\tau} \frac{1}{\sqrt{\pi}} \int_{-\frac{1+\alpha}{2}\sqrt{2\sigma^2\tau}-\frac{\ln\frac{S}{E}}{\sqrt{2\sigma^2\tau}}}^{\infty} e^{-\xi^2} d\xi$$

$$= \frac{e^{-q\tau}}{2}\left[1 - \operatorname{erf}\left(-\frac{1+\alpha}{2}\sqrt{2\sigma^2\tau} - \frac{\ln\frac{S}{E}}{\sqrt{2\sigma^2\tau}}\right)\right]$$

$$= \frac{e^{-q\tau}}{2}\left[1 + \operatorname{erf}\left(\frac{1}{\sqrt{2}}\frac{(r-q+\frac{\sigma^2}{2})\tau + \ln\frac{S}{E}}{\sigma\sqrt{\tau}}\right)\right].$$

By introducing another substitution $\xi = \frac{y}{\sqrt{2\sigma^2\tau}} - \frac{\alpha}{2}\sqrt{2\sigma^2\tau}$ we can compute the integral

$$I_2 = \frac{e^{-\beta\tau}}{\sqrt{2\sigma^2\tau}}\frac{1}{\sqrt{\pi}} \int_{-\ln\frac{S}{E}}^{\infty} e^{-\frac{y^2}{2\sigma^2\tau}+\alpha y} dy.$$

Using $\frac{\alpha^2}{2}\sigma^2 = \beta - r$ we simplify the transformed integral using the error function erf as follows:

$$I_2 = \frac{e^{-r\tau}}{2}\left[1 + \operatorname{erf}\left(\frac{1}{\sqrt{2}}\frac{(r-q-\frac{\sigma^2}{2})\tau + \ln\frac{S}{E}}{\sigma\sqrt{\tau}}\right)\right].$$

Substituting above results for the integrals I_1 and I_2 we can finally state the formula (3.6) for the call option price $V(S,0)$:

$$V(S, T-\tau) = \frac{Se^{-q\tau}}{2}\left[1 + \operatorname{erf}\left(\frac{1}{\sqrt{2}}\frac{(r-q+\frac{\sigma^2}{2})\tau + \ln\frac{S}{E}}{\sigma\sqrt{\tau}}\right)\right]$$
$$- \frac{Ee^{-r\tau}}{2}\left[1 + \operatorname{erf}\left(\frac{1}{\sqrt{2}}\frac{(r-q-\frac{\sigma^2}{2})\tau + \ln\frac{S}{E}}{\sigma\sqrt{\tau}}\right)\right].$$

Using the relations between functions $N(x)$ and $\operatorname{erf}(x)$ we finally conclude

$$V(S,t) = Se^{-q(T-t)}N(d_1) - Ee^{-r(T-t)}N(d_2), \tag{3.8}$$

where

$$d_1 = \frac{(r-q+\frac{\sigma^2}{2})(T-t) + \ln\frac{S}{E}}{\sigma\sqrt{T-t}}, \quad d_2 = d_1 - \sigma\sqrt{T-t}. \tag{3.9}$$

Expression (3.8) is called the Black–Scholes formula for pricing European call options. All parameters appearing in the formula should be known in advance. Here is a typical example of computation of an option price:

Example. The present spot price of the underlying IBM stock paying no dividends is $S = 58.5$ USD. The historical volatility of the asset price process has been estimated to the value $\sigma = 29\%$, i.e., $\sigma = 0.29$. The annualized interest rate of zero-coupon bonds is $r = 4\%$, i.e., $r = 0.04$. We make a derivative contract of the call option type for the exercise price $E = 60$ USD at the expiration time $T = 0.3$ years. Substituting these variables into the Black–Scholes formula we obtain that the option price $V = V(58.5, 0)$ is approximately 3.348 USD. Fig. 3.2 and Fig. 3.3 show the prices of call and put options $V(S, 0)$ as functions of IBM stock price S.

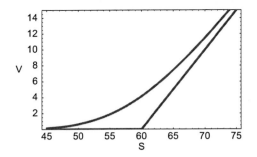 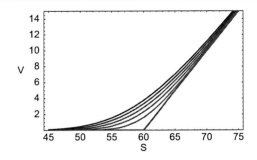

Figure 3.2. A graph of a solution for pricing the call option $V(S,0)$ together with its terminal condition $V(S,T)$ (left). Solutions $V(S,t)$ are depicted in various times $T-t$ remaining to expiration (right). Model parameters were chosen as: $E=60, \sigma=0.29, r=0.04, q=0, T=0.3$.

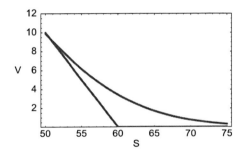 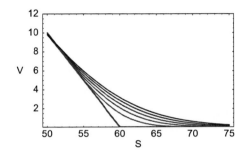

Figure 3.3. A graph of a solution for pricing the put option $V(S,0)$ and its terminal condition of $V(S,T)$ (left). Solutions $V(S,t)$ are depicted in selected times $T-t$ to expiration (right). Model parameters were chosen as: $E=60, \sigma=0.29, r=0.04, q=0, T=0.3$.

3.2. Pricing Put Options Using Call Option Prices and Forwards, Put-Call Parity

The procedure of derivation of the explicit formula for pricing a European call option from the previous section could be easily adapted also to the case of pricing a European put option. However, a more elegant way is to make use of the so-called put–call parity. The main idea behind derivation of the put–call parity is rather simple. Suppose that our portfolio consists of one call option held in the long position and one sold put option in the short position. It means that our combined option strategy for such a portfolio (denoted by V_f) has the terminal pay–off diagram

$$V_f(S,T) = V_{ec}(S,T) - V_{ep}(S,T).$$

An easy calculation yields the identity

$$V_f(S,T) = (S-E)^+ - (E-S)^+ = S - E. \tag{3.10}$$

Because of linearity of the Black–Scholes equation and uniqueness of its solution subject to a given terminal condition, we can conclude that a difference of two solutions is again a solution. Hence, for any time $t \in (0,T)$, the derivative security V_f satisfies the relationship:

$$V_f(S,t) = V_{ec}(S,t) - V_{ep}(S,t).$$

On the other hand, it should be emphasized that pricing the derivative having its pay–off diagram $V_f(S,T) = S - E$ is rather simple. In fact, it is just a difference between the asset price S and a forward contract with the expiration price E, i.e., a right, but also an obligation, to buy an asset for the expiration price E at expiration time T. As such, its price is therefore a difference of the underlying asset price S discounted by a dividend yield q and the expiration price discounted by the interest rate r, i.e.

$$V_f(S,t) = Se^{-q(T-t)} - Ee^{-r(T-t)}. \tag{3.11}$$

This property can be also verified mathematically. Substituting the function $(S,t) \mapsto Se^{-q(T-t)} - Ee^{-r(T-t)}$ into the Black–Scholes equation we easily see that it is indeed a solution satisfying the terminal condition $V_f(S,T) = S - E$. Since $V_f = V_{ec} - V_{ep}$, we obtain the relation between European call and put option in the form

$$V_{ec}(S,t) - V_{ep}(S,t) = Se^{-q(T-t)} - Ee^{-r(T-t)}, \tag{3.12}$$

which is also known as the put–call parity. Now, it follows from the put–call parity that the price of a European put option is given by the formula

$$V_{ep}(S,t) = V_{ec}(S,t) - Se^{-q(T-t)} + Ee^{-r(T-t)}, \tag{3.13}$$

which can be further simplified, by using a simple property of the cumulative distribution function of normal distribution

$$N(-d) = 1 - N(d),$$

to the form

$$V_{ep}(S,t) = Ee^{-r(T-t)}N(-d_2) - Se^{-q(T-t)}N(-d_1), \tag{3.14}$$

where the coefficients d_1, d_2 are given by (3.9).

At the end of this section, we mention an interesting symmetry between European call and put options prices given by formulae (3.8) and (3.14). Let us denote by $V_{ec}(S,t;E,r,q)$ and $V_{ep}(S,t;E,r,q)$ the functions given as in (3.8) and (3.14). Then, it follows from the relations (3.9) between coefficients d_1, d_2 and explicit formulae for pricing call and put options that the put-call symmetry

$$V_{ep}(S,t;E,r,q) = V_{ep}(E,t;S,q,r) \tag{3.15}$$

is satisfied. This symmetry can be verbally explained as a transformation of call to put option when we simultaneously exchange the underlying asset price and exercise price $S \leftrightarrow E$, as well as the interest rate and dividend yield $r \leftrightarrow q$. This relation is a straightforward consequence of explicit pricing formulae (3.8), (3.14) and transformation $d_1 \leftrightarrow -d_2$ when we exchange $S \leftrightarrow E$ and $r \leftrightarrow q$.

3.3. Pricing Combined Options Strategies: Spreads, Straddles, Condors, Butterflies and Digital Options

The aim of this section is to illustrate solutions corresponding to prices of selected option strategies. In particular, we will consider bullish spreads, bought strangles, condors and butterflies. We remind ourselves that, with regard to linearity of the Black–Scholes parabolic partial differential equation, the superposition of solutions is possible. As a consequence of this property, pricing formulae for combined option strategies, whose terminal pay-off diagrams are linear combinations of call and/or put options, turn to be the same linear combinations of corresponding call and/or put option prices given in (3.8) and (3.14) with respective exercise prices. Several illustrative graphs of the solution $V(S,0)$ for pricing a given option strategy together with terminal pay–off condition $V(S,T)$ are shown in Fig. 3.4–3.7 (left). In figures placed in the right position we furthermore show intermediate solution $S \mapsto V(S,t)$ for several times $t \in [0,T]$.

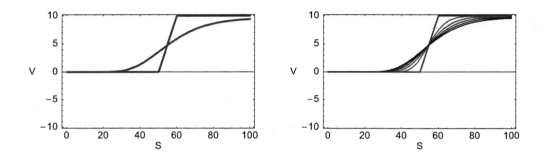

Figure 3.4. A graph of a solution for pricing the bullish spread option strategy. The parameters were chosen as: $E_1 = 50; E_2 = 60; \sigma = 0.29; r = 0.04; q = 0; T = 0.3$.

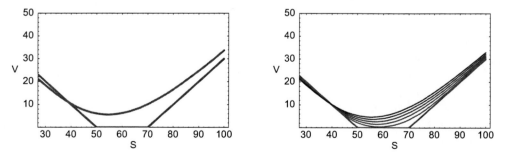

Figure 3.5. A graph of a solution for pricing the bought strangle option strategy. The parameters were chosen as: $E_1 = 50; E_2 = 70; \sigma = 0.29; r = 0.04; q = 0; T = 1$.

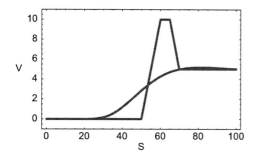 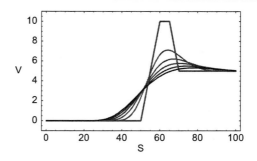

Figure 3.6. A graph of a solution for pricing the condor option strategy. The model parameters were chosen as: $E_1 = 50, E_2 = 60, E_3 = 65, E_4 = 70, \sigma = 0.29, r = 0.04, q = 0, T = 0.3$.

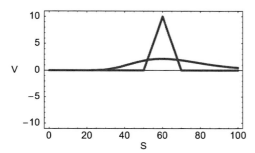 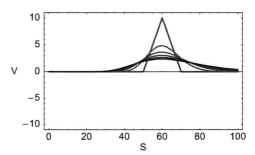

Figure 3.7. A graph of solution to pricing butterfly option strategy. The parameters were chosen as: $E_1 = 50, E_2 = E_3 = 60, E_4 = 70, \sigma = 0.29, r = 0.04, q = 0, T = 0.3$.

3.4. Comparison of Theoretical Pricing Results to Real Market Data

In this section we present practical comparison of the theoretical Black–Scholes formula for pricing options to real financial market data. As examples we have chosen the IBM and Microsoft underlying asset and their respective call option prices. By these examples we try to convince the reader that as soon as we know the underlying stock spot price and model parameters $r > 0$ (interest rate of a zero-coupon bond) and $\sigma > 0$ (volatility of the asset price) then we are in a position to price the call (respectively put) options with the exercise price E and the expiration time T.

In Fig. 3.8 we depict intraday stochastic evolution of the IBM and Microsoft underlying stock prices from May 22, 2002. The horizontal axis is depicted in the time scale of minutes. Next, in Fig. 3.9 (left) we show evolution of bid (offers to buy) and ask (offers to sell) prices of the call option with the expiration time on June 2, 2002. In the right picture we compare the computed price by the Black–Scholes formula and the mean value of bid and ask prices. The theoretical Black–Scholes values of the call option were computed for the following model parameters:

- interest rate $r = 5\%$,

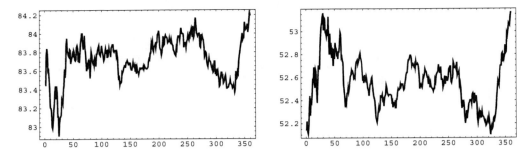

Figure 3.8. Time evolution (in the scale of minutes) of stock prices of IBM (left) and Microsoft (right) from May 22, 2002.

- expiration price $E = 70$,
- expiration time $T = 11$ days,
- volatility $\sigma = 41\%$.

We assumed that the underlying stock does not pay dividends. The comparison of real and theoretical data for the case of Microsoft is shown in Fig. 3.10. In this case we have assumed the following parameters:

- interest rate $r = 5\%$,
- expiration price $E = 35$,
- expiration time $T = 11$ days,
- volatility $\sigma = 61\%$.

It should be obvious that the theoretical Black–Scholes call option price is in a good agreement with its market value. However, it should be emphasized that the most disputable part of the computation consits in the appropriate choice of the volatility parameter σ. Notice that, in order to achieve market option values, the volatility σ had to be chosen to be much higher that it could be indicated from analysis of historical values of the underlying asset. We will address this problem in the next chapter. It is related to the phenomenon of the so–called implied volatility.

3.5. Black–Scholes Equation for Pricing Index Options

In this section we turn our attention to the problem of pricing options on underlying virtual assets represented by stock indices. In what follows, we will derive a multidimensional generalization of the Black–Scholes parabolic equation whose solution represents the price of a call or put option on a stock index.

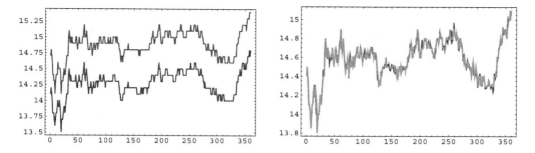

Figure 3.9. An example of comparison of theoretical Black–Scholes prices of the IBM call option price with real market call option data. Evolution of bid and ask option prices (left) and comparison of their mean with the price computed by the Black–Scholes formula (right).

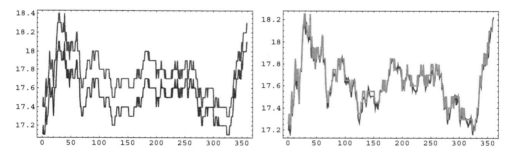

Figure 3.10. An example of comparison of theoretical computation of a call Microsoft option price with real market data. Evolution of bid and ask option prices (left) and comparison of their mean with computed price by means of the Black–Scholes formula (right).

Let us recall that the value of an index depends on n underlying asset prices $S_i, i = 1, \ldots, n$. Its value I can be defined as a weighted sum

$$I = \sum_{i=1}^{n} a_i S_i,$$

where $a_i > 0$ are prescribed positive weights corresponding to the index definition. For example, the Dow Jones Industrial Average (DJIA) is a price-weighted index. It consists of 30 stocks of great American companies representing the U.S. economy. The Index includes a wide range of companies from financial services over computer companies to retail companies. The Dow Jones Industrial Average can be calculated by the following formula:

$$DJIA = \sum_{i=1}^{30} S_i \qquad (3.16)$$

where S_i is the current market price of the i-th stock. Another popular index is the Standard & Poor's 500 Index. It is a capitalization-weighted index consisting of 500 stocks. It is intended to represent a sample of leading companies in leading industries within the U.S. economy. The formula for evaluation of the S&P 500 reads as:

$$SP = \sum_{i=1}^{500} a_i S_i, \qquad (3.17)$$

where S_i is the current market price of i-th stock and a_i represents the market capitalization of the i-th asset in the market.

The exact market definitions of the DJIA as well as Standard & Poor's 500 Index are divided by a common divisor in order to adjust the index value for events leading to no change in company's value but otherwise influencing the index, like e.g., stock splittings, etc.

In general, index derivatives are contracts to buy or sell the index at the given expiration time T. Examples of index derivatives are call and put index options. A mathematical model describing the time evolution of the derivative value is well known as the multidimensional Black-Scholes equation. Similarly, as in the case of a single underlying asset, derivation of the governing equation consists of two steps. In the first step we make use of the multidimensional variant of Itō's lemma (see Chapter 2). With this instrument, we can construct a stochastic differential equation governing the evolution of the derivative value V as a function of time t and the vector $\vec{S} = (S_1, \ldots, S_n)$ of assets prices forming the index. In the next step we construct a self-financing portfolio consisting of assets, an option written on the index and risk-free bonds.

Concerning stochastic behavior of the underlying asset prices we will suppose that each of them follows the stochastic differential equation representing the geometric Brownian motion, i.e.

$$\frac{dS_i}{S_i} = \mu_i dt + \sigma_i dZ_i, \qquad i = 1, 2, \ldots, n,$$

where μ_i and σ_i denote the expected rate of return and the volatility of the asset i, dZ_i is the differential of the Wiener process for the i-th asset. Let ρ_{ij} denote the correlation coefficient of dZ_i and dZ_j, i.e.

$$E(dZ_i dZ_j) = \rho_{ij} dt, \qquad i, j = 1, 2, \ldots, n.$$

Notice that $\rho_{ii} = 1$ for any $i = 1, \ldots, n$. Our goal is to price the index option by a function

$$V = V(S_1, S_2, \ldots, S_n, t)$$

depending on time $t \in [0, T]$ and a vector $\vec{S} = (S_1, S_2, \ldots, S_n) \in \mathbb{R}^n$ of underlying assets forming the given index. According to the multidimensional variant of Itō's lemma (see Chapter 2) we obtain the following expression for the differential dV:

$$dV = \left(\frac{\partial V}{\partial t} + \frac{1}{2} \sum_{i,j=1}^{n} \frac{\partial^2 V}{\partial S_i \partial S_j} \rho_{ij} \sigma_i \sigma_j S_i S_j \right) dt + \sum_{i=1}^{n} \frac{\partial V}{\partial S_i} dS_i. \qquad (3.18)$$

Let us construct a synthetic portfolio consisting of the amount of Q_V index options with the unit price V and amounts of Q_{S_i} assets of the type i having the unit price S_i for $i = 1, \ldots, n$. By B we denote the price of a zero coupon risk-less bond. Similarly as in the case of derivation of the Black–Scholes model for a single underlying asset, we assume the zero net investment in the portfolio, i.e.

$$\sum_{i=1}^{n} S_i Q_{S_i} + V Q_V + B = 0. \tag{3.19}$$

Furthermore, we assume that our portfolio is self-financed, meaning that

$$\sum_{i=1}^{n} S_i dQ_{S_i} + V dQ_V + \delta B = 0,$$

where dQ_{S_i}, dQ_V and δB stand for respective changes in amounts of underlying assets, options and the change of the money value in bonds needed for self-financing the portfolio. Taking a differential of (3.19) we obtain

$$\sum_{i=1}^{n} (S_i dQ_{S_i} + dS_i Q_{S_i}) + dV Q_V + V dQ_V + dB = 0.$$

The total change dB of the money value in bonds can be expressed as a sum: $dB = rBdt + \delta B$, where $r > 0$ is the risk-less interest rate on bonds. Hence

$$\sum_{i=1}^{n} dS_i Q_{S_i} + dV Q_V + rBdt = 0.$$

If we insert the differential dV given by (3.18) into the above equality and assume that the amounts of underlying assets and options satisfy the condition:

$$\frac{Q_{S_i}}{Q_V} = -\frac{\partial V}{\partial S_i},$$

for $i = 1, 2, \ldots, n$, we achieve a risk-less, delta hedged, self-financing portfolio. Moreover, the option price V fulfills the multidimensional parabolic partial differential equation of the form:

$$\frac{\partial V}{\partial t} + \frac{1}{2} \sum_{i=1}^{n} \sum_{j=1}^{n} \rho_{ij} \sigma_i \sigma_j S_i S_j \frac{\partial^2 V}{\partial S_i \partial S_j} + r \sum_{i=1}^{n} S_i \frac{\partial V}{\partial S_i} - rV = 0, \tag{3.20}$$

defined for $0 < S_i < \infty, t \in (0, T)$. Equation (3.20) is called the multidimensional Black–Scholes equation (cf. Kwok [75]). We also refer the reader for further details of derivation of the model and numerical analysis based on the additive operator splitting technique to the paper by Kilianová and Ševčovič [72].

The terminal condition at expiry $t = T$ depends on the type of an option. For example, if we consider the index call option then its pay–off diagram is given by

$$V(\vec{S}, T) = \left(\sum_{i=1}^{n} a_i S_i - E \right)^+,$$

where E is the exercise price.

In the rest of this section we focus our attention to the method how to obtain an analytical solution to the multidimensional Black-Scholes partial differential equation (3.20).

We can construct a solution $V(\vec{S},t)$ in the form of a convolution of the discounted initial condition $V(\vec{S},T)$ and the fundamental solution ψ to (3.20), i.e.

$$V(\vec{S}, T-\tau) = e^{-r\tau} \int_{\mathbb{R}^n} V(\vec{\xi},T)\psi(\vec{S}-\vec{\xi},\tau)d\vec{\xi}. \tag{3.21}$$

It is easy to verify (see e.g., [105]) that V is a solution to (3.21) if and only if the fundamental solution function ψ satisfies the parabolic equation

$$\frac{\partial \psi}{\partial \tau} = \frac{1}{2}\sum_{i=1}^{n}\sum_{j=1}^{n}\rho_{ij}\sigma_i\sigma_j S_i S_j \frac{\partial^2 \psi}{\partial S_i \partial S_j} + r\sum_{i=1}^{n} S_i \frac{\partial \psi}{\partial S_i}$$

and the initial condition $\psi(\vec{S}-\vec{\xi},0) = \delta(\vec{S}-\vec{\xi})$, where $\delta(\vec{x})$ stands for the Dirac function of a vector argument $\vec{x} \in \mathbb{R}^n$. It means that

$$\int_{\mathbb{R}^n} f(\vec{\xi})\delta(\vec{x}-\vec{\xi})d\vec{\xi} = f(\vec{x}), \quad \int_{\mathbb{R}^n} \delta(\vec{\xi})d\vec{\xi} = 1,$$

for any smooth compactly supported function $f : \mathbb{R}^n \to \mathbb{R}$.

In order to find an explicit form of the function ψ we apply a series of transformations of variables. By using the following transformation:

$$y_i = \frac{1}{\sigma_i}\left(r - \frac{\sigma_i^2}{2}\right)\tau + \frac{1}{\sigma_i}\ln S_i, \quad i = 1, 2, \ldots, n$$

and putting $\Phi(\vec{y},\tau) = \psi(\vec{S},\tau)$ we conclude that Φ is a solution to the following n-dimensional diffusion equation

$$\frac{\partial \Phi}{\partial \tau} = \frac{1}{2}\sum_{i=1}^{n}\sum_{j=1}^{n}\rho_{ij}\frac{\partial^2 \Phi}{\partial y_i \partial y_j}, \quad -\infty < y_i < \infty, \ \tau > 0. \tag{3.22}$$

Now we can transform the variable $y \in \mathbb{R}^n$ into a new variable $x \in \mathbb{R}^n$ using a linear transformation given by the $n \times n$ matrix Q such that

$$x = Qy.$$

Since the coorelation matrix $R = (\rho_{ij})$ is symmetric and positive definite there exists a matrix Q such that

$$QRQ^T = I,$$

where I is the $n \times n$ identity matrix. With this transformation we can rewrite the parabolic equation (3.22) in the standard form of a heat equation for the transformed function $\tilde{\Phi}(x,\tau) = \Phi(y,\tau) = \Phi(Q^{-1}x,\tau)$:

$$\frac{\partial \tilde{\Phi}}{\partial \tau} = \frac{1}{2}\Delta\tilde{\Phi},$$

where Δ is the so-called Laplace operator, i.e. $\Delta\tilde{\Phi} = \sum_{i=1}^{n}\frac{\partial^2 \tilde{\Phi}}{\partial x_n^2}$. A solution to the above equation is known in the explicit form (cf. [105]). It is given by

$$\tilde{\Phi}(\vec{x},\tau) = \frac{1}{(2\pi\tau)^{\frac{n}{2}}}\exp\left(-\frac{\|\vec{x}\|^2}{2\tau}\right),$$

where $||\vec{x}||^2 = \sum_{i=1}^{n} x_i^2$. By returning to original variables we finally obtain the solution $V(\vec{S}, T - \tau)$ to (3.20) given by formula (3.21) in which we have to insert and the fundamental solution ψ constructed by the procedure described above.

Although the transformation method for pricing index options lead to the explicit formula (3.21) for the option price $V(S,t)$ it should be emphasized that numerical computation of a high dimensional integral appearing in the explicit formula (3.21) is a hard task. In order to accurately compute the integral in (3.21) new numerical methods and techniques are needed. In particular, Monte–Carlo simulations are quite often used when computing high dimensional integrals. In [95] Reisinger and Wittum presented an efficient sparse grid discretization of high dimensional options that can be achieved by hierarchical approximation.

Problem Section and Exercises

1. Compute the value of a bought straddle strategy consisting of purchasing one call option and one put option written on the same underlying stock with the same exercise prices E and the same exercise time T. Perform the computation for the following model parameters: the price of stock not paying dividends is $S = 55$, the stock volatility is $\sigma = 0.4$, interest rate is $r = 0.1$, expiration of the options is in 3 months (i.e., $T = 0.25$ years), exercise price $E = 50$.

2. How does the price of a European call option depend on the interest rate r? Plot a graph of dependence of the call option price on the interest rate, when the price of the stock paying no dividends is $S = 115$, volatility of the stock is $\sigma = 0.3$, and the option expires in 6 months with the strike price is $E = 110$.

3. A contract between the writer of the buyer (holder) of an option is as follows: the option gives the holder the right to buy the stock at the exercise price E_1 at the expiration time T. However, at the moment of exercising the option the holder has to sell the writer the stock for the expiration price E_2, where $E_1 < E_2$. Compute the value of this strategy for the following data: price of the stock that does not pay dividends is $S = 65$, volatility of the stock is $\sigma = 0.5$, interest rate is $r = 0.06$, expiration time is 6 months, (i.e., $T = 0.5$ years), strike prices are $E_1 = 50$, $E_2 = 60$.

4. Compute the price of a bullish spread strategy consisting of purchasing a call option with a lower strike price and selling a call option with a higher strike price E_2 with the same expiration time. Perform the computation with the following model parameters: the price of the underlying stock paying no dividends is $S = 55$, volatility of the stock is $\sigma = 0.4$, interest rate is $r = 0.1$, expiration time is 3 months (i.e., $T = 0.25$ years), strike prices are $E_1 = 50$, $E_2 = 60$.

5. Compute the price of a bought straddle strategy consisting of purchasing call and put options with a strike price E. Perform the computation for the following data: the price of the underlying stock paying no dividends is $S = 55$, volatility of the stock is $\sigma = 0.4$, interest rate is $r = 0.05$, dividend rate of the continuously paying dividends

is $q = 0.03$, expiration time T is 3 months (i.e., $T = 0.25$ years), the strike price is $E = 60$.

6. How does the price of a European put option depend on the volatility σ of the stock price? Find the expression for the factor Vega, i.e., the derivative of the European put option price with respect to the volatility. Is it an increasing or a decreasing dependence? What is the limit of the European put option price when the volatility approaches zero?

7. Show that a graph of a European call option on the stock paying nontrivial continuous dividends always intersects its pay–off diagram for all sufficiently large values of the stock price. It means that $V_{ec}(S,t) < S - E$ for S large enough. Analyze the case of a European put option.

Chapter 4

Analysis of Dependence of Option Prices on Model Parameters

The purpose of this chapter is to study various sensitivity factors of financial derivatives. Using these factors we are able to analyze and better understand behavior of financial derivatives. In the first part of this chapter, we concentrate on the problem of estimating the historical volatility of a stochastic process describing the evolution of the underlying asset price. We discuss the computation of the so–called historical volatility and we show how to compute its value from real market data. Then we focus on a new concept of volatility represented by the so–called implied volatility. It is obtained by means of calibration of option prices to financial market data. Finally, we present and discuss other sensitivity factors yielding information about the dependence option prices with respect to various model parameters. These sensitivity factors (often called - according to the Greek alphabet - *greeks*) give us a more complete information about the behavior of derivatives prices.

4.1. Historical Volatility of Stocks

In this section we focus our attention on the methodology how to obtain an estimate of the volatility parameter σ. It characterizes the size of random fluctuations of the prices of underlying stock asset prices. The estimate of σ that can be computed from the known historical prices of the underlying asset is referred to as the historical volatility.

Assume that the underlying stock price $S = S(t)$ follows a geometric Brownian motion with a drift μ and volatility σ, i.e.

$$dS = \mu S dt + \sigma S dw. \qquad (4.1)$$

According to Itō's lemma we know that the stochastic process $X_t = \ln S_t$ is a Brownian motion with a drift $\mu - \frac{1}{2}\sigma^2$ and volatility σ. It means that

$$dX = \left(\mu - \frac{1}{2}\sigma^2\right) dt + \sigma dw.$$

Our aim is to estimate the parameter σ of the stochastic process (4.1). Suppose that we are know historical asset prices $S_{t_i}, i = 0, 1, \ldots, n$, at times $T_{start} = t_0 < t_1 < \cdots < t_n = T_{end}$, over some interval $[T_{start}, T_{end}]$. We assume time intervals are equidistant, i.e., $t_{i+1} - t_i = \tau$ for all $i = 0, 1, \ldots, n-1$. Then for the differences $X(t_{i+1}) - X(t_i)$ we have

$$X(t_{i+1}) - X(t_i) = (\mu - \frac{1}{2}\sigma^2)\tau + \sigma(w(t_{i+1}) - w(t_i)).$$

Recall that for increments of the Wiener process $w(t)$ we have $w(t_{i+1}) - w(t_i) = \Phi\sqrt{\tau}$, where $\Phi \sim N(0,1)$ is a standardized normally distributed random variable. Since $X(t_{i+1}) - X(t_i) = \ln(S(t_{i+1})/S(t_i))$, then the sample statistical dispersion of independent returns $\{\ln(S(t_{i+1})/S(t_i)), i = 0, 1, \ldots, n-1\}$ is an unbiased estimate of the parameter $\sigma^2\tau$. Thus the estimate of the historical volatility σ_{hist} can be expressed by the following formula:

$$\sigma_{hist}^2 = \frac{1}{\tau}\frac{1}{n-1}\sum_{i=0}^{n-1}\left(\ln\frac{S(t_{i+1})}{S(t_i)} - \gamma\right)^2, \tag{4.2}$$

where γ is an estimate for the mean return, i.e.

$$\gamma = \frac{1}{n}\sum_{i=0}^{n-1}\ln\frac{S(t_{i+1})}{S(t_i)}.$$

Let us emphasize that the numerical value of the estimated parameter σ depends on a selected time scale. For example, if we express all the time data (hence also the time difference τ) in years, the resulting estimate of the historical volatility is also expressed on the yearly basis, i.e., in percents *per annum* (p.a.).

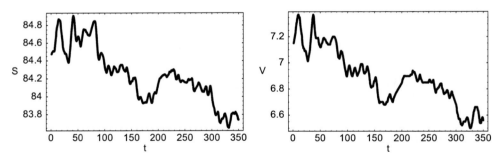

Figure 4.1. Minute behavior of prices of the IBM stock from May 21, 2002 (left) and call option prices of this stock with the strike price $E = 80$ with expiry on July 2, 2002, i.e., $T - t = 43/365$ years (right).

In Fig. 4.1 (left) we show prices of the IBM stock. We also plot call option prices with the strike strike price $E = 80$. The time series of both stock and options prices are depicted in a scale of minutes. It means that the length of the time interval $\tau = 1/(24 \times 60 \times 365)$ years. The data were smoothed by a moving average through a five minute time interval. Using the formula (4.2) we estimate the historical volatility to be $\sigma_{hist} = 0.2306$ p.a.

4.1.1. A Useful Identity for Black–Scholes Option Prices

In the remaining part of this section we first recall the formulae for pricing European call and put options. Moreover, we show some useful mathematical properties, which will be used later in analysis the derivatives prices from the point of view of their sensitivity with respect to change of model parameters.

Recall that prices V^{ec} and V^{ep} of the European call and put options are given by closed form formulae (3.8) and (3.14), i.e.

$$V^{ec}(S,t) = Se^{-q(T-t)}N(d_1) - Ee^{-r(T-t)}N(d_2), \quad (4.3)$$
$$V^{ep}(S,t) = Ee^{-r(T-t)}N(-d_2) - Se^{-q(T-t)}N(-d_1), \quad (4.4)$$

where

$$d_1 = \frac{\ln \frac{S}{E} + (r - q + \frac{\sigma^2}{2})(T-t)}{\sigma\sqrt{T-t}}, \quad d_2 = d_1 - \sigma\sqrt{T-t}. \quad (4.5)$$

Since the price of a European call and put option depends not only on the underlying stock asset price S and time t, but also on the parameters of the Black–Scholes model: E, T, r, q, σ, we can write

$$V^{ec}(S,t) = V^{ec}(S,t; E,T,r,q,\sigma),$$
$$V^{ep}(S,t) = V^{ep}(S,t; E,T,r,q,\sigma).$$

In the following lines, we derive an important identity, which will be used several times in computations of partial derivatives of a European option price with respect to the stock price, as well as to other model parameters.

We begin with computing the difference $(d_1^2 - d_2^2)/2$. Since $d_2 = d_1 - \sigma\sqrt{T-t}$, we obtain

$$\frac{d_1^2 - d_2^2}{2} = \frac{(d_1+d_2)(d_1-d_2)}{2} = \frac{2\ln\frac{S}{E} + 2(r-q)(T-t)}{\sigma\sqrt{T-t}} \cdot \frac{\sigma\sqrt{T-t}}{2}$$
$$= \ln\frac{S}{E} + (r-q)(T-t),$$

and hence

$$\frac{d_1^2}{2} = \frac{d_2^2}{2} + \ln\frac{S}{E} + (r-q)(T-t).$$

For derivative of the cumulative distribution function $N'(d)$ of the standardized normal distribution we have

$$N'(d) = \frac{1}{\sqrt{2\pi}} \exp(-d^2/2).$$

Using the above aidentity for the difference $(d_1^2 - d_2^2)/2$ we finally obtain an important identity:

$$Se^{-q(T-t)}N'(d_1) - Ee^{-r(T-t)}N'(d_2) = 0. \quad (4.6)$$

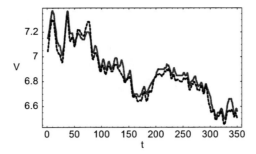

Figure 4.2. Comparison of an intraday evolution of prices of a European call option price computed from the Black–Scholes formula $V^{ec}(S_{real}(t),t;\sigma_{hist})$ (dashed line) and real market prices $V_{real}(t)$ (solid line), where $\sigma_{hist} = 0.2306$ p.a.

4.2. Implied Volatility

In Fig. 4.2 we compare the real market data of call options and options computed by the formula (4.3) for pricing European call option, where the volatility parameter σ was chosen to be the historical volatility of time evolution of the underlying stock of IBM displayed in Fig. 4.2. It should be obvious from this figure that the market price (solid line) is underestimated by the theoretical Black–Scholes value (dashed line). Since the option price is an increasing function of volatility σ, it means that the historical volatility σ_{hist} is too low for the theoretical Black–Scholes call option price V^{ec} to match the market data accurately. This observation leads us to a new concept of the so–called implied volatility.

The implied volatility $\sigma_{impl} > 0$ is such a value of the volatility parameter σ, for which the theoretical call (put) option price $V(S,t;\sigma)$ for given time t and the stock price $S = S_{real}(t)$ coincides with the market value of the option $V_{real}(t)$. It means that the task of finding the implied volatility σ_{impl} for a given option with given expiration time T and strike price E consists in solving the following inverse function problem:

$$V_{real}(t) = V(S_{real}(t),t;\sigma_{impl}). \qquad (4.7)$$

Next we have to discuss the problem of existence and uniqueness of implied volatility $\sigma_{impl} > 0$, which satisfies (4.7). In the first step we show that the price of a call or put option is an increasing function of the volatility σ. Intuitively this is an obvious fact, since with increasing volatility of the underlying stock, the role, and consequently also the price, of a hedging instrument such as a call or put option increases. Analytically we prove this property by differentiating the functions V^{ec} and V^{ep} with respect to the parameter σ and to show that this derivative is always positive. From the formula (4.3) we deduce

$$\begin{aligned}\frac{\partial V^{ec}}{\partial \sigma} &= Se^{-q(T-t)}N'(d_1)\frac{\partial d_1}{\partial \sigma} - Ee^{-r(T-t)}N'(d_2)\frac{\partial d_2}{\partial \sigma} \\ &= \left(Se^{-q(T-t)}N'(d_1) - Ee^{-r(T-t)}N'(d_2)\right)\frac{\partial d_1}{\partial \sigma} \\ &\quad + Ee^{-r(T-t)}N'(d_2)\sqrt{T-t},\end{aligned}$$

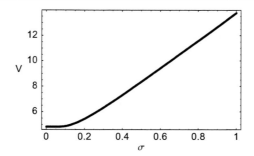

Figure 4.3. A price of European call option as a function of the volatility σ for model parameters $E = 80, S = 84.45, r = 0.04, q = 0, T - t = 43/365$.

because $d_2 = d_1 - \sigma\sqrt{T-t}$. With regards to the fundamental formula (4.6), the term $Se^{-q(T-t)}N'(d_1) - Ee^{-r(T-t)}N'(d_2)$ vanishes and so we obtain

$$\frac{\partial V^{ec}}{\partial \sigma} = Ee^{-r(T-t)}N'(d_2)\sqrt{T-t}.$$

In the same way, for a European put option we obtain:

$$\frac{\partial V^{ep}}{\partial \sigma} = Ee^{-r(T-t)}N'(-d_2)\sqrt{T-t}.$$

Since $N'(-d_2) = N'(d_2) = \exp(-d_2^2/2)/\sqrt{2\pi}$, we finally conclude that

$$\frac{\partial V^{ec}}{\partial \sigma} = \frac{\partial V^{ep}}{\partial \sigma} = Ee^{-r(T-t)}N'(d_2)\sqrt{T-t}. \qquad (4.8)$$

Recall that equality (4.8) for a put option also easily follows from the put–call parity.

It means that $\frac{\partial V^{ec}}{\partial \sigma} = \frac{\partial V^{ep}}{\partial \sigma} > 0$, and hence the price of the European call and put options is an increasing function with respect to the volatility parameter $\sigma > 0$. Graphical dependence of the option price on the volatility for the model parameters $E = 80, r = 0.04, q = 0$ and time to expiry $T - t = 43/365$ years is shown in Fig. 4.3.

In the next step we determine the interval for option prices, in which the existence of a solution σ_{impl} to equation (4.7) can be guaranteed. The basis for construction of this interval follows from the computation of the limit of an option price as $\sigma \to 0$ and $\sigma \to \infty$. Clearly, it holds

$$\lim_{\sigma \to 0} d_1 = \lim_{\sigma \to 0} d_2 = \begin{cases} -\infty & \text{for } \ln(S/E) + (r-q)(T-t) < 0, \\ +\infty & \text{for } \ln(S/E) + (r-q)(T-t) > 0. \end{cases}$$

Since for the distribution function N we have $N(-\infty) = 0, N(+\infty) = 1$, we obtain

$$\lim_{\sigma \to 0} V^{ec}(S,t;\sigma) = \max(Se^{-q(T-t)} - Ee^{-r(T-t)}, 0),$$

$$\lim_{\sigma \to 0} V^{ep}(S,t;\sigma) = \max(Ee^{-r(T-t)} - Se^{-q(T-t)}, 0). \qquad (4.9)$$

On the other hand, $\lim_{\sigma \to \infty} d_1 = +\infty, \lim_{\sigma \to \infty} d_2 = -\infty$, and hence

$$\lim_{\sigma \to \infty} V^{ec}(S,t;\sigma) = Se^{-q(T-t)},$$

$$\lim_{\sigma \to \infty} V^{ep}(S,t;\sigma) = Ee^{-r(T-t)}. \qquad (4.10)$$

Using the already proved fact that the option price is an increasing function of volatility σ, we have shown the following proposition:

Theorem 4.1. *If a market price of a European call option V_{real}^{ec} satisfies the inequalities*

$$(S_{real}e^{-q(T-t)} - Ee^{-r(T-t)})^+ < V_{real}^{ec} < S_{real}e^{-q(T-t)},$$

where S_{real} is a market price of the underlying stock, then there exists the unique implied volatility $\sigma_{impl}^{ec} > 0$ such that

$$V_{real}^{ec} = V^{ec}(S_{real}, t; \sigma_{impl}^{ec}).$$

If a market price of a European put option V_{real}^{ep} satisfies the inequalities

$$(Ee^{-r(T-t)} - S_{real}e^{-q(T-t)})^+ < V_{real}^{ep} < Ee^{-r(T-t)},$$

then there exists the unique implied volatility $\sigma_{impl}^{ep} > 0$ such that

$$V_{real}^{ep} = V^{ep}(S_{real}, t; \sigma_{impl}^{ep}).$$

Notice that the interval of values for which the market price can belong to, is not very restrictive. For example, for a call option the upper bound $S_{real}e^{-q(T-t)}$ is higher than the value of the terminal pay–off diagram $(S-E)^+$ of a call option, in the neighborhood of which we can expect the option price. An analogous consideration can be done in the case of restrictions on the put option price.

An example of a graphical solution of the equation $V_{real} = V^{ec}(S_{real}, t; \sigma_{impl})$ for the case of finding the implied volatility of the call option on IBM stock is shown in Fig. 4.4.

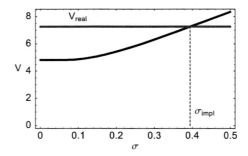

Figure 4.4. A graphical solution of the equation $V_{real} = V^{ec}(S_{real}, t; \sigma_{impl})$ for the market values of the stock price $S_{real} = 84.45$ and the call option $V_{real} = 7.25$ computed for model parameters $E = 80, r = 0.04, q = 0, T - t = 43/365$.

Another way of computing the implied volatility can be based on considering a longer time interval of option price. In this case it is not possible to expect that it would be possible to find one common value of the implied volatility. We can, however, look for an implied volatility by minimization of the sum of squares of differences between option's market

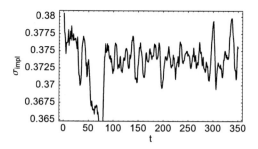

Figure 4.5. A plot of the implied volatility σ_{impl} for the option on IBM stock from 21st May 2002 (tick marks represent scale of minutes). The average implied volatility is $\bar{\sigma}_{impl} = 0.3733$ p.a.

prices and their theoretical values obtained from the Black–Scholes formula. This idea leads us to a problem of minimizing the function $U : (0,\infty) \to (0,\infty)$ defined as:

$$U(\sigma) = \left(\frac{1}{m} \sum_{i=1}^{m} |V_{real}(t_i) - V(S_{real}(t_i), t_i; \sigma)|^2 \right)^{\frac{1}{2}},$$

where $V(S,t;\sigma)$ is the price of a European call (put) option, $S_{real}(t)$ is the real market underlying stock price at time t and $V_{real}(t)$ is the real market price of a call (put) option at time t. The parameter σ corresponds to the volatility of a stochastic process describing the underlying stock price evolution. The argument of a minimum of this function can be taken to be an estimate of the time averaged implied volatility σ_{impl}^{ta} based on the time series of underlying stock and option prices, i.e.

$$\sigma_{impl}^{ta} = \arg\min_{\sigma > 0} U(\sigma).$$

Notice that in the case of $m = 1$, the minimum (with a zero optimal value) of the function U is attained exactly at the value $\sigma_{impl} > 0$ corresponding to the solution of equation (4.7) at the time t.

In Fig. 4.7 we choose the volatility parameter $\sigma_{impl} = 0.3733$ obtained by minimizing the function U over the whole intraday time interval of 360 minutes, i.e., $m = 360$. Comparing with results based on the usage of the historical volatility shown in Fig. 4.2 it is now clear that using the implied volatility results in a more accurate matching between theoretical and time averaged real market data.

4.3. Volatility Smile

In this section we briefly mention an interesting phenomenon closely related to the concept of implied volatility. At given time t we have only single underlying asset price listed. On the other hand, several values of options with the same maturity T but written on different strike prices $E_1, E_2, \ldots, E_k$ are listed at the given time t. Hence, for every option it is possible

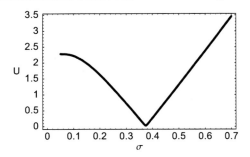

Figure 4.6. A graph of the function $U = U(\sigma)$ for a call option written on the IBM stock from May 21, 2002. A value of the parameter σ, in which the function U attains its minimum, is $\sigma_{impl} = 0.3734$ p.a.

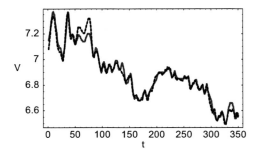

Figure 4.7. A comparison of values of a European call option $V^{ec}(S_{real}(t), t; \sigma_{impl})$ (dashed line) and market prices $V_{real}(t)$ (solid line) in the case we take the time averaged implied volatility $\sigma^{ta}_{impl} = 0.3733$ p.a.

to compute the value of implied volatility corresponding to the underlying asset price S and strike price $E_i, i = 1, \ldots, k$. It is often the case that the computed implied volatility need not be necessarily the same for all options. This is rather a rare situation. We often find examples, when dependence of the implied volatility on the ratio S/E of asset to strike price is a convex function in the neighborhood of the value $S/E \approx 1$. In Fig. 4.8 we have a practical example of this phenomenon for the case of call options on IBM stocks from May 21, 2002 with maturity July 2, 2002 and strike prices $E = 65, 70, 75, \ldots, 150$. In the neighborhood of $S/E \approx 1$ the implied volatility attains its global minimum and the function is convex in the neighborhood of 1. A slang name *volatility smile* comes from the shape of a graph implied volatility reminding us a smile in the neighborhood of $S/E \approx 1$.

4.4. Delta of an Option

The basic sensitivity factor, which is often evaluated when analyzing the market data, is dependence of a change of a derivative price with respect to a change of the price of the underlying asset stock. In the infinitesimal form this factor can be written as a partial

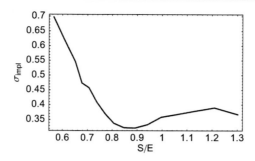

Figure 4.8. A volatility smile of the implied volatility for IBM options from May 21, 2002 with the maturity July 2, 2002.

derivative:
$$\Delta = \frac{\partial V}{\partial S}. \tag{4.11}$$

The importance of the factor Δ is especially in its close relation with construction of the risk-neutral portfolio consisting of amount of Q_S underlying stocks with a unit price S, and Q_V options with a unit price V written on the stock and the riskless bonds. When deriving the Black–Scholes partial differential equation, in order to eliminate the stochastic random part of the portfolio we came to the conclusion (2.15). It determines the ratio of amounts of stocks and options in the portfolio, i.e.

$$\frac{Q_S}{Q_V} = -\frac{\partial V}{\partial S} = -\Delta.$$

In means that the value of the sensitivity parameter Δ yields the ratio between the stocks and options leading to a risk-neutral portfolio. For example, if the factor Δ for call option equals 0.7, then in order to obtain a risk-neutral portfolio containing $Q_S = 7$ stock, we need to have $Q_V = -10$ call options on that stock.

For European call and put options we are able to derive an explicit formulae for the factor Δ. We differentiate the functions V^{ec} and V^{ep} with respect to S. For a call option, it follows from (4.3), the relationship $\partial d_1/\partial S = \partial d_2/\partial S$ and the identity (4.6) that:

$$\Delta^{ec} = \frac{\partial V^{ec}}{\partial S} = Se^{-q(T-t)}N'(d_1)\frac{\partial d_1}{\partial S} - Ee^{-r(T-t)}N'(d_2)\frac{\partial d_2}{\partial S}$$
$$+ e^{-q(T-t)}N(d_1) = e^{-q(T-t)}N(d_1).$$

Similarly, for a European put option we obtain:

$$\Delta^{ep} = \frac{\partial V^{ep}}{\partial S} = Se^{-q(T-t)}N'(-d_1)\frac{\partial d_1}{\partial S} - Ee^{-r(T-t)}N'(-d_2)\frac{\partial d_2}{\partial S}$$
$$- e^{-q(T-t)}N(-d_1) = -e^{-q(T-t)}N(-d_1).$$

In summary, we derived the following formulae:

$$\begin{aligned} \Delta^{ec} &= e^{-q(T-t)}N(d_1), \\ \Delta^{ep} &= -e^{-q(T-t)}N(-d_1). \end{aligned} \tag{4.12}$$

A graph of dependence of the factor Δ on the stock price S for parameters $E = 80, r = 0.04, q = 0, T - t = 43/365$ is shown in Fig. 4.9.

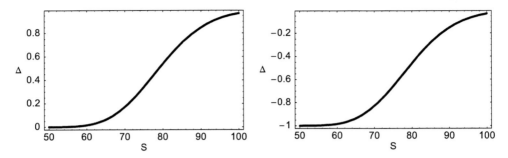

Figure 4.9. Dependence of the factor $\Delta = \frac{\partial V}{\partial S}$ of a call (left) and put option (right) on the stock price S.

Experimentally we can obtain the parameter Δ from the real market data in such a way that we approximate the derivative $\frac{\partial V}{\partial S}$ at the time t_i by a ratio of differences of the option price with respect to the changes of underlying asset prices, i.e.,

$$\Delta_{t_i} = \frac{\partial V}{\partial S}(S_{t_i}, t_i) \approx \frac{V_{t_i} - V_{t_{i-1}}}{S_{t_i} - S_{t_{i-1}}}. \quad (4.13)$$

However, it is necessary to emphasize that the direct use of the formula (4.13) and taking into account non-smoothed data from the option market would often lead to unusable results, since small price differences $S_{t_i} - S_{t_{i-1}}$ in the numerator of the right hand side of the equality (4.13) will result in high values of the parameter Δ. Hence it is necessary to use pre-smoothed time series of option and underlying stock asset prices, as well as smoothing the resulting values of parameter Δ. To smooth the time series data of S it is possible to use a simple arithmetic averaging of the time series values for some specified period. Fig. 4.10 depicts pre-smooth prices IBM stock and option prices (compare with nonsmoothed data in Fig. 4.1!). In this example, we have used the arithmetic average with a length of 60 minutes. A practical computation of the factor Δ by the formula (4.13) and usage of these smoothed time series of underlying stock and option prices is depicted in Fig. 4.11 (left).

The factor Δ can be experimentally determined also by using the explicit formula for the price of a European call (put) option (see (4.3) and (4.4)). In Fig. 4.11 (right) we show the time series of the factor Δ computed as $\Delta_{t_i}^{ec} = \frac{\partial V^{ec}}{\partial S}(S_{t_i}, t_i; \sigma_{impl})$. Notice that in Fig. 4.11 (right) we show the behavior of the factor Δ^{ec} scaled to the interval $0.67 - 0.72$. These values correspond to the dashed line in the left.

4.5. Gamma of an Option

Not less important sensitivity factor, that we often evaluate when analyzing the market data, is represented by dependence of change of the factor Δ itself on change of the underlying asset stock price. This factor, denoted as Gamma of the option Γ, can be written in a

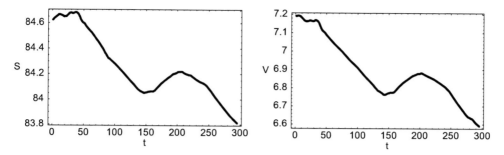

Figure 4.10. Pre-smoothed time series of IBM stock asset price from May 21, 2002 (left) and the time series of pre-smoothed call option with the strike price $E = 80$ with expiration on July 21, 2002, i.e., $T - t = 43/365$. To smooth the time series we have used the arithmetic average with the length 60 minutes.

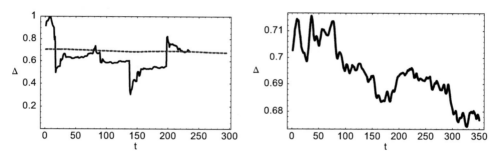

Figure 4.11. Left: pre-smoothed time dependence of the function Δ computed from the IBM stock prices from May 21, 2002 using the smoothed data from the Fig. 4.10. To smooth the function Δ we have again used the arithmetic average with the length 60 minutes. The dashed line in the left and in the zoomed graph (right) corresponds to the value $\Delta^{ec}(t) = \frac{\partial V^{ec}}{\partial S}(S_{real}(t), t; \sigma_{impl})$.

differential form as follows:

$$\Gamma = \frac{\partial \Delta}{\partial S} = \frac{\partial^2 V}{\partial S^2}. \qquad (4.14)$$

Since the factor Δ is the derivative of the option price with respect to the underlying stock price, the factor Γ is, in fact, the second derivative of the option price with V respect to the underlying asset price S.

The sensitivity factor Γ indicates the magnitude in change of the factor Δ. Recall that the factor Δ represents a ratio of the amount of stocks and amount of options in the risk-neutral delta hedged portfolio consisting of stocks, options and riskless bonds. Hence, in the case when the factor Δ on the option market changes its value it can be interpreted in such a way, that the options are being sold (purchased). It means that any increase in the value of the factor Γ can indicate a movement in the volume of sold (purchased) options of a given type.

Similarly as for the factor Δ, we are able to derive an explicit formula for the factor Γ

for European call and put options. Differentiating formulae (4.12) for the factor Δ we obtain

$$\Gamma^{ec} = \Gamma^{ep} = \frac{\partial \Delta^{ec}}{\partial S} = e^{-q(T-t)} N'(d_1) \frac{\partial d_1}{\partial S}$$
$$= e^{-q(T-t)} \frac{\exp(-\frac{1}{2}d_1^2)}{\sigma \sqrt{2\pi(T-t)} S}. \qquad (4.15)$$

A graph of dependence of the factor Γ on the underlying stock asset price S for parameters $E = 80, r = 0.04, q = 0, T - t = 43/365$ is shown in Fig. 4.12 (left). In the right figure, we can see evolution of dependence of the factor Γ on S for various times to expiry. The smaller is the time to expiry $T - t$, the higher and narrower is a graph of the factor Γ, having its maximum near the strike price E.

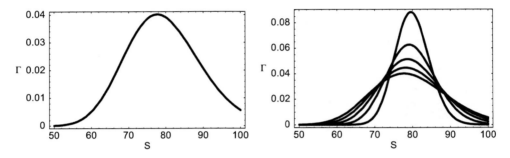

Figure 4.12. Dependence of the factor $\Gamma^{ec} = \Gamma^{ep} = \frac{\partial^2 V}{\partial S^2}$ for call and put options on the underlying asset stock price S for time T to expiry (left) and several graphs of the factor Gamma for different times $T - t$ to expiry (right).

4.6. Other Sensitivity Factors: Theta, Vega, Rho

4.6.1. Sensitivity with Respect to a Change in the Interest Rate – Factor Rho

Factor P (from the capital Greek letter *Rho*) shows the sensitivity of the derivative price with respect to a change of the interest rate of a riskless bond $r > 0$. Factor P is therefore given as a derivative

$$P = \frac{\partial V}{\partial r}.$$

An analytical expression of the factor P can be obtained by differentiating the explicit formulae for prices of European call and put options (4.3) and (4.4). Using the fundamental identity (4.6) and the fact that $d_2 = d_1 - \sigma\sqrt{T-t}$ we obtain:

$$P^{ec} = \frac{\partial V^{ec}}{\partial r} = Se^{-q(T-t)} N'(d_1) \frac{\partial d_1}{\partial r} - Ee^{-r(T-t)} N'(d_2) \frac{\partial d_2}{\partial r}$$
$$+ E(T-t)e^{-r(T-t)} N(d_2) = E(T-t)e^{-r(T-t)} N(d_2),$$
$$P^{ep} = \frac{\partial V^{er}}{\partial r} = -Ee^{-r(T-t)} N'(-d_2) \frac{\partial d_2}{\partial r} + Se^{-q(T-t)} N'(-d_1) \frac{\partial d_1}{\partial r}$$
$$- E(T-t)e^{-r(T-t)} N(-d_2) = -E(T-t)e^{-r(T-t)} N(-d_2).$$

In Fig. 4.13 we show the factor P for call and put options as a function of the underlying stock price S. It is important to note that always $P^{ec} > 0$ and $P^{ep} < 0$, i.e., the price of a European call option increases with increasing interest rate and the price of a European put option decreases with respect to r.

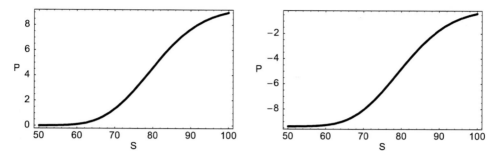

Figure 4.13. Dependence of the factor $P = \frac{\partial V}{\partial r}$ of a call (left) and a put option (right) on the stock price S.

4.6.2. Sensitivity to the Time to Expiration – Factor Theta

The factor Θ represents sensitivity of a derivative price with respect to the expiration time of the derivative T and it is defined as the derivative $\Theta = -\frac{\partial V}{\partial T}$. Since the price of a derivative at time $t \in [0, T]$ depends only on the difference $T - t$, the factor Θ can be expressed as:

$$\Theta = \frac{\partial V}{\partial t}.$$

An analytical expression of the factor Θ can be again obtained by differentiating the explicit formulae for the prices of European call and put options (4.3) and (4.4) with respect to time t. Since $d_2 = d_1 - \sigma\sqrt{T-t}$, we obtain $\partial d_2/\partial t = \partial d_1/\partial t + \sigma/(2\sqrt{T-t})$. After some rearrangements, we finally obtain:

$$\Theta^{ec} = \frac{\partial V^{ec}}{\partial t} = Sqe^{-q(T-t)}N(d_1) - Ere^{-r(T-t)}N(d_2)$$
$$- \frac{E\sigma}{2\sqrt{T-t}}e^{-r(T-t)}N'(d_2),$$

$$\Theta^{ep} = \frac{\partial V^{ep}}{\partial t} = Ere^{-r(T-t)}N(-d_2) - Sqe^{-q(T-t)}N(-d_1)$$
$$- \frac{E\sigma}{2\sqrt{T-t}}e^{-r(T-t)}N'(-d_2).$$

In Fig. 4.14 we plot the factor Θ for call and put options as a function of the underlying stock asset price S. A price of a European option on stock paying no dividends ($q = 0$) is always a decreasing function of the time t, and the limit, as $t \to T$, is given by the pay–off diagram of the call option. It means that $\Theta^{ec} < 0$, if $q = 0$. On the other hand, for the European put option, its value for the zero stock price $S = 0$ is always equal to $Ee^{-r(T-t)}$, and hence $V^{ep}(0,t) < E$. It means that the function $V^{ep}(0,t)$ increases as $t \to T$, and hence $\Theta^{ep} > 0$ for $S \approx 0$. However, for the value $S > E$ the price of put option decreases as $t \to T$, i.e., $\Theta^{ep} < 0$. This phenomenon can be seen also in Fig. 4.14 (right).

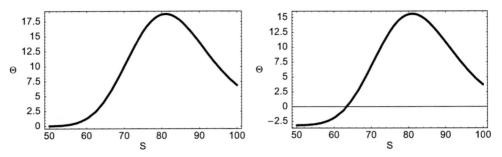

Figure 4.14. Dependence of the factor $\Theta = \frac{\partial V}{\partial t}$ of a call (left) and put option (right) on the stock price S for the strike price $E = 80$.

4.6.3. Sensitivity to a Change in Volatility – Factor Vega

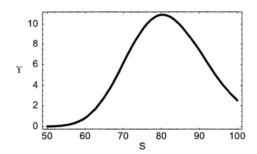

Figure 4.15. Dependence of the factor $\Upsilon = \frac{\partial V}{\partial \sigma}$ of a call (left) and put option (right) on the stock price S where the parameter $E = 80$.

The factor Vega Υ[1] describes sensitivity of a derivative security with respect to change if the volatility of the underlying asset price and hence it is defined as a derivative

$$\Upsilon = \frac{\partial V}{\partial \sigma}.$$

An analytical expression of the factor Υ has been already derived in the formula (4.8). Thus

$$\Upsilon^{ec} = \Upsilon^{ep} = E e^{-r(T-t)} N'(d_2) \sqrt{T-t}. \tag{4.16}$$

It means that $\Upsilon^{ec} = \Upsilon^{ep} > 0$ and hence the price of a European call or put option is always an increasing function of the volatility parameter $\sigma > 0$. A graph of dependence of the option price on the volatility for model parameters $E = 80, r = 0.04, q = 0$ and time to expiry $T - t = 43/365$ years is shown in Fig. 4.15.

---
[1] Vega is not a Greek letter

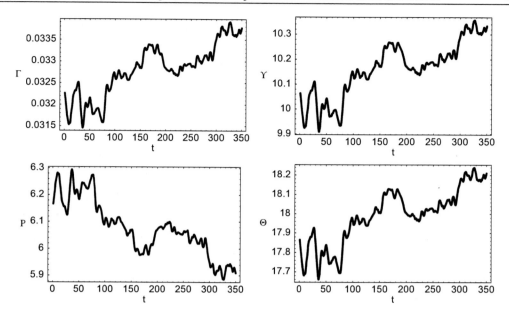

Figure 4.16. Time series (time scale of minutes) of the factors $\Gamma^{ec}_{t_i}, \Upsilon^{ec}_{t_i}, P^{ec}_{t_i}, \Theta^{ec}_{t_i}$ for the IBM call option from May 21, 2002 with the strike price $E = 80$ and time to expiry $T - t = 43/365$ years.

Problem Section and Exercises

1. The price of the IBM stock is 68.86 USD. The price of the European call option with the strike price 70 USD and expiration in 2 months is 3.5 USD. The riskless interest rate is 2% p.a.

    a) Compute the implied volatility σ_{impl} from the stock and option prices.

    b) Repeat the computation of the implied volatility, but as a basis for the computation consider the call option with the strike price 75 USD, expiration in 2 months, whose market price is 1 USD.

    c) Discuss the results obtained in parts a) and b) from the point of view of investor's exposure to risk. Which option do you consider to be more favorable for an investor?

2. How prices of European call and put options depend on the strike price? Is it an increasing or a decreasing dependence?

3. The present market value of an IBM stock is 64 USD. The value of the European put option with expiration price 70 USD and expiration time 6 months is 9.5 USD. The riskless interest rate is 2% p.a. Compute the implied volatility σ_{impl} from the stock and option prices.

4. Construct a graph of dependence of the sensitivity factor Γ on the stock price S and time to expiry $t \in (0, T)$. Display the results for the European call option on stock

paying nod dividends for $\sigma = 0.4$, strike price $E = 50$, time to expiry 4 months and riskless interest rate $r = 0.05$.

5. How prices of European call and put options depend on the interest rate r? Plot a graph of dependence of call and put option prices on the interest rate, assuming the price of the underlying stock is $S = 115$, volatility of the stock $\sigma = 0.3$, time to expiry is 6 months and the strike prices of both call and put options are $E = 110$.

6. Derive an explicit formula and plot dependence of the factor Lambda λ on underlying asset price S. The factor λ is defined as the logarithmic derivative of the option price with respect to the stock price, i.e., $\lambda = \frac{1}{V}\frac{\partial V}{\partial S}$.

7. Derive an explicit formula and plot dependence of the factor Vanna on the underlysing asset price S. The factor Vanna measures sensitivity of the factor Δ with respect to change in the volatility σ, i.e., Vanna $= \frac{\partial^2 V}{\partial S \partial \sigma}$.

8. Derive an explicit formula and plot dependence of the factor Speed on the underlysing asset price S. The factor Speed measures sensitivity of the factor Γ with respect to change in the underlying asset price S, i.e., Speed $= \frac{\partial^3 V}{\partial S^3}$.

Chapter 5

Option Pricing under Transaction Costs

The classical Black–Scholes theory presented in the previous chapters is capable of pricing options and other derivative securities over moderate time intervals in which transaction costs are negligible. However, if transaction costs due to e.g., bid–ask spreads of the underlying asset are taken into account then the classical Black–Scholes theory is no longer applicable. In order to maintain the delta hedge one has to make frequent portfolio adjustments yielding thus a substantial increase in transaction costs. Our purpose is to present a systematic way how to modify assumptions of the classical Black–Scholes theory in order to take into account nontrivial transaction costs. In this chapter we will derive the so-called Leland model for pricing derivative securities under transaction costs.

5.1. Leland Model, Hoggard, Wilmott and Whalley Model

In the paper [78] H. Leland generalized the Black–Scholes model for pricing plain vanilla call and put options for the case there are nontrivial transaction costs arising from maintaining the delta hedged portfolio by buying or selling underlying assets. The model has been further generalized to more complex options strategies by Hoggard, Whalley and Wilmott in [64].

Similarly as in the classical Black–Scholes theory, we assume that the underlying asset price $S = S_t, t \geq 0$, follows a geometric Brownian motion with a drift μ and volatility $\sigma > 0$. To simplify derivation, we will assume that the asset pays no dividends. It means that the stochastic differential equation governing the underlying asset price has the form:

$$dS = \mu S dt + \sigma S dW \qquad (5.1)$$

where dW denotes the differential of a standard Wiener process. This assumption is usually made when deriving the classical Black–Scholes equation (see e.g., [65, 75]).

Following derivation of the Black–Scholes equation we will construct a synthesized portfolio Π consisting of a one long-positioned option with a price V and δ underlying

assets with a unit price S:
$$\Pi = V + \delta S. \tag{5.2}$$

It means that the option is in the long position and we maintain delta hedged portfolio by buying and selling underlying assets. Since the holder of a long positioned option bears transaction costs the price of such an option should be less than the one corresponding to the case there are no transaction costs.

We recall that the key step in the Black–Scholes theory is to examine the differential of equation (5.2). The term $V + \delta S$ appearing in (5.2) can be differentiated by using Itō's formula whereas the change $\Delta \Pi_t = \Pi_{t+\Delta t} - \Pi_t$ should be equal to the change of a riskless bond with a risk-free interest rate $r > 0$ over the time interval Δt, i.e

$$\Delta \Pi = r \Pi \Delta t. \tag{5.3}$$

In reality, when selling and buying underlying assets lead to nontrivial transaction costs, a new term measuring transaction costs should be added to (5.3). More precisely, the change $\Delta \Pi$ of the portfolio Π is composed of two parts:

$$\Delta \Pi = \Delta(V + \delta S) - \Delta TC,$$

where the term ΔTC stands for the transaction costs over the time interval of the length Δt. It means that the balance equation reads as follows:

$$r \Pi \Delta t = \Delta \Pi = \Delta(V + \delta S) - \Delta TC. \tag{5.4}$$

Our next goal is to show how the transaction cost part ΔTC depends on other model parameters, e.g., σ, S, V, and derivatives of V. In practice, we have to adjust our portfolio by frequent buying and selling of assets. In the presence of nontrivial transaction costs, continuous portfolio adjustments may lead to infinite total transaction costs. A natural way how to consider transaction costs within the framework of the Black–Scholes theory is to follow the well known Leland approach [78] extended by Hoggard, Whalley and Wilmott (cf. [78, 64, 75]. In what follows, we recall crucial lines of the Hoggard, Whalley and Wilmott derivation of Leland's model in order to show how to incorporate the effect of transaction costs into the governing equation. More precisely, we will derive the coefficient of transaction costs ΔTC occurring in (5.4).

Henceforth, we denote by S_{ask} and S_{bid} the so-called ask and bid prices of the underlying asset, i.e., the market price offers for selling and buying assets, respectively. Furthermore, let us denote by C the round trip transaction cost per unit dollar of transaction. By this we mean that the ask and bid underlying asset prices can be expressed as follows:

$$S_{ask} = S(1+C/2), \quad S_{bid} = S(1-C/2), \quad \text{where} \quad S = \frac{S_{ask} + S_{bid}}{2}.$$

Hence the parameter C represents relative transaction costs for buying or selling one stock. Clearly,

$$C = 2 \frac{S_{ask} - S_{bid}}{S_{ask} + S_{bid}}. \tag{5.5}$$

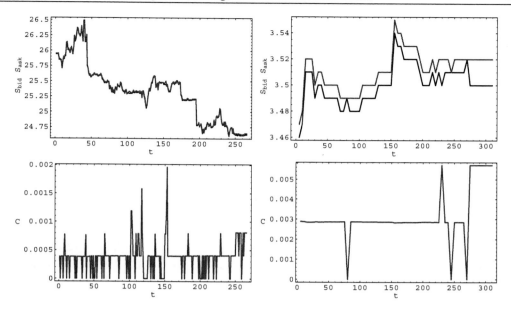

Figure 5.1. Intraday behavior (time scale of minutes) of bid–ask spreads for Microsoft asset prices, April 19, 2003 (left) and Sun Inc., May 17, 2003 (right). The estimated parameter C (below).

It means that transaction costs ΔTC are given by the value $C|k|S/2$ where k is the number of sold assets ($k < 0$) or bought assets ($k > 0$). Thus

$$\Delta TC = C|k|S/2.$$

Clearly, the number k of bought or sold assets depends on the change $k = \Delta\delta$ over one time step Δt. Hence $\Delta TC = C|\Delta\delta|S/2$. We suppose that the portfolio is maintained by following the δ-hedging strategy, i.e.

$$\delta = -\frac{\partial V}{\partial S}$$

(see Chapter 2). Recall that the underlying asset price fulfills

$$\Delta S = \mu S \Delta t + \sigma S \Delta W,$$

where $\Delta W = W(t + \Delta t) - W(t)$ is the increment of the Wiener process. Therefore, in the lowest order approximation in Δt, we obtain

$$\Delta\delta = -\sigma S \frac{\partial^2 V}{\partial S^2} \Delta W.$$

Since W is the Wiener process we have

$$E(|\Delta W|) = \sqrt{2/\pi}\sqrt{\Delta t}.$$

For a situation when the time lag Δt is small compared to $T-t$, Leland in [78] suggested to take the simple approximation: $|\Delta W| \approx E(|\Delta W|)$ (see also [64]). Hence,

$$\Delta TC = r_{TC} S \Delta t \qquad (5.6)$$

where the coefficient r_{TC} of the rate of transaction costs is given by the formula:

$$r_{TC} = \frac{C\sigma}{\sqrt{2\pi}} S \left|\frac{\partial^2 V}{\partial S^2}\right| \frac{1}{\sqrt{\Delta t}} \qquad (5.7)$$

(cf. [64]). Clearly, by increasing the time-lag Δt between portfolio adjustments we can decrease transaction costs. Now, using Itō's lemma (see Chapter 2) applied to the function V we obtain

$$\Delta \Pi = \Delta V + \delta \Delta S - \Delta TC = \left(\frac{\partial V}{\partial t} + \frac{1}{2}\sigma^2 S^2 \frac{\partial^2 V}{\partial S^2}\right)\Delta t + \left(\frac{\partial V}{\partial S} + \delta\right)\Delta S - r_{TC} S \Delta t.$$

Comparing the above expression for $\Delta \Pi$ with equation (5.3) and taking into fact that $\delta = -\frac{\partial V}{\partial S}$ we finally end up with the equation

$$\frac{\partial V}{\partial t} + \frac{1}{2}\sigma^2 S^2 \frac{\partial^2 V}{\partial S^2}\left(1 - \text{sign}\left(\frac{\partial^2 V}{\partial S^2}\right)\sqrt{\frac{2}{\pi}}\frac{C}{\sigma\sqrt{\Delta t}}\right) + r\left(S\frac{\partial V}{\partial S} - V\right) = 0.$$

The above partial differential equation can rewritten in a compact form

$$\frac{\partial V}{\partial t} + \frac{1}{2}\sigma^2 S^2 \left(1 - \text{Le sign}\left(\frac{\partial^2 V}{\partial S^2}\right)\right)\frac{\partial^2 V}{\partial S^2} + rS\frac{\partial V}{\partial S} - rV = 0, \qquad (5.8)$$

where Le is the so-called Leland number defined as follows:

$$\text{Le} = \sqrt{\frac{2}{\pi}}\frac{C}{\sigma\sqrt{\Delta t}}. \qquad (5.9)$$

The Leland number depends on the round trip transaction cost coefficient C, the time lag between two consecutive portfolio adjustments Δt and on the volatility $\sigma > 0$ of the underlying asset price.

A solution $V = V(S,t)$ the Leland equation (5.8) is defined for $(S,t) \in (0,\infty) \times (0,T)$. The terminal condition $V(S,T)$ at expiration time $t = T$ corresponds to type of an option. For example,

$$\begin{aligned} V(S,T) &= (S-E)^+ \quad \text{for call option,} \\ V(S,T) &= (E-S)^+ \quad \text{for put option.} \end{aligned}$$

In the case when we are pricing plain vanilla call or put options the Leland equation becomes a linear equation with the adjusted volatility. This is due to convexity of the solution profile $S \mapsto V(S,t)$ of a call and puts options. Therefore the second derivative $\frac{\partial^2 V}{\partial S^2}$ is always positive and so for its sign we obtain: $\text{sign}(\frac{\partial^2 V}{\partial S^2}) = 1$. In this case the Leland equation reduces to the Black–Scholes equation with a constant volatility $\hat{\sigma}$ given by

$$\hat{\sigma}^2 = \sigma^2(1 - \text{Le}).$$

Since a solution to the Black–Scholes equation is an increasing function with respect to the volatility σ (see Chapter 4) we conclude that, for any positive Leland number Le > 0, the following inequalities are satisfied:

$$V^{ec}_{lel}(S,t) < V^{ec}_{bs}(S,t), \quad V^{ep}_{lel}(S,t) < V^{ep}_{bs}(S,t) \quad \text{for any } S > 0, t \in (0,T),$$

where $V^{ec}_{bs}(S,t)$, respectively $V^{ep}_{bs}(S,t)$ is a solution to the Black–Scholes equation for pricing plain vanilla call, respectively put option and $V^{ec}_{lel}(S,t)$, respectively $V^{ep}_{lel}(S,t)$ is a solution to the Leland equation (5.8) with the same parameters $\sigma > 0, r > 0, E, T$.

Let us emphasize that the governing equation (5.8) is a backward parabolic equation satisfying a terminal condition if and only if the diffusion coefficient

$$1 - \text{Le sign}\left(\frac{\partial^2 V}{\partial S^2}\right) > 0$$

is positive. For call or put options it yields the restriction for the Leland number

$$0 \leq \text{Le} < 1.$$

This condition is achieved by taking moderate time intervals Δt between two consecutive portfolio adjustments. On the other hand, if $\Delta t \ll 1$ is very small, or the round trip transaction cost C is very large then the above condition is clearly violated and the Leland approach in modeling transaction costs is doubtful.

Notice that Leland in [78] claimed that, in the presence of transaction costs, a call option can be perfectly hedged using the Black–Scholes delta hedging with a modified volatility. Kabanov and Safarian [69] have shown failure of Leland's statement and they proved that the limiting hedging error in Leland's strategy is equal to zero only in the case the level of transaction costs tends to zero (sufficiently fast) in the limit when the time lag between two consecutive portfolio adjustments goes to zero. On the other hand, they have shown that the plain vanilla option is always under-priced (i.e., the hedging error is negative) in such a limit (see also Grandits and Schachinger [57]). Nevertheless, in the derivation of the Leland model we have only used Leland's approximation $|\Delta W| \approx E(|\Delta W|) = \sqrt{2/\pi}\sqrt{\Delta t}$ which holds only for sufficiently small $0 < \Delta t \ll 1$. Such an approximation can be justified by the following reasoning (see e.g., [66]). Let us consider another approximation of $|\Delta W|$ in the form $\phi(|\Delta W|) \approx E(\phi(|\Delta W|))$ where ϕ is a smooth increasing convex function with the property $\phi(0) = 0$. For instance, if we take $\phi(x) = x^2$ then the approximation of $|\Delta W|$ has the form $|\Delta W| \approx \sqrt{E(|\Delta W|^2)} = \sqrt{\Delta t}$. For a general function ϕ we obtain an approximation

$$|\Delta W| \approx \phi^{-1}(E(\phi(|\Delta W|))).$$

If we insert such an approximation into the formula for $r_{TC} \approx \frac{C S \sigma \Gamma}{2\Delta t}|\Delta W|$ (see (5.7)) we obtain

$$r^{\phi}_{TC} = \frac{b}{\Delta t}\phi^{-1}(E(\phi(|\Delta W|))),$$

where $b = C\sigma S|\Gamma|/2$. The coefficient of transaction costs r^{ϕ}_{TC} now also depends on the way we approximate $|\Delta W|$ expressed through the function ϕ. By Jensen's inequality (see

Exercise 7) applied to a convex increasing function ϕ we have $E(\phi(|\Delta W|)) \geq \phi(E(|\Delta W|))$ and thus

$$r_{TC}^{\phi} \geq b\frac{E(|\Delta W|)}{\Delta t} = r_{TC},$$

where r_{TC} is given by (5.7). In other words, Leland's choice of approximation $|\Delta W| \approx E(|\Delta W|)$ yields lowest possible transaction costs among all admissible approximations of $|\Delta W|$ (cf. Jandačka and Ševčovič [66])..

5.2. Modeling Option Bid–Ask Spreads by Using Leland's Model

In real market quotes data sets, there are listed two different option prices $V_{bid} < V_{ask}$ called bid and ask price representing thus offers for buying and selling options, respectively. In the Leland model presented in the previous section, a holder of a long positioned option bears transaction costs for maintaining the delta hedged portfolio by buying and selling assets. It turned out that the price of an option under the presence of transaction costs is always less that the option price on asset not paying transaction costs, i.e., $V_{lel} < V_{bs}$. From the point of view of a perspective holder who wants to buy a long positioned option, the Leland price V_{lel} can be therefore identified with the bid price of the option.

If we want to derive a pricing equation for a short positioned call option then we have to take into account that the pay–off diagram $V^{sp}(S,T) = -(S-E)^+$ is a concave function and so does the solution $V^{sp}(S,t)$. Hence, sign$\left(\frac{\partial^2 V^{sp}}{\partial S^2}\right) = -1$. The same conclusion is true for the put option. In this case the governing Leland equation changes slightly. Namely, in front of the Leland coefficient Le there is a reversed sign. It means that the Leland equation modeling higher ask option prices reads as follows:

$$\frac{\partial V}{\partial t} + \frac{1}{2}\sigma^2 S^2 \left(1 + \text{Le sign}\left(\frac{\partial^2 V}{\partial S^2}\right)\right)\frac{\partial^2 V}{\partial S^2} + rS\frac{\partial V}{\partial S} - rV = 0, \qquad (5.10)$$

where Le is again the Leland number defined as in (5.9). A solution $V = V(S,t)$ is again defined on the domain $(S,t) \in (0,\infty) \times (0,T)$. It represents the higher ask option price, i.e., the offer to sell the option. It is subject to the terminal pay–off diagram representing a chosen type of an option, e.g., a call or put option.

In Fig. 5.2 we plot solutions $S \mapsto V(S,t)$ of the Leland model for call (left) and put (right) options. The lower curve represents the bid price of an option whereas the upper curve corresponds to its ask price. The curve in the middle is the solution obtained by means of the classical Black–Scholes model, i.e., Le $= 0$. In Fig. 5.3 we depict differences (spreads) between ask and bide prices of a call option as a function of the underlying asset price S for various values of the Leland number Le. Spreads for a put option are the same because of the put–call parity (see Chapter 3).

The Leland model for pricing options under transaction costs can be used in order to capture different bid and ask prices of an option. Therefore, it can be also used for calibration of model parameters obtained from time series of bid and ask prices of option and underlying asset prices. Indeed, suppose that at the t we know the market underlying asset

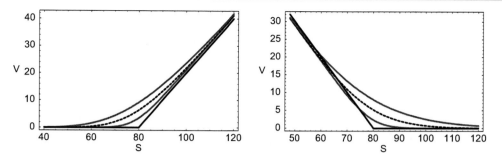

Figure 5.2. Graphs of a solution $S \mapsto V(S,t)$ of the Leland model for call (left) and put (right) options. Lower curve represents the option bid price, the middle curve is the Black–Scholes price, the upper curve is the option ask price. The parameters of the model were chose as: $E = 80, \sigma = 0.3, \mathrm{Le} = 0.15, r = 0.04, D = 0, T - t = 43/365$.

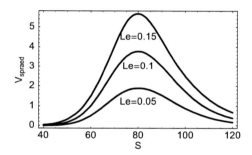

Figure 5.3. A graphical description of the bid–ask spreads between ask and bid call option prices for various Leland numbers $\mathrm{Le} = 0.15, 0.1, 0.05$. Model parameters were chosen as: $E = 80, \sigma = 0.3, r = 0.04, T - t = 43/365$.

price S_{real}, the bid and ask prices V_{real}^{bid} and V_{real}^{ask} of a call option with maturity T and expiration price E. We furthermore suppose that the interest rate $r > 0$ is also known. The coefficient C of a round trip transaction costs is given by (5.5). At the given time t it can be computed from bid and ask prices of the underlying asset where $S_{real} = (S_{ask} + S_{bid})/2$. Then we can compute the implied parameters σ and Le from the Leland model for the bid price (5.8) and ask price (5.10). Indeed, they are solutions of the system of two equations:

$$\sigma^2(1+\mathrm{Le}) = \sigma_{ask}^2, \quad \sigma^2(1-\mathrm{Le}) = \sigma_{bid}^2,$$

where σ_{ask} and σ_{bid} are uniquely determined implied volatilities of the classical Black–Scholes model (see Chapter 4), i.e.

$$V_{real}^{ask} = V_{bs}^{ec}(S_{real}, t; \sigma_{ask}), \quad V_{real}^{bid} = V_{bs}^{ec}(S_{real}, t; \sigma_{bid}).$$

In Fig. 5.4 we plot a time evolution of the asset price of Microsoft from May 21, 2002. We also show bid–ask call options prices and their arithmetic averaged prices. The option were written on the expiration price $E = 50$ with the maturity July 2, 2002. In Fig. 5.5

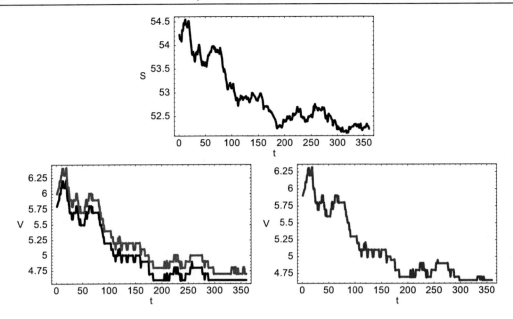

Figure 5.4. A time evolution (time scale of minutes) of the asset price of Microsoft from May 21, 2002 (upper). Bottom-left: evolution of bid–ask call options prices with expiration price $E = 50$ and maturity July 2, 2002 ($T - t = 43/365$). Bottom-right: evolution of the average $(V^{ask} + V^{bid})/2$ of call option prices.

we show time evolution (time scale of minutes) of the implied Leland volatility and Leland number. We also plot the implied time Δt between two consecutive portfolio adjustmens which can be computed from the formula (5.9) as follows:

$$\Delta t = \frac{2}{\pi} \frac{C^2}{\sigma^2 \mathrm{Le}^2}.$$

The coefficient C of a round trip transaction cost was chosen $C = 0.0004$. It corresponds to the level 0.02 of bid–ask spreads in the underlying asset price (see Fig. 5.1). It follows from intraday computations of Δt during May 21, 2002, that the optimal time between two consecutive portfolio adjustments is around the value of 50 minutes.

Furthermore, important conclusions can be made when comparing implied Leland quantities (i.e., the volatility, Leland number and time interval Δ) computed from prices of options with different maturities. In Fig. 5.6 we present such a comparison. Clearly, the time Δt between consecutive transaction is bigger for an option with longer expiration when compared to the one expiring sooner.

Problem Section and Exercises

1. The bid and ask prices of the asset of IBM are USD 118 and USD 119. It is known from the data that the historical volatility σ_{hist} of the underlying asset price was $\sigma =$

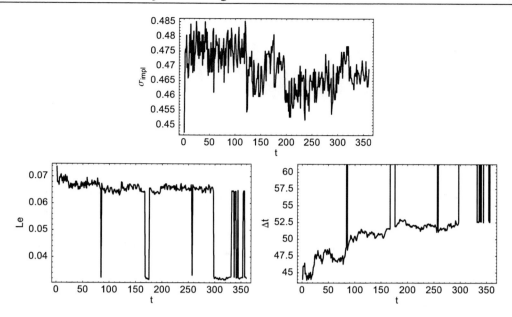

Figure 5.5. Top: the implied Leland volatility. Bottom: the implied Leland number (left) and implied transaction time Δt in the scale of minutes (right).

0.2. The price of a call option with expiration price USD 115 and maturity $T = 1/2$ (half year) is USD 6. The interest rate of a zero coupon bond is $r = 0.02$.

   a) Consider the classical Black–Scholes model. Compute the implied volatility from the option price when the asset price is the arithmetic average of the bid and ask asset prices.

   b) In the Leland model for pricing options under transaction cost, compute the coefficient C of a round trip transaction cost for a long positioned call option. Determine the time step Δt between two consecutive transactions.

2. Present a detailed derivation of the Leland model for the case of a short positioned option (see (5.10))

3. Show that for a normally distributed random variable $\Phi \sim N(0,1)$ we have $E(|\Phi|) = \sqrt{2/\pi}$.

4. How the price of a call or put long positioned option depends on the time Δt between two consecutive portfolio adjustmens? Is it an increasing or decreasing function?

5. How the bid–ask spread for a call or put long positioned option depends on the Leland number Le and the time Δt between two consecutive portfolio adjustmens? Find the value S of the underlying asset price for which the spread maximum is attained.

6. In the Leland model, by using the put–call parity show that the difference (spread) between ask and bid prices of call and put options is the same.

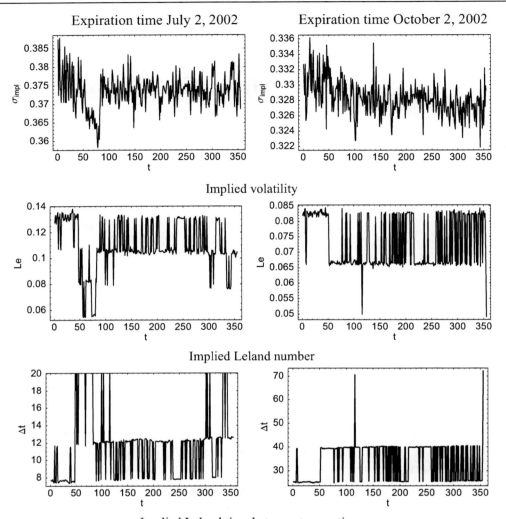

Figure 5.6. A comparison of intraday behavior of the implied volatility computed from call option prices on the stock of IBM from May 21, 2002. The expiration price was $E = 80$ and expiration time July 2, 2002 (top-left) ($T - t = 43/365$) and October 2, 2002 (top-right). In the middle and bottom row we plot implied Leland numbers and implied Leland times between transactions for both maturities.

7. Let ϕ be a smooth convex function. Prove the Jensen inequality $\phi(\sum_{i=1}^{n} f_i x_i) \leq \sum_{i=1}^{n} f_i \phi(x_i)$ for a real numbers x_i and weights $f_1, \ldots, f_n \geq 0$ such that $\sum_{i=1}^{n} f_i = 1$. With help of a discrete Jensen's inequality prove that $E(\phi(X)) \geq \phi(E(X))$ for a continuous random variable X. Notice that $E(\phi(X)) = \int_{\mathbb{R}} \phi(x) f(x) dx$, where f is a nonnegative density distribution function of X such that $\int_{\mathbb{R}} f(x) dx = 1$.

Chapter 6

Modeling and Pricing Exotic Financial Derivatives

The main purpose of this chapter is to provide a swift introduction into the subject of pricing the so-called exotic options. Pay–off diagrams of these options often depend on the entire history of an underlying asset price. They are also referred to as path–dependent options. For instance, there are Asian options, barrier options, lookback options, Parisian and Russian options. One of the reasons for introducing path-dependent options was the strong need for protecting individual call or put option holders from speculative attacks of large traders on the underlying asset price close to expiry. Furthermore, path-dependent options play an important role since they are quite common in currency and commodity markets, e.g., oil trading. In the second part of this chapter, we present an overview of pricing methods for other types of options including, in particular, binary options and compound options. Although their pay–off diagrams do not depend on the history of the underlying asset, pricing of such options is much more tricky and involved when compared to plain vanilla options.

6.1. Asian Options

Asian options represent a financial derivative depending not only on the underlying asset spot value but also on the entire time evolution of an asset in the predetermined time interval. They belong to a class of the so-called path–dependent options. In the case of Asian options, the terminal pay–off diagram depends on the terminal underlying asset price as well as on the historical average of asset values.

Classification of Asian options is more complicated because of their dependence on the way how the averaged underlying asset price enters their pay–off diagrams. We can distinguish the following basic types of Asian options:

1. According to the way how we average the underlying asset price we can define either arithmetically or geometrically averaged Asian options. In the case of discrete averaging, the average A_{t_n} at the time t_n can be defined as follows:

$$\begin{array}{cc} \text{arithmetic average} & \text{geometric average} \\ A_{t_n} = \frac{1}{n}\sum_{i=1}^{n} S_{t_i}, & \ln A_{t_n} = \frac{1}{n}\sum_{i=1}^{n} \ln S_{t_i}, \end{array} \quad (6.1)$$

where $t_1 < t_2 < \cdots < t_n$, is a division of the time interval $[0, t_n]$ such that $t_{i+1} - t_i = 1/n$ for any i.

In the case of continuous averaging, the average A_t at the time t can be defined by means of integrals as follows:

$$\begin{array}{cc} \text{arithmetic average} & \text{geometric average} \\ A_t = \frac{1}{t}\int_0^t S_\tau d\tau, & \ln A_t = \frac{1}{t}\int_0^t \ln S_\tau d\tau. \end{array} \quad (6.2)$$

2. According to the way how the averaged asset price enters the pay–off diagram we can distinguish

- Average rate

$$\begin{array}{cc} \text{call option} & \text{put option} \\ V^{arc}(S,A,T) = (A-E)^+, & V^{arp}(S,A,T) = (E-A)^+. \end{array} \quad (6.3)$$

- Average strike

$$\begin{array}{cc} \text{call option} & \text{put option} \\ V^{asc}(S,A,T) = (S-A)^+, & V^{asp}(S,A,T) = (A-S)^+, \end{array} \quad (6.4)$$

where E is the predetermined strike price (specified only for average rate options) and $S = S_T, A = A_T$ stand for the spot and averaged underlying asset prices at the time of expiration T.

For example, we can consider an Asian arithmetically averaged strike call option, or Asian geometrically averaged rate put option, etc. In summary, there are eight subclasses of Asian options. Let us also notice that the discrete form of a geometric average (6.1) can be expressed in a standard notation as follows:

$$A_{t_n} = \left(\prod_{i=1}^{n} S_{t_i}\right)^{\frac{1}{n}}, \quad (6.5)$$

representing the geometric average of the set of positive numbers - asset prices: $S_{t_1}, S_{t_2}, \ldots, S_{t_n}$.

6.1.1. A Partial Differential Equation for Pricing Asian Options

In this part, we focus our attention on derivation of the partial differential equation for pricing Asian average strike call (put) options. Similarly as in the Black–Scholes theory, we suppose the underlying asset price S is stochastic and it follows the geometric Brownian motion

$$dS = (\mu - q)Sdt + \sigma Sdw,$$

where w is the Wiener process, μ is a drift, $q \geq 0$ is a continuous dividend yield and σ is a volatility of the underlying asset price process.

The price V of an Asian option is a now function depending not only on the underlying asset spot price S and time t but also on the average A of the underlying asset price over the interval $[0,t]$. It means that $V = V(S,A,t)$. Since both the price $S = S_t$ as well as its average $A = A_t$ are stochastic variables so does their derivative, i.e., the option price $V_t = V(S_t, A_t, t)$. Before calculating the differential dV we establish the relationship between the differential of the average A and time t. In the case of arithmetic averaging we obtain

$$\frac{dA}{dt} = -\frac{1}{t^2}\int_0^t S_\tau d\tau + \frac{1}{t}S_t = \frac{S_t - A_t}{t}.$$

On the other hand, for the case of geometric averaging, we have

$$\frac{1}{A}\frac{dA}{dt} = \frac{d\ln A}{dt} = -\frac{1}{t^2}\int_0^t \ln S_\tau d\tau + \frac{1}{t}\ln S_t = \frac{\ln S_t - \ln A_t}{t}.$$

In both cases, we can conclude that the differential dA is a stochastic variable with the leading order term of the order dt. It means that

$$\text{arithmetic average} \qquad \text{geometric average}$$

$$dA = \frac{S-A}{t}dt, \qquad dA = A\frac{\ln S - \ln A}{t}dt. \qquad (6.6)$$

For both types of averaging we can express the differential dA in the form:

$$dA = Af\left(\frac{S}{A},t\right)dt, \qquad (6.7)$$

where the function f has the form: $f(x,t) = (x-1)/t$ for arithmetic averaging and $f(x,t) = (\ln x)/t$ for the case of geometric averaging.

Applying Itō's lemma 2.2 for the function $V = V(S,A,t)$ and taking into account the fact that dA is of the same order as dt, we end up with the following stochastic differential equation for the option price V:

$$\begin{aligned}dV &= \frac{\partial V}{\partial S}dS + \frac{\partial V}{\partial A}dA + \left(\frac{\partial V}{\partial t} + \frac{\sigma^2}{2}S^2\frac{\partial^2 V}{\partial S^2}\right)dt \\ &= \frac{\partial V}{\partial S}dS + \left(\frac{\partial V}{\partial t} + \frac{\sigma^2}{2}S^2\frac{\partial^2 V}{\partial S^2} + \frac{\partial V}{\partial A}Af\left(\frac{S}{A},t\right)\right)dt.\end{aligned}$$

Next steps in derivation of a partial differential equation for the option price $V = V(S,A,t)$ are almost identical with derivation of the Black–Scholes equation for plain vanilla options discussed in Chapter 3. Indeed, by construction of a self-financed delta hedged portfolio with zero net investments, we can derive the following parabolic partial differential equation for pricing Asian options:

$$\frac{\partial V}{\partial t} + \frac{\sigma^2}{2}S^2\frac{\partial^2 V}{\partial S^2} + (r-q)S\frac{\partial V}{\partial S} + Af\left(\frac{S}{A},t\right)\frac{\partial V}{\partial A} - rV = 0. \qquad (6.8)$$

Its solution $V = V(S,A,t)$ is defined on the domain $(S,A,t) \in (0,\infty) \times (0,\infty) \times (0,T)$ and satisfies the terminal pay–off diagram depending whether we are pricing call or put option. In the case of an arithmetically averaged strike call option we have

$$V(S,A,T) = (S-A)^+, \qquad (6.9)$$

and the function f appearing in equation (6.8) has the form $f(x,t) = (x-1)/t$. In the case of geometric averaging we have to take $f(x,t) = (\ln x)/t$. In the case of the average rate Asian call option the terminal pay–off diagram has the form:

$$V(S,A,T) = (A-E)^+, \qquad (6.10)$$

where E is a given expiration price at expiration time T.

6.1.2. Dimension Reduction Method and Numerical Approximation of a Solution

Let us emphasize that the partial differential equation (6.8) is not a regular parabolic equation because it does not contain the second derivative of V with respect to A. Therefore its numerical approximation may lead to complications with stability of a numerical scheme. However, in the case of an averaged strike option we can perform a dimension reduction transformation of variables such that the resulting equation is indeed a parabolic equation. This reduction is possible because of the homothetic structure of the equation and the function f. More precisely, let us consider the following change of variables:

$$V(S,A,t) = AW(x,t), \quad \text{where } x = \frac{S}{A}, \; x \in (0,\infty). \qquad (6.11)$$

If we insert these transformed variables into equation (6.8) and the terminal condition (6.9) then, after straightforward algebraic manipulations we conclude that the transformed function $W = W(x,t)$ is a solution to the following partial differential equation of parabolic type:

$$\frac{\partial W}{\partial t} + \frac{\sigma^2}{2}x^2\frac{\partial^2 W}{\partial x^2} + (r-q)x\frac{\partial W}{\partial x} + f(x,t)\left(W - x\frac{\partial W}{\partial x}\right) - rW = 0, \qquad (6.12)$$

where a solution $W = W(x,t)$ is defined on the domain $(x,t) \in (0,\infty) \times (0,T)$. The terminal pay–off diagram for an Asian averaged strike call or put option for the transformed function W has the form:

$$W^{asc}(x,T) = \max(x-1,0), \text{ respectively, } W^{asp}(x,T) = \max(1-x,0). \qquad (6.13)$$

In Fig. 6.1 we depict a solution W to the partial differential equation (6.12). In Fig. 6.2 we show 3D and contour plots of the option price $V(S,A,t)$ as a function of the underlying asset spot price S and averaged strike asset price A at a given time $t = 0.1$. The function is expressed through the transformed function W, i.e., $V(S,A,t) = AW(S/A,t)$ (see (6.11)). We also present a short numerical code written in Mathematica language for computation of the averaged strike Asian call option (see Table 6.1).

Finally, let us mention that in the case of a geometrically averaged rate call or put option we can explicitly compute the Asian option price (see Exercises 6 and 7).

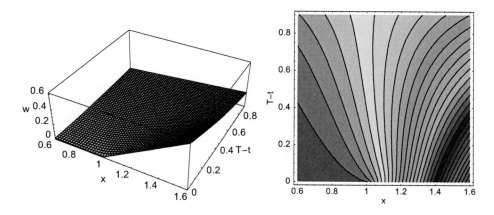

Figure 6.1. A 3D plot of a solution $W(x,t)$ to equation (6.12) (left) and its contour plot (right). Model parameters were chosen as: $\sigma = 0.4, r = 0.04, q = 0, T = 1$.

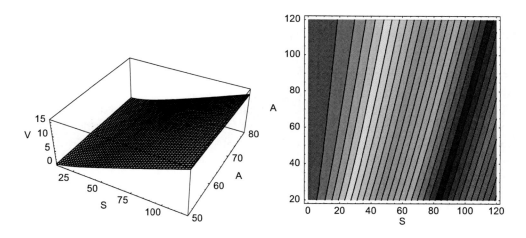

Figure 6.2. A 3D plot of the price of an Asian averaged strike call option $V(S,A,t) = AW(S/A,t)$ at the time $t = 0.1$ (i.e., $T - t = 0.9$) (left). A contour plot of the function $V(S,A,t)$ (right). The model parameters are the same as in Fig. 6.1.

6.2. Barrier Options

Barrier options can be considered as examples of simple path-dependent options whose price depend on the history of the underlying asset price. Barrier options are either call or put options with the possibility of their early exercising in cases when the underlying asset price hits from above (or from below) the predetermined barrier value B. In the case of early exercising the writer of a barrier option contract pays the option holder the prescribed amount of money called a rebate. The rebate may be set to zero as well. After that time the barrier option contract is no longer valid.

Table 6.1. The Mathematica source code for pricing of an Asian arithmetically averaged call option.

```
sigma=0.4; r=0.04; d=0; T=1; t=0.9; xmax=8;

PayOff[x_] := If[x - 1 > 0, x - 1, 0];

solution = NDSolve[{
  D[w[x, tau],tau] == (sigma^2/2) x^2 D[w[x, tau], x,x]
  + (r - d)*x * D[w[x, tau], x]
  + ((x - 1)/(T - tau))*(w[x, tau] - x*D[w[x, tau], x])
    - r*w[x, tau],
        w[x, 0] == PayOff[x],
        w[0, tau] == 0,
        w[xmax, tau] == PayOff[xmax]},
      w, {tau, 0, t}, {x, 0, xmax}
      ];

w[x_, tau_] := Evaluate[w[x, tau] /. solution[[1]] ];
V[tau_, S_, A_] := A w[S/A, tau];
Plot3D[ V[t, S, A], {S, 10, 120}, {A, 50, 80}];
```

"Out" barrier options are active from the beginning. However, they become null and void in the event a given knock-out barrier B price is attained. On the other hand, "in" barrier options are void at the beginning of a contract. They become active in the time a given knock-in barrier B price is attained. Based on this terminology, we can distinguish four types of barrier options:

- Down-and-out barrier option: in the beginning the underlying asset price is above the predetermined barrier B. If the underlying asset price moves down and hits the barrier B from above then the option contract becomes void. The writer of an option pays the holder a predetermined rebate $R \geq 0$.

- Up-and-out barrier option: analogously as in the case of a down-and-out option, the underlying asset price is below the predetermined barrier B in the beginning. If the underlying asset price moves up and hits the barrier B from below then the option contract becomes void. Again, the writer of an option pays the holder a predetermined rebate $R \geq 0$.

- Up-and-in barrier option: the underlying asset price is below the barrier B in the beginning. The option becomes active in the event when underlying asset moves up and hits the barrier from below.

- Down-and-in barrier option: the underlying asset price is above the barrier B in the beginning. The option becomes active in the event when underlying asset moves down and hits the barrier from above.

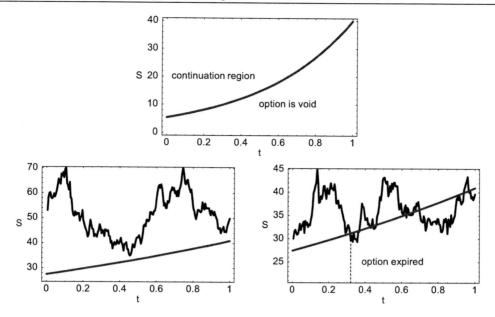

Figure 6.3. An example of an exponential barrier function $B(t)$ for a down-and-out barrier option (upper). In the left figure we show an example of evolution of the asset for which the barrier is inactive for the entire interval $[0, T]$. An example of activation of the barrier is shown in the right picture.

In the rest of this section, we will analyze the case of a down-and-out barrier call or put option. The remaining cases can be treated analogously. Suppose we are given the barrier B expressed in terms of a function depending time $t \in [0, T]$, i.e., $B = B(t)$. The terminal condition of a down-and-out call or put option is given by the pay–off diagram of a call or put option. In the case, the underlying asset price moves down and hits the barrier $B(t)$ at some time $t \in (0, T)$ then the option contract is void and the holder receives the predetermined rebate $R(t) \geq 0$. Notice that the predetermined rebate can be also prescribed to zero.

As a typical example of a predetermined barrier function one can consider the exponential barrier function of the form:

$$B(t) = bEe^{-\alpha(T-t)}, \tag{6.14}$$

where $0 < b \leq 1, \alpha \geq 0$, are given constants. In Fig. 6.3 we plot examples of evolution of the underlying asset price and the case of activation of the down-and-out barrier option. At the time of activation the holder receives predeterminded rebate $R(t) \geq 0$. As a typical rebate function one can consider a function of the form

$$R(t) = E\left(1 - e^{-\beta(T-t)}\right), \tag{6.15}$$

where $\beta \geq 0$. Notice that such a rebate function $R(t) \geq 0$ satisfies a natural terminal condition $R(T) = 0$ at expiry $t = T$.

Since the down-and-out option is active only in the time dependent domain of underlying asset prices $S > B(t)$ we can price such an option by means of the Black–Scholes theory only in the region in which $S > B(t)$. It means the option price $V = V(S,t)$ satisfies the following parabolic equation:

$$\frac{\partial V}{\partial t} + \frac{1}{2}\sigma^2 S^2 \frac{\partial^2 V}{\partial S^2} + (r-q)S\frac{\partial V}{\partial S} - rV = 0, \qquad (6.16)$$

for $t \in (0,T), S > B(t)$. At the boundary of the computational domain $S = B(t)$ the price $V(S,t)$ is given by the predetermined rebate value $R(t)$. It means that the following boundary condition

$$V(B(t),t) = R(t) \qquad (6.17)$$

has to be satisfied for any $t \in (0,T)$.

Finally, depending on type of an option, we have to prescribe the terminal condition given by the pay–off diagram of a call or put option, i.e.

$$V^{bc}(S,T) = (S-E)^+, \quad \text{respectively,} \quad V^{bp}(S,T) = (E-S)^+, \qquad (6.18)$$

where E is the expiration price at the expiration time T.

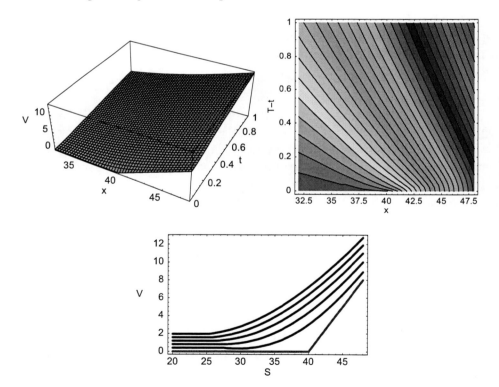

Figure 6.4. A 3D graph of a solution to equation (6.20) for pricing a down-and-out barrier call option (upper left). A contour plot graph of level sets of the function W (upper right). Option prices $S \mapsto V(S,t)$ for various times t (bottom).

6.2.1. Numerical and Analytical Solutions to the Partial Differential Equation for Pricing Barrier Options

Table 6.2. The Mathematica source code for pricing a down-and-out barrier call option.

```
b = 0.7; alfa = 0.1; beta = 0.05; X = 40;
sigma = 0.4; r = 0.04; q = 0; T = 1;

xmax = 2;

Bariera[t_] := X b Exp[-alfa (T - t)];
Rabat[t_]   := X (1 - Exp[-beta(T - t)]);

PayOff[x_] := X*If[b Exp[x] - 1 > 0, b Exp[x] - 1, 0];

solution = NDSolve[{
    D[w[x, tau], tau] == (sigma^2/2)D[w[x, tau], x, x]
       + (r - q - sigma^2/2 - alfa )* D[w[x, tau], x]
       - r *w[x, tau] ,

       w[x, 0]    == PayOff[x],
       w[0, tau]  == Rabat[T - tau],
       w[xmax, tau] == PayOff[xmax] },

    w, {tau, 0, T}, {x, 0, xmax}
    ];

w[x_, tau_] := Evaluate[w[x, tau] /. solution[[1]] ];
Plot3D[w[x, tau], {x, 0, xmax}, {tau, 0, T}];

V[S_, tau_] :=
   If[S > Bariera[T - tau],
       w[ Log[S/Bariera[T - tau]], tau],
       Rabat[T - tau]
   ];

Plot[
{V(S,0.2 T],V(S,0.4 T], V(S,0.6 T], V(S,0.8 T], V(S,T]},
{S,20,50}];
```

Equation (6.16) can be transformed by the following simple transformation

$$V(S,t) = W(x,t), \quad \text{where } x = \ln\left(\frac{S}{B(t)}\right), \quad x \in (0,\infty). \qquad (6.19)$$

It is easy to verify that the transformed function W is then a solution of the partial differen-

tial equation:
$$\frac{\partial W}{\partial t} + \frac{\sigma^2}{2}\frac{\partial^2 W}{\partial x^2} + \left(r - q - \frac{\sigma^2}{2} - \alpha\right)\frac{\partial W}{\partial x} - rW = 0 \qquad (6.20)$$
defined on a fixed domain $(x,t) \in (0,\infty) \times (0,T)$. The boundary condition (6.17) and terminal condition (6.18) are transformed as follows:
$$W(0,t) = R(t), \qquad (6.21)$$
$$W^{bc}(x,T) = E\max(be^x - 1, 0), \text{ respectively, } W^{bp}(x,T) = E\max(1 - be^x, 0). \qquad (6.22)$$

In Fig. 6.4 we depict a graph of a solution to the partial differential equation (6.20) with a rebate function defined as in (6.15). We again present a simple Mathematica code for computation of the price of a down-and-out barrier call option (see Table 6.2).

Notice that equation (6.20) together with the terminal condition (6.22) and boundary condition (6.21) can be solved analytically. Its explicit solution obtained by the transformation method based on the so-called Duhamel integral is presented in the exercise section. As far as practical implementation is concerned, the explicit formula is however rather involved and its numerical computation turns to be equally hard when compared to the numerical solution of (6.20).

6.3. Binary Options

Binary options (also referred to as digital options) represent very simple form of exotic options. The pay–off diagram of a binary option is given by
$$V^{bin}(S,T) = \begin{cases} 1, & \text{if } S \in [E_1, E_2], \\ 0, & \text{otherwise,} \end{cases}$$
where $0 < E_1 < E_2$ are prescribed values (see Fig. 2.13). It is easy to derive the following explicit solution to a binary option:
$$V^{bin}(S,t) = e^{-r(T-t)}(N(d_2^{E_1}) - N(d_2^{E_2})),$$
where
$$d_2^E = \frac{\ln(S/E) + (r - q - \sigma^2/2)(T-t)}{\sigma\sqrt{T-t}}.$$

The above formula follows from the fact that the pay–off diagram can be obtained as a limit, when $\varepsilon \to 0^+$, of a combination of four plain vanilla call options written on the different exercise prices $E_1, E_1 + \varepsilon, E_2, E_2 + \varepsilon$:
$$V^{bin}(S,T) \approx \frac{(-V^{ec}(S,T;E_1 + \varepsilon) + V^{ec}(S,T;E_1))}{\varepsilon}$$
$$- \frac{(-V^{ec}(S,T;E_2 + \varepsilon) + V^{ec}(S,T;E_2))}{\varepsilon}.$$

Hence
$$V^{bin}(S,t) = -\frac{\partial V^{ec}}{\partial E}(S,t;E_1) + \frac{\partial V^{ec}}{\partial E}(S,t;E_2)$$
and the pricing formula for a binary option follows from the identity (4.6). A graphical description of the price $V^{bin}(S,t)$ of a binary option is depicted in Fig. 6.5.

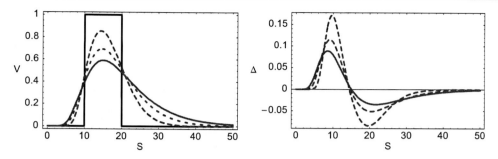

Figure 6.5. Left: a plot of the profile $S \mapsto V^{bin}(S,t)$ of a binary option for times $t = T$ (the pay-off diagram), $t = 2T/3, t = T/3$ (dashed lines), $t = 0$ (solid line) computed for the model parameters: $E_1 = 10, E_2 = 20, r = 0.04, q = 0.02, T = 1$. Right: a plot of the corresponding factor $\Delta = \frac{\partial V^{bin}}{\partial S}$.

6.4. Compound Options

A compound option is a financial derivative on the underlying asset represented again by another option. We restrict ourselves to the class of plain call or put options written on underlying call or put options. We can distinguish four types of plain compound options:

1. a call option on a call option,
2. a put option on a call option,
3. a call option on a put option,
4. a put option on a put option.

The compound options were introduced and studied by Geske in [55] in the context of pricing of an option on the underlying stock asset where the stock asset of a firm itself is priced as an option written on the underlying value of the firm.

In this section, we will present the methodology of pricing a compound option on a particular example of a call option written on another call option. The other cases of compound options can be treated analogously.

Let us consider a compound call option A with the expiration time T_A and expiration price E_A. Its underlying asset is another call option B written on the underlying stock asset S paying no dividends ($q = 0$) and having the expiration time $T_B > T_A$ and expiration price E_B. Hence the terminal pay-off diagram of such a call on call option has the form:

$$V_A(S, T_A) = (V_B(S, T_A) - E_A)^+.$$

We suppose that the underlying stock asset process S for the option B follows a geometric Brownian motion of the form:

$$dS = \mu S dt + \sigma S dw,$$

with some drift μ and volatility σ. Then a price of the underlying asset for a compound option, i.e., the call option B at the time $t < T_B$ is known and it is given by the explicit formula (4.3),

$$V_B(S,t) = SN(d_1^B) - E_B e^{-r(T_B-t)} N(d_2^B), \quad 0 \le t < T_B, \tag{6.23}$$

where

$$d_1^B = \frac{\ln\left(\frac{S}{E_B}\right) + (r + \frac{\sigma^2}{2})(T_B - t)}{\sigma\sqrt{T_B - t}}, \qquad d_2^B = d_1^B - \sigma\sqrt{T_B - t}.$$

Moreover, $V_B = V_B(S,t)$ is a solution to the Black–Scholes equation

$$\frac{\partial V_B}{\partial t} + \frac{1}{2}\sigma^2 S^2 \frac{\partial^2 V_B}{\partial S^2} + rS\frac{\partial V_B}{\partial S} - rV_B = 0. \tag{6.24}$$

Now, by using Itō's lemma (2.2), we can derive the stochastic differential equation for the stochastic process $V_B(S_t,t)$:

$$dV_B = \tilde{\mu}V_B dt + \tilde{\sigma}V_B dw, \tag{6.25}$$

where

$$\tilde{\mu}(S,t) = \frac{1}{V_B}\left(\frac{\partial V_B}{\partial t} + \frac{1}{2}\sigma^2 S^2 \frac{\partial^2 V_B}{\partial S^2} + \mu S \frac{\partial V_B}{\partial S}\right), \quad \tilde{\sigma}(S,t) = \sigma \frac{S}{V_B}\frac{\partial V_B}{\partial S}$$

and $V_B = V_B(S,t)$. The compound option A, considered now as a derivative of the underlying asset V_B, is therefore a solution to the Black–Scholes equation with variable volatility $\tilde{\sigma}(S,t)$, i.e.

$$\frac{\partial V_A}{\partial t} + \frac{1}{2}\tilde{\sigma}^2(S,t) V_B^2 \frac{\partial^2 V_A}{\partial V_B^2} + rV_B \frac{\partial V_A}{\partial V_B} - rV_A = 0. \tag{6.26}$$

In what follows, we will derive a PDE for the compound option price $V_A = V_A(S,t)$ considered as a function of the stock asset price S. To this end, we calculate the partial derivatives:

$$\frac{\partial V_A}{\partial S} = \frac{\partial V_A}{\partial V_B}\frac{\partial V_B}{\partial S}, \qquad \frac{\partial^2 V_A}{\partial S^2} = \frac{\partial^2 V_A}{\partial V_B^2}\left(\frac{\partial V_B}{\partial S}\right)^2 + \frac{\partial V_A}{\partial V_B}\frac{\partial^2 V_B}{\partial S^2}.$$

Inserting partial derivatives of V_A with respect to V_B into (6.26), taking into account the expression for $\tilde{\sigma}(S,t)$ and using the fact that the function V_B is a solution to (6.24) we conclude $V_A = V_A(S,t)$ is a solution to the Black–Scholes equation with a right hand side:

$$\frac{\partial V_A}{\partial t} + \frac{1}{2}\sigma^2 S^2 \frac{\partial^2 V_A}{\partial S^2} + rS\frac{\partial V_A}{\partial S} - rV_A = c(S,t)\frac{\partial V_A}{\partial S}, \quad \text{where } c(S,t) = -\frac{\frac{\partial V_B}{\partial t}}{\frac{\partial V_B}{\partial S}}, \tag{6.27}$$

satisfying the terminal pay–off diagram $V_A(S,T) = (V_B(S,T_A) - E_A)^+$. The explicit solution to the above parabolic equation has been derived by Geske in the seminal paper [55]. It is a straightforward but very long verification that the unique solution to (6.27) is given by the following formula:

$$V_A(S,t) = SM(d_1^A, d_1^B, \rho) - E_B e^{-r(T_B-t)} M(d_2^A, d_2^B, \rho) - E_A e^{-r(T_A-t)} N(d_2^A). \tag{6.28}$$

Here

$$M(x,y,,\rho) = \frac{1}{2\pi\sqrt{1-\rho^2}} \int_{-\infty}^{x} \int_{-\infty}^{y} e^{-\frac{1}{2(1-\rho^2)}(\xi^2 - 2\rho\xi\eta + \eta^2)} d\eta d\xi$$

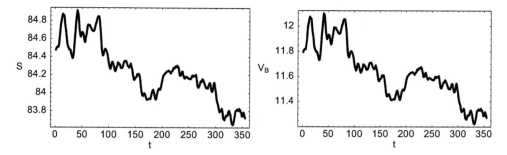

Figure 6.6. A graph of the IBM stock price from May 21, 2002 plotted with one minute time stepping (left). The price V_B of a call option with $E_B = 80, T_B = 1, \sigma = 0.23, r = 0.04, q = 0$ (right).

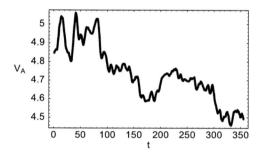

Figure 6.7. The price V_A of a compound call option with $E_A = 10, T_A = 0.5, E_B = 80, T_B = 1, \sigma = 0.23, r = 0.04$.

denotes the bivariate normal distribution function with a correlation coefficient

$$\rho = \sqrt{(T_A - t)/(T_B - t)} \in [0, 1)$$

and

$$d_1^A = \frac{\ln\left(\frac{S}{S^*}\right) + (r + \frac{\sigma^2}{2})(T_A - t)}{\sigma\sqrt{T_A - t}}, \quad d_2^A = d_1^A - \sigma\sqrt{T_A - t},$$

where $S = S^*$ is the unique root of the equation

$$V_B(S^*, T_A) = E_A.$$

In Fig. 6.6 we plot the intraday behavior of the IBM stock from May 21, 2002. We also plot the prices of a call option V_B which are considered as the underlying asset for the compound call option V_A. The behavior of the compound option V_A is depicted in Fig. 6.7.

6.5. Lookback Options

Lookback options represent financial derivatives whose price depend not only on the underlying asset spot price but also on the maximum value of the underlying asset process during the prescribed time period. We can distinguish two types of lookback options:

- floating maximum strike lookback put options,
- floating maximum rate lookback call options.

Notice that it is meaningless to consider the case of a call lookback option on a floating maximum strike because the terminal underlying stock price S_T is always less or equal to the floating strike price $M_T = \max_{t \in [0,T]} S_t$. Hence the price of such a call option is always zero.

Analogously, if we consider minimum value of the underlying asset process during the prescribed time period we can distinguish:

- floating minimum strike lookback call options,
- floating minimum rate lookback put options.

Again, it is meaningless to consider the case of a put lookback option on a floating minimum strike because the terminal underlying stock price S_T is always greater or equal to the floating minimum strike price $m_T = \min_{t \in [0,T]} S_t$ and therefore a price of such a put option is always zero. Notice that for floating maximum or minimum rate options we consider both call as well as put options.

In the rest of this section we will present a method how to derive an explicit formula for pricing lookback options. We concentrate on derivation of the pricing formula for the floating maximum strike lookback put option. The procedure of derivation in remaining cases is similar and therefore omitted.

A floating maximum strike lookback put option has the exercise price M given by a maximum of the underlying asset price over the entire time interval $[0,T]$, i.e.

$$M = \max_{t \in [0,T]} S_t.$$

It means that the terminal pay–off of such a lookback put option is given by:

$$V^{lp}(S,M,T) = (M-S)^+, \qquad (6.29)$$

where $S = S_T$ is the price of an underlying asset at the expiration time T.

Let us emphasize that the maximal value of a positive function S_τ taken over the interval $[0,t]$ can be computed as a limit

$$M_t = \max_{0 \le \tau \le t} S_\tau = \lim_{p \to \infty} A_{p,t}, \qquad \text{where } A_{p,t} = \left(\frac{1}{t} \int_0^t S_\tau^p d\tau\right)^{\frac{1}{p}}$$

(see Exercise 5 below). Therefore we can approximate the maximum M_t by the p-integral average $A_{p,t}$, where $p \gg 1$ is large enough. Notice that the differential $dA_{p,t}$ of the p-integral average can be easily computed as follows:

$$\frac{dA_{p,t}}{A_{p,t}} = f_p(x,t), \quad x = \frac{S_t}{A_{p,t}}, \qquad \text{where } f_p(x,t) = \frac{x^p - 1}{pt}.$$

In the limit $p \to \infty$, the floating maximum strike lookback put option can be therefore viewed as an Asian put option with a floating strike given by the p-integral average $A_{p,t}$. By introducing the following change of variables

$$V^{lp}(S,A,t) = AW(x,t), \quad \text{where } x = \frac{S}{A}, \quad x \in (0,\infty),$$

we end up with the parabolic equation for the transformed function W

$$\frac{\partial W}{\partial t} + \frac{\sigma^2}{2} x^2 \frac{\partial^2 W}{\partial x^2} + rx\frac{\partial W}{\partial x} + f_p(x,t)\left(W - x\frac{\partial W}{\partial x}\right) - rW = 0, \tag{6.30}$$

defined for $0 < x < \infty, 0 < t < T$. Clearly, for $0 < x < 1$ we have $\lim_{p\to\infty} f_p(x,t) = 0$. Hence, in the limit $p \to \infty$, the Black–Scholes equation for W reads as follows:

$$\frac{\partial W}{\partial t} + \frac{\sigma^2}{2} x^2 \frac{\partial^2 W}{\partial x^2} + rx\frac{\partial W}{\partial x} - rW = 0, \quad 0 < x < 1, \; 0 < t < T. \tag{6.31}$$

On the other hand, if $1 < x$ we have $\lim_{p\to\infty} f_p(x,t) = \infty$. By dividing equation (6.30) by the term f_p and passing to the limit as $p \to \infty$ we deduce that

$$W(x,t) - x\frac{\partial W}{\partial x} = 0, \quad 1 < x, \; 0 < t < T.$$

By taking the limit $x \to 1^+$, from the latter equation we can deduce the boundary condition for the solution $W(x,t)$ at $x = 1$, i.e.

$$W(1,t) = \frac{\partial W}{\partial x}(1,t), \quad 0 < t < T. \tag{6.32}$$

A solution W is subject to the terminal pay–off diagram of a put option, i.e., $W(x,T) = (1-x)^+$. Since $0 < x < 1$ we simply have

$$W(x,T) = 1 - x, \quad 0 < x < 1. \tag{6.33}$$

After long, but very straightforward algebraic computations (see also Exercises 8,9), one can verify that the solution W to equation (6.30) satisfying the boundary and initial conditions (6.32) and (6.33) can be explicitly found in the form:

$$W(x,t) = W^{ep}(x,t) + \frac{\sigma^2}{2r}\left[xN(d_1) - e^{-r(T-t)}x^{1-\frac{2r}{\sigma^2}}N(d_3)\right], \tag{6.34}$$

where $W^{ep}(x,t)$ is the price of an European put option with a unit strike price, i.e.

$$W^{ep}(x,t) = e^{-r(T-t)}N(-d_2) - xN(-d_1)$$

and

$$d_1 = \frac{\ln x + (r + \sigma^2/2)(T-t)}{\sigma\sqrt{T-t}}, \quad d_2 = d_1 - \sigma\sqrt{T-t}, \quad d_3 = d_1 - \frac{2r}{\sigma}\sqrt{T-t}.$$

Problem Section and Exercises

1. Using transformation of variables and the method of the Duhamel integral applied to equation (6.20) derive the following explicit formula:

$$w(x,t) = E \frac{e^{\lambda x + Q(T-t)}}{\sigma\sqrt{2\pi(T-t)}} \int_{-\ln b}^{\infty} \left(be^{-(\lambda-1)\xi} - e^{-\lambda\xi}\right)$$
$$\times \left(e^{-(x-\xi)^2/(2\sigma^2(T-t))} - e^{-(x+\xi)^2/(2\sigma^2(T-t))}\right) d\xi$$
$$+ \frac{xe^{\lambda x}}{\sigma\sqrt{2\pi}} \int_0^{T-t} \frac{e^{Qs}e^{-x^2/(2\sigma^2 s)}}{s^{\frac{3}{2}}} R(T-t-s)\, ds,$$

for the price of a down-and-out barrier call option with a given rebate function $R(t)$. Here $\lambda = \frac{1}{2} + \frac{\alpha-r}{\sigma^2}, Q = -\frac{\lambda^2\sigma^2}{2} - r$.

2. Compute numerically the price of an Asian arithmetically averaged strike call option for model parameters: $\sigma = 0.4, r = 0.04, q = 0, T = 1, t = 0.9$. Compare the price of such an Asian call option with the price of the plain vanilla call option with a strike price E equal to the averaged price A. Which price is higher?

3. Following the derivation of the price of a compound call on a call derive formulae for all three remaining cases of compound options.

4. Derive the pricing formula for the asset-or-nothing option (see Chapter 2) with the pay–off diagram $V(S,T) = S$ if $S > E$ and $V(S,T) = 0$, otherwise.

5. Show that the maximum of positive numbers $S_{t_1}, S_{t_2}, \ldots, S_{t_n}$ can be expressed as the limit

$$\max_{i=1,\ldots,n} S_{t_i} = \lim_{p\to\infty} \left(\frac{1}{n}\sum_{i=1}^n S_{t_i}^p\right)^{\frac{1}{p}}.$$

For a continuous function $S_\tau, \tau \in [0,T]$, show that

$$\max_{\tau\in[0,t]} S_\tau = \lim_{p\to\infty} \left(\frac{1}{t}\int_0^t S_\tau^p d\tau\right)^{\frac{1}{p}}.$$

6. Assume the underlying stock price follows the geometric Brownian motion $dS = \mu S dt + \sigma S dw$. Show that the geometric average defined by $\ln A_t = \frac{1}{t}\int_0^t \ln S_\tau d\tau$ is again a solution to the SDE of the geometric Brownian motion. More precisely, show that

$$dA = \tilde{\mu} A dt + \tilde{\sigma} A dw,$$

where $\tilde{\mu} = \frac{1}{2}(\mu - \frac{\sigma^2}{2} + \tilde{\sigma}^2)$ and $\tilde{\sigma} = \frac{\sigma}{\sqrt{3}}$.

7. With regard to the previous example, find an explicit solution for pricing the Asian style geometrically averaged rate call option.

8. Prove the identity
$$N'(d_2) - x^{1-\frac{2r}{\sigma^2}} N'(d_3) = 0,$$
where
$$d_1 = \frac{\ln x + (r+\sigma^2/2)(T-t)}{\sigma\sqrt{T-t}}, \quad d_2 = d_1 - \sigma\sqrt{T-t}, \quad d_3 = d_1 - \frac{2r}{\sigma}\sqrt{T-t}.$$

With help of this identity show that the function $W(x,t)$ defined as in (6.34) is indeed a solution to the parabolic equation (6.31). Notice that the European put option price W^{ep} solves this equation as well.

9. Show that $d_3 = -d_2$ for $x = 1$, where d_2, d_3 are defined as in the previous example. Using this fact prove that the function $W(x,t)$ defined as in (6.34) satisfies the boundary condition (6.32).

Chapter 7

Short Interest Rate Modeling

In the following two chapters we will deal with modeling of interest rates and their simple derivatives. In this chapter, we present analysis of the so-called short rate models for description of the instantaneous (or short) interest rate. The instantaneous interest rate will be described by a solution to the stochastic differential equation. We will discuss the role of model parameters. With regards to the coefficients of the governing stochastic differential equation, we will distinguish between various types of interest rate models. In the modeling of short rate processes, the important role is played by knowledge of the probability distribution of the short rate stochastic process. Therefore we present the mathematical framework of the Fokker-Planck equation describing the time dependent probability distribution function for a stochastic process satisfying the prescribed stochastic differential equation. At the end of the chapter, we will show how the information about this distribution can be used in order to calibrate the interest rate models.

In Fig. 7.1 we present an example of the time evolution of the instantaneous (short) interest rate of the term structure BRIBOR (BRatislava Interbank Offered Rate) from Slovakia. It is obvious from the displayed behavior that the evolution of the short interest rate can be considered as a stochastic process fluctuating within the interval of 2% up to 6% p.a. In contrast to stochastic behavior of stock prices (see Chapter 2), the short interest rate process exhibits the so-called mean reversion property. Loosely speaking, by this we mean that the interest rate is fluctuating around long term interest rate θ (3.5% in our example) and upward and downward deviations from this value are reverted towards θ with some speed $\kappa > 0$.

7.1. One-Factor Interest Rate Models

In one-factor models it is assumed that the instantaneous interest rate r is characterized by the solution to the stochastic differential equation that has a general form

$$dr = \mu(t,r)dt + \sigma(t,r)dw. \qquad (7.1)$$

The deterministic part of the process $\mu(t,r)$ represents the trend (or drift) of the short rate behavior whereas the volatility $\sigma(t,r)$ measures the size of random fluctuations that are modulated around the deterministic (trend) part.

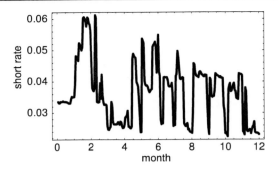

Figure 7.1. Time evolution of the short interest rate from the BRIBOR term structure in the year 2007.

A usual choice of the drift function is a linear decreasing function $\mu(t,r) = \kappa(\theta - r)$, where κ, θ are positive constants. The parameter θ is called the long term limiting interest rate and κ is called the speed (or rate) of mean reversion. It measures a speed of return of the short rate towards the limiting interest rate θ. Stochastic processes with such a specific form of the deterministic part are also referred to as the Ornstein–Uhlenbeck mean reversion processes. The essence of the drift of the form $\mu(t,r) = \kappa(\theta - r)$ is that the mean value of the short rate process is attracted towards the limit value θ with a speed of attraction proportional to the parameter κ. Indeed, for the mean value $E(r_t)$ it holds:

$$dE(r_t) = E(dr_t) = \kappa(\theta - E(r_t))dt + E(\sigma(t,r)dw) = \kappa(\theta - E(r_t))dt.$$

It follows from the construction of Itō's integral (see Chapter 2) that

$$E(\sigma(t,r)dw) = 0.$$

Now it is easy to verify that a solution to the differential equation

$$\frac{d}{dt}E(r_t) = \kappa(\theta - E(r_t))$$

is uniquely given by

$$E(r_t) = \theta + (E(r_0) - \theta)e^{-\kappa t}. \tag{7.2}$$

Clearly, it means that $\lim_{t \to \infty} E(r_t) = \theta$.

In the rest of this section we restrict our attention on the class of Orstein–Uhlenbeck processes of mean reversion type. Depending on the form of volatility $\sigma(r,t)$ we distinguish particular the short interest rate models. Assuming a constant volatility, i.e., $\sigma(t,r) = \sigma$, we obtain the so-called Vasicek model

$$dr = \kappa(\theta - r)dt + \sigma dw. \tag{7.3}$$

It was derived by O. A. Vasicek in [120]. The Vasicek model is one of the first and most simple models for short interest rates. The analysis of the stochastic properties of the short rate process driven by Vasicek's model is relatively simple. It will turn out (see the next section) that the random variable r at a given time t is just a normally distributed random variable with mean and variance depending on time. However, one of principal disadvantages of the Vasicek model is the fact, that it allows negative interest rates. Indeed, even if the short rate is close to zero, its volatility is always equal to the given constant σ and therefore the process r_t can become negative.

The Cox–Ingersoll–Ross model (abbreviated as the CIR model) derived by the authors in [31], eliminates the possibility of negative short rate for certain range of model parameters. The model still preserves the assumption on the drift $\kappa(\theta - r)$ in the form of a mean-reversion process. Unlike the Vasicek model, the volatility of the CIR process is no longer assumed to be constant, but it is proportional to the square root of r, i.e.

$$dr = \kappa(\theta - r)dt + \sigma\sqrt{r}dw. \tag{7.4}$$

A stochastic process for a random variable r of the form (7.4) where the differential of the Wiener process dw is multiplied by the square root of r is also referred to as the Bessel square root process. It means that, for small values r of interest rates, the whole volatility term $\sigma\sqrt{r}$ is also as well. Furthermore, if the zero value of r is attained then the volatility of r is vanishing as well. The further evolution of r is therefore deterministic and is described just by the drift function, which is positive for $r = 0$. It means that the interest rate increases and again attains positive values. Moreover, if coefficients of the CIR process satisfy the inequality

$$2\kappa\theta \geq \sigma^2,$$

then the process r attains a zero value with zero probability. Intuitively, this property can be seen by applying the transformation $x = \ln r$. The value $r = 0$ then corresponds to $x = -\infty$ for the new variable x. From Itō's lemma it follows that

$$dx = (e^{-x}(\kappa\theta - \sigma^2/2) - \kappa)dt + e^{-x/2}\sigma dw.$$

Now, if $2\kappa\theta < \sigma^2$ then for $x \to -\infty$ the drift of the above stochastic process tends to $-\infty$. The process x is therefore driven to towards $-\infty$ without possibility of reversion. More precisely, for the mean value $E(x)$ of the process x we have

$$dE(x) = (E(e^{-x})(\kappa\theta - \sigma^2/2) - \kappa)dt$$

because $E(e^{-x/2}\sigma dw) = 0$ (see Chapter 2). Using Jensen's inequality $E(e^{-x}) > e^{-E(x)}$ applied for the convex function $x \mapsto e^{-x}$, we deduce

$$\frac{d}{dt}E(x) < (\kappa\theta - \sigma^2/2)e^{-E(x)}.$$

Hence, $e^{E(x_t)} \leq e^{E(x_0)} - (\sigma^2/2 - \kappa\theta)t$ and so $E(x_t) \to -\infty$ as $t \to T_{max} := r_0/(\sigma^2/2 - \kappa\theta)$, where $r_0 = e^{x_0} = e^{E(x_0)}$. On the other hand, the condition $2\kappa\theta \geq \sigma^2$ clearly eliminates such an undesirable behavior.

Notice that there are several other models that are generalizations of the previously mentioned Vasicek and CIR models. For example, another popular model for the short rate process was proposed by Chan, Karolyi, Longstaff and Sanders in [23]. Their model is given by a solution to the SDE:

$$dr = \kappa(\theta - r)dt + \sigma r^\gamma dw, \tag{7.5}$$

Table 7.1. An overview of one-factor short rate models having a form of the Chan, Karolyi, Longstaff and Sanders (CKLS) process.

Model	Stochastic equation for r
CKLS	$dr = \kappa(\theta - r)dt + \sigma r^\gamma dw$
Vasicek	$dr = \kappa(\theta - r)dt + \sigma dw$
Dothan	$dr = \sigma r dw$
Brennan–Schwartz	$dr = \kappa(\theta - r)dt + \sigma r dw$
Cox–Ingersoll–Ross	$dr = \kappa(\theta - r)dt + \sigma \sqrt{r} dw$
Cox–Ross	$dr = \beta r dt + \sigma r^\gamma dw$

where $\gamma \geq 0$ is a constant. A brief and incomplete overview of other interest rate models is presented in Table 7.1.

Let us recall that there are also other possibilities how to design a short rate model with the property of nonnegativity of the short rate. For example, if we use a process of the type (7.3) for an auxiliary process x_t and we define the short rate by $r_t = e^{x_t}$ then resulting model is called the exponential Vasicek model. It has a SDE representation:

$$d \ln r = \kappa(\theta - \ln r)dt + \sigma dw.$$

The short rate r_t at a time t is almost surely a positive random variable. Since the variable $x = \ln r$ is a solution to (7.3), by using Itō's lemma, we deduce that the exponential Vasicek process r satisfies the SDE:

$$dr = \kappa(\theta + \sigma^2/2 - \ln r)r dt + \sigma r dw.$$

It has the same volatility term $\sigma r dw$ as the Brennan-Schwartz process (see Table 7.1). The drift term is however a nonlinear function of r. Taking the expected value $E(r_t)$ of the exponential Vasicek process at a time t we conclude

$$\frac{d}{dt}E(r_t) = \kappa\left((\theta + \sigma^2/2)E(r_t) - E(r_t \ln r_t)\right)$$

because $E(\sigma r_t dw_t) = 0$. Since the function $(0, \infty) \ni r \mapsto r \ln r$ is convex we have, by Jensen's inequality, $E(r_t \ln r_t) > E(r_t) \ln E(r_t)$. Solving the differential inequality

$$\frac{d}{dt}E(r_t) < \kappa\left((\theta + \sigma^2/2)E(r_t) - E(r_t) \ln E(r_t)\right)$$

yields the upper estimate for the expected value $E(r_t)$:

$$\ln E(r_t) < \ln E(r_0)e^{-\kappa t} + (\theta + \sigma^2/2)\left(1 - e^{-\kappa t}\right).$$

On the other hand, as the auxiliary function $x = \ln r$ is a solution to (7.3) we can deduce

$$E(\ln r_t) = E(\ln r_0)e^{-\kappa t} + \theta\left(1 - e^{-\kappa t}\right)$$

(see (7.2) with r replaced by x variable). Now, as the function $(0, \infty) \ni r \mapsto \ln r$ is concave, again by Jensen's inequality, we obtain the lower estimate $E(\ln r_t) < \ln E(r_t)$. Combining the above upper and lower inequalities for $\ln E(r_t)$ we obtain, for the limit of the expected value $E(r_t)$ of the exponential Vasicek model, the following inequality bounds:

$$\theta \leq \lim_{t \to \infty} \ln E(r_t) \leq \theta + \frac{\sigma^2}{2}. \tag{7.6}$$

We end this section by recalling the model of Black and Karasinski. If the parameters κ, θ, σ are not constants, but they are allowed to be functions of time, we obtain the governing SDE for the short rate process:

$$dr = \kappa(t)(\theta(t) - r)dt + \sigma(t)r^\gamma dw.$$

We refer the reader for detailed survey of qualitative and quantitative properties of various short rate models to the books of Brigo and Mercurio [20] and Kwok [75].

7.2. The Density of Itō's Stochastic Process and the Fokker–Plank Equation

In this section we address the question what is the probability distribution of the random variable x_t, which is a solution to the general stochastic Itō's process of the form:

$$dx = \mu(x,t)dt + \sigma(x,t)dw, \tag{7.7}$$

having a drift $\mu(x,t)$ and volatility $\sigma(x,t)$. Denote the cumulative distribution function $G = G(x,t) = \text{Prob}(x_t < x | x_0)$ of the process $x_t, t \geq 0$, which satisfies the stochastic differential equation (7.7) and starts almost surely from the initial condition x_0. Then the cumulative distribution function G can be computed from the probability density function $g = \partial G/\partial x$, which is a solution to the so-called Fokker–Planck equation:

$$\frac{\partial g}{\partial t} = \frac{1}{2}\frac{\partial^2}{\partial x^2}(\sigma^2 g) - \frac{\partial}{\partial x}(\mu g), \quad g(x,0) = \delta(x - x_0). \tag{7.8}$$

By $\delta(x - x_0)$ we have denoted Dirac delta function located at a point x_0. Recall that the Dirac delta function is not a function in a classical sense, but it is a distribution. For our needs we can represent it as a function satisfying

$$\delta(x - x_0) = \begin{cases} 0, & \text{if } x \neq x_0, \\ +\infty, & \text{if } x = x_0 \end{cases} \quad \text{and} \quad \int_{-\infty}^{\infty} \delta(x - x_0)dx = 1. \tag{7.9}$$

In our case it means that for the distribution function $G(x,0)$ at the time $t = 0$ we have

$$G(x,0) = \int_{-\infty}^{x} \delta(\xi - x_0)d\xi = \begin{cases} 0, & \text{if } x < x_0, \\ 1, & \text{if } x \geq x_0, \end{cases}$$

i.e., the process x_t is at time $t = 0$ almost surely at the point x_0.

An intuitive proof of the fact that the density g satisfies the Fokker-Planck equation can be done in a following way: Let $V = V(x,t)$ be a smooth function with a compact support,

i.e., $V \in C_0^\infty(\mathbb{R} \times (0,T))$. It means that for any such a function V there exists a compact set $[-R,R] \times [\alpha, \beta] \subset \mathbb{R} \times (0,T)$ with the property $V = 0$ outside of this set. According to Itō's lemma we have
$$dV = \left(\frac{\partial V}{\partial t} + \frac{\sigma^2}{2}\frac{\partial^2 V}{\partial x^2} + \mu\frac{\partial V}{\partial x}\right) dt + \sigma\frac{\partial V}{\partial x}dw.$$
Since the differential of the Wiener process $dw_{t_i} \approx w_{t_{i+1}} - w_{t_i}$ and the variable $\sigma(x_i, t_i)\frac{\partial V}{\partial x}(x_i, t_i)$ are at the time t_i independent, we obtain $E_t\left(\sigma(.,t)\frac{\partial V}{\partial x}(.,t)dw_t\right) = 0$, where E_t is the expected value with respect to the density $g(.,t)$. Then
$$dE_t(V(.,t)) = E_t(dV(.,t)) = E_t\left(\frac{\partial V}{\partial t} + \frac{\sigma^2}{2}\frac{\partial^2 V}{\partial x^2} + \mu\frac{\partial V}{\partial x}\right) dt.$$

Notice that the function $V(x,t) \in C_0^\infty$ is zero for $t = 0$ and $t = T$ and for $|x| > R$, where $R > 0$ is sufficiently large. Using this fact and integration by parts in x and t-variables, we obtain
$$\begin{aligned}
0 &= \int_0^T \frac{d}{dt} E_t(V(.,t)) dt = \int_0^T E_t\left(\frac{\partial V}{\partial t} + \frac{\sigma^2}{2}\frac{\partial^2 V}{\partial x^2} + \mu\frac{\partial V}{\partial x}\right) dt \\
&= \int_0^T \int_\mathbb{R} \left(\frac{\partial V}{\partial t} + \frac{\sigma^2}{2}\frac{\partial^2 V}{\partial x^2} + \mu\frac{\partial V}{\partial x}\right) g(x,t)\, dx dt \\
&= \int_0^T \int_\mathbb{R} V(x,t)\left(-\frac{\partial g}{\partial t} + \frac{1}{2}\frac{\partial^2}{\partial x^2}(\sigma^2 g) - \frac{\partial}{\partial x}(\mu g)\right) dx dt.
\end{aligned}$$

Recall that the function $V \in C_0^\infty(\mathbb{R} \times (0,T))$ was arbitrary. Hence the differential expression for the function g in the brackets has to be identically equal to zero. This way we have shown that the function g satisfies the Fokker–Planck equation (7.8). As we have already noted, the initial value $g(x,0) = \delta(x - x_0)$ is equivalent to the fact that the process x_t almost surely starts at the point x_0 in the initial time $t = 0$.

7.2.1. Multidimensional Version of the Fokker-Planck Equation

In this part we briefly present derivation of the multidimensional generalization of the Fokker–Planck equation for vector random processes.

Let $\{\vec{x}_t, t \geq 0\}$ be a vector stochastic process where $\vec{x} = (x_1, x_2, \ldots, x_n)^T$ and such that its increments $dx_i, i = 1, \ldots, n$, satisfy the system of stochastic differential equations
$$dx_i = \mu_i(\vec{x},t)dt + \sum_{k=1}^n \sigma_{ik}(\vec{x},t)dw_k.$$

Here $\vec{w} = (w_1, w_2, \ldots, w_n)^T$ is a vector of Wiener processes having mutually independent increments
$$E(dw_i dw_j) = 0, \text{ for } i \neq j, \quad E((dw_i)^2) = dt.$$
It can be rewritten in a vector form of the system of SDEs
$$d\vec{x} = \vec{\mu}(\vec{x},t)dt + K(\vec{x},t)d\vec{w},$$

where K is an $n \times n$ matrix

$$K(\vec{x},t) = (\sigma_{ij}(\vec{x},t))_{i,j=1,\ldots,n}.$$

Now, let us suppose that $f = f(\vec{x},t) = f(x_1, x_2, \ldots, x_n, t) : \mathbb{R}^n \times [0,T] \to \mathbb{R}$ is at least a C^2 smooth function. The multidimensional version of Itō's lemma (see Chapter 2) yields the SDE for the composite function $f = f(\vec{x},t)$:

$$df = \left(\frac{\partial f}{\partial t} + \frac{1}{2} K : \nabla_x^2 f K\right) dt + \nabla_x f \, d\vec{x},$$

where

$$K : \nabla_x^2 f K = \sum_{i,j=1}^{n} \frac{\partial^2 f}{\partial x_i \partial x_j} \sum_{k=1}^{n} \sigma_{ik} \sigma_{jk}.$$

Now, for the joint density distribution function $g(x_1, x_2, \ldots, x_n, t)$,

$$g(x_1, x_2, \ldots, x_n, t) = P(x_1(t) = x_1, x_2(t) = x_2, \ldots, x_n(t) = x_n, t)$$

conditioned to the initial condition state $x_1^0, x_2^0, \ldots, x_n^0$ one can follow the same procedure as in the scalar case to obtain the parabolic partial differential equation

$$\frac{\partial g}{\partial t} + \nabla \cdot (\vec{\mu} g) = \frac{1}{2} \sum_{i,j=1}^{n} \frac{\partial^2}{\partial x_i \partial x_j} (a_{ij} g), \quad \text{where } a_{ij} = \sum_{k=1}^{n} \sigma_{ik} \sigma_{jk}.$$

$$g(\vec{x}, 0) = \delta(\vec{x} - \vec{x}^0).$$

The above parabolic partial differential equation for the multidimensional density function g is refereed to as the multidimensional Fokker–Planck equation.

As an example, let us derive the multidimensional Fokker–Planck equation for a system of uncorrelated SDEs

$$\begin{aligned} dx_1 &= \mu_1(\vec{x},t)dt + \bar{\sigma}_1 dw_1, \\ dx_2 &= \mu_2(\vec{x},t)dt + \bar{\sigma}_2 dw_2, \\ &\vdots \\ dx_n &= \mu_n(\vec{x},t)dt + \bar{\sigma}_n dw_n, \end{aligned}$$

with mutually independent increments of Wiener processes

$$E(dw_i dw_j) = 0, \text{ for } i \neq j, \quad E((dw_i)^2) = dt.$$

The corresponding Fokker–Planck equation reads as follows:

$$\frac{\partial g}{\partial t} + \nabla \cdot (\vec{\mu} g) = \frac{1}{2} \sum_{i=1}^{n} \frac{\partial^2}{\partial x_i^2} (\bar{\sigma}_i^2 g).$$

This is a scalar parabolic advection–diffusion equation for g.

In the rest of this section we analyze the probabilistic distribution corresponding to short interest rate SDEs discussed in the first section of this chapter. Using the technique based on the solution to the Fokker–Planck equation we will be able to derive the probability distribution of the random variable x_t for the time $t \to \infty$.

As a first example of application of the Fokker-Planck equation we will find a distribution of the Vasicek stochastic process r_t at a time t when conditioned by the given initial value r_0 at time $t = 0$. The Fokker-Planck equation for this process reads as follows:

$$\frac{\partial f}{\partial t} = \frac{\sigma^2}{2} \frac{\partial^2 f}{\partial r^2} - \frac{\partial}{\partial r}(\kappa(\theta - r)f) \tag{7.10}$$

with the initial condition $f(r,0) = \delta(r - r_0)$. Its solution is the density function $f(r,t)$ of the random variable r_t subject almost surely to the initial condition r_0. Instead of the density function f, we will construct the characteristic function $\phi(s,t)$ of the random variable. Recall that the characteristic function $\phi(s)$ of a random variable X is defined as the expected value $E(e^{isX})$, where i is an imaginary unit, $i^2 = -1$. Hence

$$\phi(s,t) = E\left(e^{isr_t}\right) = \int_{-\infty}^{\infty} e^{isr} f(r,t)\, dr.$$

We also recall that the characteristic function of the normal distribution $N(\mu, \sigma^2)$ is $\exp\left(i\mu s - \frac{\sigma^2}{2} s^2\right)$.

Now we multiply equation (7.10) by the expression e^{isr} and integrate it with respect to r from $-\infty$ to ∞. We obtain

$$\int_{-\infty}^{\infty} e^{isr} \frac{\partial f}{\partial t} dr = \frac{\sigma^2}{2} \int_{-\infty}^{\infty} e^{isr} \frac{\partial^2 f}{\partial r^2} dr - \kappa\theta \int_{-\infty}^{\infty} e^{isr} \frac{\partial f}{\partial r} dr + \kappa \int_{-\infty}^{\infty} r e^{isr} \frac{\partial f}{\partial r} dr + \kappa \int_{-\infty}^{\infty} e^{isr} f dr.$$

For the density function f we have $f(r,t) \to 0$ and $\frac{\partial f}{\partial r}(r,t) \to 0$ for $r \to \pm\infty$. Hence we can use

$$\int_{-\infty}^{\infty} e^{isr} \frac{\partial f}{\partial t} dr = \frac{\partial \phi}{\partial t}, \quad \int_{-\infty}^{\infty} e^{isr} \frac{\partial f}{\partial r} dr = -is\phi,$$

$$\int_{-\infty}^{\infty} e^{isr} \frac{\partial^2 f}{\partial r^2} dr = -s^2 \phi, \quad \int_{-\infty}^{\infty} r e^{isr} \frac{\partial f}{\partial r} dr = -\left(\phi + s \frac{\partial \phi}{\partial s}\right).$$

As a consequence, we obtain the equation

$$\frac{\partial \phi}{\partial t} + \kappa s \frac{\partial \phi}{\partial s} = -\left(\frac{\sigma^2}{2} s^2 - \kappa\theta i s\right) \phi,$$

which is a quasilinear partial differential equation of the first order. Using method of characteristics (see e.g., [105]) we can construct its general solution ϕ given in the implicit form

$$F\left(se^{-\kappa t}, \ln \phi + \frac{\sigma^2}{4\kappa} s^2 - \theta i s\right) = 0. \tag{7.11}$$

For the initial time $t = 0$ we have

$$\phi(s,0) = \int_{-\infty}^{\infty} e^{isr} \delta(r - r_0) dr = e^{isr_0}.$$

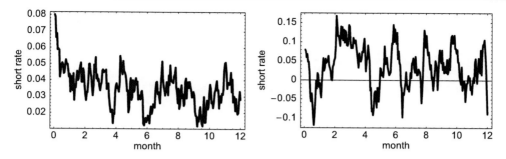

Figure 7.2. Simulation of one year data of the short rate evolution according to the Vasicek model. Left: parameters estimated from data in 2007 using the maximum likelihood method, right: 5-time higher volatility.

By substituting it into (7.11) we find implicit relation, i.e., the function F:

$$F(\xi_1, \xi_2) = ir_0\xi_1 + \frac{\sigma^2}{4\kappa}\xi_1^2 - \theta i \xi_1 - \xi_2.$$

It means that the solution ϕ is given by

$$\phi(s,t) = \exp\left(-\frac{s^2}{2}\left[\frac{\sigma^2}{2\kappa}(1-e^{-2\kappa t})\right] + is\left[\theta(1-e^{-\kappa t}) + r_0 e^{-\kappa t}\right]\right).$$

Clearly, this function $\phi(s,t)$ is the characteristic function of the normal distribution $N(\bar{r}_t, \bar{\sigma}_t^2)$, where

$$\bar{r}_t = \theta(1-e^{-\kappa t}) + r_0 e^{-\kappa t}, \quad \bar{\sigma}_t^2 = \frac{\sigma^2}{2\kappa}(1-e^{-2\kappa t}).$$

In this way we have shown that the expected value is a weighted average of the initial value r_0 and the limit value θ. Note that the weight of the starting values r_0 is decreasing with the strength of the mean-reversion of the process, i.e., with the increasing parameter κ.

Considering the limit for $t \to \infty$ we obtain the limiting distribution of the short rate:

$$r_\infty \sim N\left(\theta, \frac{\sigma^2}{2\kappa}\right).$$

Another way how to deduce the limit density is based on a solution to the stationary Fokker-Planck equation. Indeed, in the case when we expect that the density function $g(x,t)$ of a random variable x_t is stabilized at some limiting density $\tilde{g}(x)$ as $t \to \infty$, i.e., $\lim_{t\to\infty} g(x,t) = \tilde{g}(x)$ we obtain $\frac{\partial g}{\partial t}(x,t) \to 0$ for $t \to \infty$. It means that the limiting density $\tilde{g}(x)$ has to satisfy the stationary Fokker–Planck equation:

$$\frac{1}{2}\frac{\partial^2}{\partial x^2}(\sigma^2 \tilde{g}) - \frac{\partial}{\partial x}(\mu \tilde{g}) = 0 \qquad (7.12)$$

and the normalization condition on the density distribution, i.e., $\int_\mathbb{R} \tilde{g}(x)dx = 1$.

In the case of the Vasicek model, the stationary Fokker–Planck equation reads as follows:

$$\frac{\sigma^2}{2}\frac{\partial^2 \tilde{g}}{\partial r^2} - \frac{\partial}{\partial r}(\kappa(\theta - r)\tilde{g}) = 0.$$

Its solution can be easily found in the explicit form

$$\tilde{g}(r) = C\exp\left(-\frac{\kappa}{\sigma^2}(r-\theta)^2\right), \quad C = \sqrt{\frac{\kappa}{\pi\sigma^2}}, \qquad (7.13)$$

where $C > 0$ is a normalization constant, such that $\tilde{g}$ is probability density, i.e., $\int_{\mathbb{R}} \tilde{g}(r)dr = 1$. Notice that the density (7.13) is a density of a random variable with a normal distribution having the expected value θ and variance $\frac{\sigma^2}{2\kappa}$.

As an another example we consider the Cox–Ingersoll–Ross model for a short interest rate r_t. Similarly as in the case of the Vasicek model, we will compute the limiting density function of the distribution of r_t for $t \to \infty$. In the CIR model we assume that the process for the short rate r satisfies the stochastic differential equation of the Bessel square root process (7.4). The corresponding stationary Fokker–Planck equation has the form

$$\frac{\sigma^2}{2}\frac{\partial^2}{\partial r^2}(r\tilde{g}) - \frac{\partial}{\partial r}(\kappa(\theta - r)\tilde{g}) = 0. \qquad (7.14)$$

Suppose that the parameters of the process satisfy

$$\frac{2\kappa\theta}{\sigma^2} > 1.$$

Then, by integrating equation (7.14), we can find the explicit solution

$$\tilde{g}(r) = \begin{cases} \frac{1}{C}r^{\frac{2\kappa\theta}{\sigma^2}-1}\exp\left(-2\kappa r/\sigma^2\right), & r > 0, \\ 0, & r \leq 0, \end{cases} \qquad (7.15)$$

where $C = \left(\frac{\sigma^2}{2\kappa}\right)^{\frac{2\kappa\theta}{\sigma^2}} \Gamma\left(\frac{2\kappa\theta}{\sigma^2}\right)$ is the normalization constant and Γ is the Gamma function. In this case the limiting density (7.15) is the density of a random variable with a Gamma distribution. The limiting random variable r does not attain the zero value almost surely. This is one of the main reasons why the Cox–Ingersoll–Ross model is considered to be more realistic when compared to the Vasicek model of the short rate.

In Fig. 7.3 (left) we plotted the limiting distribution of the short rate that evolves according to the Vasicek process. The densities correspond to model parameters by which we generated simulations shown in Fig. 7.2. We can see that, for very large value of the volatility, we obtain a distribution which has negative values with a high probability. In Fig. 7.3 (right) we consider the parameters of the process as they were estimated from the real data. Suppose that the current level of the short rate is 0.02. We computed the distribution of the short rate in the future. Further examples of these distributions for 1 day and 1 week are also depicted in the figure. Together with these densities, the density of the limiting distribution is drawn. In Fig. 7.4 we compare the limiting distributions of the short rate according to the Vasicek a CIR models. The parameters of the both models were estimated from the same market data.

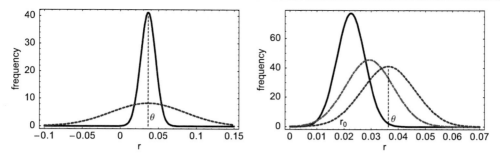

Figure 7.3. Left: limiting distribution of the short rate in the Vasicek model with parameters $\theta = 0.0364, \kappa = 41.9624, \sigma = 0.0888$ and with 5-times larger volatility. Right: a distribution of the short rate in the Vasicek model with parameters estimated from data and initial value $r_0 = 0.02$ - in 1 day, 1 week, limit distribution (dotted line).

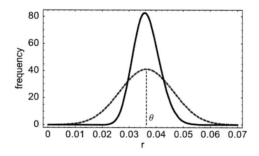

Figure 7.4. Limiting distribution of the short rate in the Vasicek model with parameters $\theta = 0.0364, \kappa = 41.9624, \sigma = 0.0888$ and CIR model with $\theta = 0.0364, \kappa = 44.9889, \sigma = 0.4917$

7.3. Two- and Multi-factor Interest Rate Models

The basic idea of two-factor interest rate models consits in the assumption that the short rate r is a function of two factors which we can denote in general as x, y. It means that $r = r(x,y)$ and the stochastic equation for the factor x can also depend on the factor y. This class of models contains the models in which:

- The short rate is a sum of two factors x and y. These factors can be interpreted as a systematic and a speculative component of the short rate. In this case $r(x,y) = x+y$;

- A stochastic equation for the short rate depends on other stochastic variables. For example r is a short rate in a certain country and we assume that it also depends on the European short interest rate r_e. Hence $x = r$, $y = r_e$ and $r(x,y) = x$;

- Some of the parameters from the one-factor model is not constant, but stochastic. Examples are represented by stochastic volatility models, where $r = x$ and the volatility of the process r is described by another stochastic factor y.

In a general case of a two-factor model we assume that the factors x, y satisfy the following system of stochastic differential equations:

$$dx = \mu_x(x,y)dt + \sigma_x(x,y)dw_1,$$
$$dy = \mu_y(x,y)dt + \sigma_y(x,y)dw_2,$$

where the correlation of differentials dw_1 and dw_2 of the Wiener processes w_1 and w_2 is a constant ρ, i.e., $E(dw_1 dw_2) = \rho dt$.

Let us present a couple of examples. In the two-factor Vasicek and CIR models the short rate is considered to be a sum of two independent factors. Each of them evolves according to a stochastic differential equation of the same form as is the equation for the short rate in the corresponding one-factor model. It means that $r = r_1 + r_2$, where r_1 and r_2 are given by the following processes:

- in the two-factor Vasicek model

$$dr_1 = \kappa_1(\theta_1 - r_1)dt + \sigma_1 dw_1,$$
$$dr_2 = \kappa_2(\theta_2 - r_2)dt + \sigma_2 dw_2, \qquad (7.16)$$

- in the two-factor CIR model

$$dr_1 = \kappa_1(\theta_1 - r_1)dt + \sigma_1\sqrt{r_1}dw_1,$$
$$dr_2 = \kappa_2(\theta_2 - r_2)dt + \sigma_2\sqrt{r_2}dw_2. \qquad (7.17)$$

In the convergence model proposed by Corzo and Schwartz in [27], the European short rate is modelled by using one-factor stochastic differential equation. It is assumed that the evolution of the domestic rate is stochastic and it depends on the stochastic European short rate also. The European rate is modelled by the one-factor Vasicek model

$$dr_e = c(d - r_e)dt + \sigma_e dw_e.$$

The process for the domestic short rate[1] is given by the stochastic differential equation

$$dr_d = (a + b(r_e - r_d))dt + \sigma_d dw_d,$$

which can be written as

$$dr_d = b\left(\frac{a}{b} + r_e - r_d\right)dt + \sigma_d dw_2. \qquad (7.18)$$

As it can be seen from the form of the process (7.18), the domestic rate r_d is pulled towards the value $\frac{a}{b} + r_e$. The differentials of the Wiener processes w_1 a w_2 can be correlated with a constant correlation ρ, i.e., $E(dw_1 dw_2) = \rho dt$.

In stochastic volatility models the first factor is the short rate itself and the second factor defines its volatility. There are several possibilities for choosing this second factor. One of the them is the so-called Fong–Vasicek model. In this model we consider

$$dr = \kappa_1(\theta_1 - r)dt + \sqrt{y}dw_1,$$
$$dy = \kappa_2(\theta_2 - y)dt + v\sqrt{y}dw_2.$$

---

[1] In the model of Corzo and Schwartz it is Spanish interest rate before adoption of EURO currency.

It means that the stochastic volatility of the short rate r is the square root of the second stochastic factor y. In this model the increments of the Wiener processes w_1 and w_2 can be correlated, $E(dw_1 dw_2) = \rho dt$.

7.4. Calibration of Short Rate Models

In this section we present one of the possible approaches to the calibration of the short rate models of the form (7.5). Firstly, we derive estimates of the parameters κ, θ, σ of the Vasicek model using the maximum likelihood methodology. Then, we generalize this method for the CIR stochastic process.

7.4.1. Maximum Likelihood Method for Estimation of the Parameters in the Vasicek and CIR Models

We already computed the conditional distribution of the short rate in the Vasicek model by solving a Fokker-Planck partial differential equation. Hence we know the distribution of $r_{t+\Delta t}$ when conditioned to the initial state r_t at time t. Here $\Delta t > 0$ is a time step evaluated on a yearly basis, e.g., $\Delta t = 1/365$. Now we derive this result in another way, which will turn to be useful when estimating models with nonconstant volatility (see Chapter 13).

We multiply the equation for the short rate $dr_s = \kappa(\theta - r)ds + \sigma dw_s$ by the term $e^{\kappa s}$. Using Itō's lemma for $f(s,t) = e^{\kappa s} r$ we obtain the expression for the differential

$$d(e^{\kappa s} r_s) = \kappa \theta e^{\kappa s} ds + \sigma e^{\kappa s} dw_s.$$

Integrating it from the time t to time $t + \Delta t$ we obtain

$$\begin{aligned}
e^{\kappa(t+\Delta t)} r_{t+\Delta t} - e^{\kappa t} r_t &= \kappa \theta \int_t^{t+\Delta t} e^{\kappa s} ds + \sigma \int_t^{t+\Delta t} e^{\kappa s} dw_s \\
&= (e^{\kappa(t+\Delta t)} - e^{\kappa t}) \theta + \sigma \int_t^{t+\Delta t} e^{\kappa s} dw_s.
\end{aligned}$$

Therefore $r_{t+\Delta t} = e^{-\kappa \Delta t} r_t + (1 - e^{-\kappa \Delta t}) \theta + \sigma e^{-\kappa(t+\Delta t)} \int_t^{t+\Delta t} e^{\kappa s} dw_s$. The conditional distribution $r_{t+\Delta t}$ conditioned to the state r_t at time t is a normal distribution. Its first two moments can be evaluated as

$$\begin{aligned}
E(r_{t+\Delta t} | r_t) &= e^{-\kappa \Delta t} r_t + (1 - e^{-\kappa \Delta t}) \theta, \\
Var(r_{t+\Delta t} | r_t) &= \sigma^2 e^{-2\kappa(t+\Delta t)} Var\left(\int_t^{t+\Delta t} e^{\kappa s} dw_s\right) \\
&= \sigma^2 e^{-2\kappa(t+\Delta t)} E\left(\left[\int_t^{t+\Delta t} e^{\kappa s} dw_s\right]^2\right) \\
&= \sigma^2 e^{-2\kappa(t+\Delta t)} \int_t^{t+\Delta t} (e^{\kappa s})^2 ds = \frac{\sigma^2}{2\kappa} (1 - e^{-2\kappa \Delta t}),
\end{aligned}$$

where we have used Itō's isometry from Chapter 2. Hence

$$r_{t+\Delta t} | r_t \sim N\left(e^{-\kappa \Delta t} r_t + (1 - e^{-\kappa \Delta t}) \theta, \; \frac{\sigma^2}{2\kappa}(1 - e^{-2\kappa \Delta t})\right).$$

Let us consider that the statistical data for the short rate are: $r_0, r_{\Delta t}, \ldots, r_{N\Delta t}$ evaluated at times: $0, \Delta t, \ldots, N\Delta t$. We define

$$v_t^2 = \frac{\sigma^2}{2\kappa}\left(1 - e^{-\kappa \Delta t}\right), \quad \varepsilon_t = r_t - \theta\left(1 - e^{-\kappa \Delta t}\right) - e^{-\kappa \Delta t} r_{t-\Delta t}. \tag{7.19}$$

Then the likelihood function L of the random vector ε is a product of normal distributions (see [96])

$$f(\varepsilon_t) = \frac{1}{\sqrt{2\pi v_t^2}} e^{-\frac{\varepsilon_t^2}{2 v_t^2}}.$$

Up to an additive constant not influencing the optimum, the logarithm of likelihood function L can be written as

$$\ln L = -\frac{1}{2} \sum_{t=2}^{N} \ln v_t^2 + \frac{\varepsilon_t^2}{v_t^2}. \tag{7.20}$$

Maximizing this function we obtain the estimates of the parameters κ, θ, σ^2.

Table 7.2. August 2003 European short rates in percent p.a.

PRIBOR	BUBOR	BRIBOR	EURIBOR
5.1	9.42	5.21	2.08
5.16	9.41	5.15	2.07
5.12	9.38	5.27	2.07
5.07	9.36	6.51	2.06
4.94	9.21	6.48	2.06
5.1	8.74	6.41	2.06
5.2	8.73	6.45	2.06
5.28	8.75	6.38	2.06
5.45	9.19	5.73	2.06
5.7	9.33	5.35	2.06
5.34	9.94	5.295	2.07
5.22	10.00	5.575	2.07
5.16	9.99	5.65	2.07
5	10.33	5.22	2.09
4.95	10.50	5.14	2.15
4.65	10.53	5.16	2.45
4.28	10.51	5.12	2.1
4.23	9.82	5.13	2.1
4.08	10.48	6.9	2.09
5	10.57	7.385	2.09

Problem Section and Exercises

1. Consider the Vasicek model with parameters $\theta = 0.0364, \kappa = 41.9624, \sigma = 0.0888$ (estimated from BRIBOR overnight rates from the year 2007). Find the density function of the short rate in a month, if its current value is 3.5% p.a. What is its expected value and its standard deviation? Find 95% confidence interval for its value.

2. Solving the stationary Fokker-Planck equation find the limiting distribution of CKLS model of short rate, in which

$$dr = \kappa(\theta - r)dt + \sigma r^\gamma dw,$$

for $\gamma > 0, \gamma \neq 1/2$.

3. By using the maximum likelihood method, estimate the parameters κ, θ, σ of the Vasicek and CIR models for the short rate data of several different European countries: PRIBOR – the Czech Republic, BUBOR – Hungary, BRIBOR – Slovakia and EURIBOR – Eurozone, which are collected in Table 7.2. Compare the value of the long term interest rate θ for both models.

Chapter 8

Pricing of Interest Rate Derivatives

In this chapter, we investigate methods for pricing derivatives of the short interest rate. At present, these derivatives belong among the category of most traded derivatives in financial markets. In this category of financial derivatives there are bonds, swaps, caps, floors, options on bonds and others. When analyzing interest rate derivatives (e.g., bonds), it is necessary to consider not only the underlying asset itself (e.g., the interest rate) but also the entire yield curve describing the term structure of interest rates for different maturities. Interest rate derivatives are characterized by the property that their pay–off diagrams depend on value of the short rate. A contract, according to which we obtain at the specified time T the specified amount of money X, is called a zero-coupon bond. The value X is called a par value and time T is called the maturity. If the bond pays coupons at specified time intervals, the bond is called a coupon bond and the payments are called coupons. Since a coupon bond is equivalent to a portfolio of zero coupon bonds (cf. Kwok [75], Dewynne et al. [122] or Melicherčík [83]), we will only investigate pricing of zero-coupon bonds.

8.1. Bonds and Term Structures of Interest Rates

A zero-coupon bond is one of the basic derivative contracts based on the interest rate as an underlying asset. A bond is an agreement to pay a certain amount today against the promise to obtain a higher amount in the future. The maturity of the bond is usually denoted by T. The price of a bond $P(t,T)$ is therefore a function of the current time t and the maturity T. Naturally, it depends also on the underlying asset - the interest rate. The most simple derivative is therefore a bond which pays its holder a unit amount of money at the maturity T. Bond prices determine time structure of interest rates $R(t,T)$, which are given by the formula

$$P(t,T) = e^{-R(t,T)(T-t)},$$

where $P(t,T)$ is the bond price, $R(t,T)$ is the interest rate at present day t with maturity at expiration day T. It means that the term structure of the interest rate can be computed from bond prices as follows:

$$R(t,T) = -\frac{\ln P(t,T)}{T-t}.$$

Table 8.1. Quarterly descriptive statistics for different European term structures of interest rates. The mean and standard deviation are given for the short rate (o.n.) and yield on a bond with one year maturity in percents p.a.

		1/4 2003		2/4 2003		3/4 2003		4/4 2003	
		Mean	STD	Mean	STD	Mean	STD	Mean	STD
BRIBOR	o.n.	5.75	1.041	6.27	1.279	5.63	0.802	5.48	0.992
	1y	5.48	0.205	5.42	0.208	5.80	0.066	5.50	0.033
WIBOR	o.n.	6.65	0.761	5.76	0.359	5.22	0.438	5.17	0.438
	1y	5.95	0.138	5.19	0.255	4.97	0.053	5.79	0.380
BUBOR	o.n.	5.42	1.813	7.08	0.879	9.58	0.547	10.52	1.213
	1y	6.57	0.433	6.76	0.773	8.80	0.207	10.02	1.334
PRIBOR	o.n.	2.52	0.107	2.44	0.045	2.06	0.135	1.94	0.032
	1y	2.43	0.130	2.33	0.084	2.13	0.063	2.19	0.061
EURIBOR	o.n.	2.77	0.188	2.44	0.199	2.07	0.120	2.02	0.169
	1y	2.54	0.140	2.23	0.189	2.20	0.106	2.36	0.081
EUROLIB	o.n.	2.79	0.196	2.47	0.196	2.08	0.101	2.02	0.139
	1y	2.54	0.139	2.23	0.187	2.20	0.105	2.35	0.085

The instantaneous interest rate (or short rate) r_t at time t is then the interest rate $R(t,T)$ with instantaneous maturity $T = t$, i.e.

$$r_t = R(t,t).$$

In Fig. 8.1 we plot four sample term structures of interest rates for various countries from 5/27/2008. It is important to note that the term structure can have different behavior, as far as the increasing or decreasing property of yields as functions of maturity, are concerned.

The data sets contained in Table 8.1 are taken from the paper Ševčovič and Urbánová-Csajková [102]. One can see descriptive statistics for short rates and yields of bonds with one year maturity for different European countries. The base for the term structure of interest rates in the countries of Eurozone is EURIBOR and the overnight (or short rate) EONIA. Recall that the short rate substitute called EONIA is computed as a weighted average from 48 European banks. Term structures of interest rates in Central European countries are based on the first letters of their capital cities: Bratislava – BRIBOR, Prague – PRIBOR, Warsaw – WIBOR, Budapest – BUBOR. Among widely known term structures there are: London LIBOR, European EURIBOR and EUROLIBOR. In typical cases, these term structures consist of the instantaneous interest rate (called also overnight or short rate) and interest rates with maturities from 1, 2, 3 weeks up to 1, 3, 6, 12 months.

8.1.1. One–Factor Equilibrium Models, Vasicek and CIR Models

We begin our analysis of one-factor equilibrium models by the description of the short rate stochastic process. Recall that the short rate plays a role of an underlying asset for its derivative which is, for instance, a zero-coupon bond. For this description we will use a general one-factor model (7.1) from the previous chapter, which can be expressed as a SDE

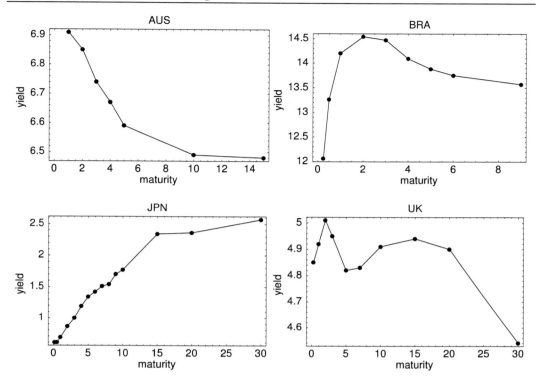

Figure 8.1. The term structure of interest rates of government bonds yields in percent from the day $t = 5/27/2008$, as a function of the maturity T (in years) for various countries: Australia, Brazil, Japan and United Kingdom.

for the short rate $r = r_t$:
$$dr = \mu(t,r)dt + \sigma(t,r)dw.$$

In what follows, we will derive a partial differential equation for the bond price P with a maturity at time T. This price depends on time t, maturity T as well as on the value of the short rate process r, i.e., $P = P(r,t,T)$. From Itō's formula (see Chapter 2) we conclude

$$\begin{aligned} dP &= \left(\frac{\partial P}{\partial t} + \mu \frac{\partial P}{\partial r} + \frac{\sigma^2}{2} \frac{\partial^2 P}{\partial r^2} \right) dt + \sigma \frac{\partial P}{\partial r} dw \\ &= \mu_B(t,r)dt + \sigma_B(t,r)dw, \end{aligned} \quad (8.1)$$

where $\mu_B(r,t)$ and $\sigma_B(r,t)$ denote the drift and volatility of the bond price stochastic process. Next we construct a portfolio consisting of bonds with two different maturities. It contains one bond with maturity T_1 and Δ bonds with maturity T_2. Its value π is therefore

$$\pi = P(r,t,T_1) + \Delta P(r,t,T_2). \quad (8.2)$$

We can express the change $d\pi$ of its value as follows:

$$\begin{aligned} d\pi &= dP(r,t,T_1) + \Delta dP(r,t,T_2) \\ &= (\mu_B(r,t,T_1) + \Delta \mu_B(r,t,T_2))\, dt + (\sigma_B(r,t,T_1) + \Delta \sigma_B(r,t,T_2))\, dw. \end{aligned}$$

If we choose the ratio of the number of the bonds Δ such that

$$\Delta = -\frac{\sigma_B(t,r,T_1)}{\sigma_B(t,r,T_2)}, \qquad (8.3)$$

then the stochastic part in (8.3) is eliminated and we obtain a riskless portfolio having the deterministic part only.

$$d\pi = \left(\mu_B(t,r,T_1) - \frac{\sigma_B(t,r,T_1)}{\sigma_B(t,r,T_2)}\mu_B(t,r,T_2)\right)dt.$$

In order to avoid a possibility of arbitrage opportunities, the yield of this portfolio should be equal to the instantaneous riskless interest rate r, i.e. $d\pi = r\pi dt$. Hence

$$\mu_B(t,r,T_1) - \frac{\sigma_B(t,r,T_1)}{\sigma_B(t,r,T_2)}\mu_B(t,r,T_2) = r\pi.$$

Substituting the value of the portfolio π from (8.2) and (8.3) we obtain

$$\mu_B(t,r,T_1) - \frac{\sigma_B(t,r,T_1)}{\sigma_B(t,r,T_2)}\mu_B(t,r,T_2)$$

$$= r\left(P(t,r,T_1) - \frac{\sigma_B(t,r,T_1)}{\sigma_B(t,r,T_2)}P(t,r,T_2)\right).$$

From the above equality it follows that

$$\frac{\mu_B(t,r,T_1) - rP(t,r,T_1)}{\sigma_B(t,r,T_1)} = \frac{\mu_B(t,r,T_2) - rP(t,r,T_2)}{\sigma_B(t,r,T_2)}.$$

Since the maturities T_1 and T_2 were chosen arbitrarily we conclude that the above expression does not depend on the maturity, i.e., there must exists a function $\lambda(r,t)$ such that the following identity holds:

$$\lambda(r,t) = \frac{\mu_B(r,t,T) - rP(r,t,T)}{\sigma_B(r,t,T)} \qquad (8.4)$$

for every maturity T. This function λ is called a market price of risk since it measures increase of the yield $\mu_B(r,t,T) - rP(r,t,T)$ on a bond with respect to the risk expressed by the volatility $\sigma_B(r,t,T)$ of the bond. Substituting the functions μ_B and σ_B into (8.4) we finally obtain a partial differential equation for the bond price $P(r,t,T)$:

$$\frac{\partial P}{\partial t} + (\mu(r,t) - \lambda(r,t)\sigma(r,t))\frac{\partial P}{\partial r} + \frac{\sigma^2(r,t)}{2}\frac{\partial^2 P}{\partial r^2} - rP = 0. \qquad (8.5)$$

At the time of maturity the value of the bond is equal to unity, regardless of the current value of the short rate. Hence the price $P(r,t,T)$ has to satisfy the terminal condition

$$P(r,T,T) = 1, \quad \text{for all } r > 0.$$

Vasicek Model

One of the first term structure models was proposed by mathematician and statistician O. A. Vasicek. The basis for this models is a description of the short rate with the model (7.3), i.e.

$$dr = \kappa(\theta - r)dt + \sigma dw.$$

We will look for a bond price as a function of the instantaneous (short) interest rate r and the time τ remaining to maturity, i.e., $\tau = T - t$. Therefore we can seek the bond price P in the form of a function of two variables: r and $\tau = T - t$. It means $P = P(r, \tau)$. For a constant market price of risk λ it satisfies the partial differential equation

$$-\frac{\partial P}{\partial \tau} + (\kappa(\theta - r) - \lambda\sigma)\frac{\partial P}{\partial r} + \frac{\sigma^2}{2}\frac{\partial^2 P}{\partial r^2} - rP = 0$$

and initial condition $P(r, 0) = 1$, for all $r > 0$. In what follows, we will show that there is an exact solution to the above bond pricing equation. We will search the solution in the form

$$P(r, \tau) = A(\tau)e^{-B(\tau)r},$$

where the time dependent functions A, B satisfy the initial conditions $A(0) = 1$, $B(0) = 0$. Next we compute all the partial derivatives appearing in the bond pricing PDE:

$$\frac{\partial P}{\partial \tau} = e^{-Br}(\dot{A} - A\dot{B}r),$$

$$\frac{\partial P}{\partial r} = -BAe^{-Br}, \quad \frac{\partial^2 P}{\partial r^2} = B^2 A e^{-Br}.$$

Here we have denoted by

$$\dot{\phi} = \frac{d}{d\tau}\phi$$

the time derivative of a function $\phi = \phi(\tau)$. Substituting them into the governing partial differential equation we obtain:

$$(\dot{A} - A\dot{B}r) + \frac{\sigma^2}{2}B^2 A - (\kappa(\theta - r) - \lambda\sigma)AB - rA = 0.$$

Collecting all the terms containing r and those that do not contain r we obtain:

$$\left(-\dot{A} + \frac{\sigma^2}{2}AB^2 - (\kappa\theta - \lambda\sigma)AB\right) + rA\left(\dot{B} + \kappa B - 1\right) = 0.$$

To satisfy this equation for all r, both parentheses have to be identically zero. This way we obtain a system of ODEs for the function A, B:

$$-\dot{A} + \frac{\sigma^2}{2}AB^2 - (\kappa\theta - \lambda\sigma)AB = 0,$$
$$\dot{B} + \kappa B - 1 = 0.$$

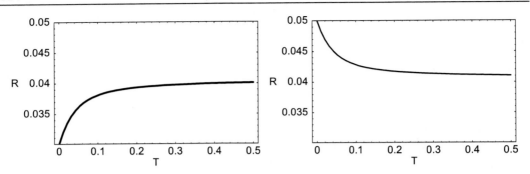

Figure 8.2. The term structure of interest rates $R(r,t,T)$ computed from the Vasicek model for two values of the short rate ($r = 0.03$ and $r = 0.05$) at a given time t.

The ordinary differential equation for B is linear and so its solution satisfying the initial condition $B(0) = 0$ can be easily constructed. We have

$$B(\tau) = \frac{1 - e^{-\kappa\tau}}{\kappa}. \tag{8.6}$$

Knowing the function B, we are yet able to compute the remaining function A. Integrating the equation for A yields

$$\ln A = \int \frac{d\ln A}{d\tau} = \int \frac{\sigma^2}{2} B^2 - (\kappa\theta - \lambda\sigma) B d\tau.$$

By substituting the function $B(\tau)$, evaluating the integral and using the initial condition $A(0) = 1$ we deduce

$$\ln A(\tau) = \left[\frac{1}{\kappa}(1 - e^{-\kappa\tau}) - \tau\right] R_\infty - \frac{\sigma^2}{4\kappa^3}(1 - e^{-\kappa\tau})^2, \tag{8.7}$$

where

$$R_\infty = \theta - \frac{\lambda\sigma}{\kappa} - \frac{\sigma^2}{2\kappa^2}.$$

If we use the notation $R(r,t,t+\tau)$ for the term structure of interest rates at time t with a maturity at time $t + \tau$, in which the short rate is r, then, for Vasicek model, the term structure $R(r,t,t+\tau)$ can be expressed as follows:

$$\begin{aligned} R(r,t,t+\tau) &= -\frac{\ln P(r,\tau)}{\tau} \\ &= \left(1 - \frac{1 - e^{-\kappa\tau}}{\kappa\tau}\right) R_\infty + \frac{\sigma^2}{4\kappa^3}\frac{(1 - e^{-\kappa\tau})^2}{\tau} + \frac{1 - e^{-\kappa\tau}}{\kappa\tau} r. \end{aligned}$$

Furthermore, the following limit $\lim_{\tau \to \infty} R(r,t,t+\tau) = R_\infty$ holds.

Cox–Ingersoll–Ross Model

Another popular one-factor model describing the bond price as a derivative of the short rate is the so-called Cox–Ingersoll–Ross model, abbreviated as the CIR model. It is based on the description (7.4) of the short rate model, i.e.

$$dr = \kappa(\theta - r)dt + \sigma\sqrt{r}dw.$$

Recall that in the case the parameters of the short rate process satisfy the inequality $2\kappa\theta/\sigma^2 > 1$, then the random variable r_t representing the short rate at time t is positive almost surely. We also remind ourselves that the Vasicek model does not possess this property since the statistical distribution of the short rate r_t is a normal distribution allowing for negative values of r_t. As usual, we express the bond price as a function of the variable τ, denoting time to maturity i.e., $\tau = T - t$. For a specific choice of the market price of risk $\lambda\sqrt{r}$ the partial differential equation for the bond price $P = P(r,\tau)$ takes the form

$$-\frac{\partial P}{\partial \tau} + (\kappa(\theta - r) - \lambda\sigma r)\frac{\partial P}{\partial r} + \frac{\sigma^2}{2}r\frac{\partial^2 P}{\partial r^2} - rP = 0, \tag{8.8}$$

with the initial condition $P(r,0) = 1$, for every $r > 0$. We can again look for a solution in the the form

$$P(r,\tau) = A(\tau)e^{-B(\tau)r},$$

where the functions A, B satisfy the initial conditions $A(0) = 1$ and $B(0) = 0$. Inserting this form of a solution into the governing (8.8) we obtain the equality

$$(-\dot{A} - \kappa\theta AB) + \left(\dot{B} + (\kappa + \lambda\sigma)B + \frac{\sigma^2}{2}B^2 - 1\right)Ar = 0,$$

which has to be satisfied for all $r > 0$ and $\tau > 0$. Similarly as in the case of the Vasicek model we obtain the system of ODEs

$$\dot{A} + \kappa\theta AB = 0,$$
$$\dot{B} + (\kappa + \lambda\sigma)B + \frac{\sigma^2}{2}B^2 - 1 = 0.$$

Together with the conditions $A(0) = 1$ and $B(0) = 0$ it represents an initial problem for the system of ordinary differential equations for unknown functions A and B. From the second equation we can compute the function B. Indeed, it is a separable ordinary differential equation and therefore we obtain

$$\frac{dB}{-\frac{\sigma^2}{2}B^2 - (\kappa + \lambda\sigma)B + 1} = d\tau. \tag{8.9}$$

In order to simplify further notation we let denote

$$\psi = \kappa + \lambda\sigma, \quad \phi = \sqrt{\psi^2 + 2\sigma^2} = \sqrt{(\kappa + \lambda\sigma)^2 + 2\sigma^2}. \tag{8.10}$$

Then

$$\frac{1}{-\frac{\sigma^2}{2}B^2 - (\kappa+\lambda\sigma)B + 1} = -\frac{2}{\sigma^2} \frac{1}{\left(B + \frac{\psi+\phi}{\sigma^2}\right)\left(B + \frac{\psi-\phi}{\sigma^2}\right)}$$

$$= -\frac{1}{\phi}\left(\frac{1}{B + \frac{\psi-\phi}{\sigma^2}} - \frac{1}{B + \frac{\psi+\phi}{\sigma^2}}\right).$$

Integrating the equation (8.9) we obtain

$$-\frac{1}{\phi}\ln\left(\frac{B + \frac{\psi-\phi}{\sigma^2}}{B + \frac{\psi+\phi}{\sigma^2}}\right) = \tau + c_1,$$

where c_1 is a constant. We write in the form

$$\frac{B(\tau) + \frac{\psi+\phi}{\sigma^2}}{B(\tau) + \frac{\psi-\phi}{\sigma^2}} = e^{\phi\tau} c_2,$$

from which we compute the constant c_2 using the initial condition $B(0) = 0$ at time $\tau = 0$, i.e., $c_2 = \frac{\psi+\phi}{\psi-\phi}$. We obtain the expression

$$B(\tau) = \frac{\frac{\psi+\phi}{\sigma^2}\left(e^{\phi\tau} - 1\right)}{1 - \frac{\psi+\phi}{\psi-\phi}e^{\phi\tau}},$$

which can be written in the following form commonly known in the literature:

$$B(\tau) = \frac{2\left(e^{\phi\tau} - 1\right)}{(\psi+\phi)\left(e^{\phi\tau} - 1\right) + 2\phi}. \tag{8.11}$$

Since the function B is already known the function A can be easily computed by integrating the equation $\dot{A} + \kappa\theta AB = 0$:

$$\ln A(\tau) = -\int_0^\tau \kappa\theta B(s)ds,$$

from which, after computing the integral and straightforward algebraic manipulations, we obtain

$$A(\tau) = \left(\frac{2\phi e^{(\phi+\psi)\tau/2}}{(\phi+\psi)(e^{\phi\tau} - 1) + 2\phi}\right)^{\frac{2\kappa\theta}{\sigma^2}}. \tag{8.12}$$

Similarly as in the case of the Vasicek model, the term structure of interest rates $R(r,t,t+\tau) = -\frac{\ln P(r,\tau)}{\tau}$ a linear function of the short rate r. We have

$$R(r,t,t+\tau) = -\frac{1}{\tau}\frac{2\kappa\theta}{\sigma^2}\ln\left(\frac{2\phi e^{(\phi+\psi)\tau/2}}{(\phi+\psi)(e^{\phi\tau} - 1) + 2\phi}\right)$$

$$+ \frac{1}{\tau}\frac{2\left(e^{\phi\tau} - 1\right)}{(\psi+\phi)(e^{\phi\tau} - 1) + 2\phi}r. \tag{8.13}$$

The limit of the term structure as time to maturity $\tau = T - t$ approaches infinity does not depend on the short rate r. More precisely, we have

$$\lim_{\tau \to \infty} \frac{1}{\tau} \frac{2\left(e^{\phi\tau} - 1\right)}{(\psi + \phi)\left(e^{\phi\tau} - 1\right) + 2\phi} = 0,$$

$$\lim_{\tau \to \infty} \frac{1}{\tau} \ln\left(\frac{2\phi e^{(\phi+\psi)\tau/2}}{(\phi+\psi)(e^{\phi\tau} - 1) + 2\phi}\right) = -\frac{\sigma^2}{\phi + \psi}.$$

We have used the l'Hospital rule to compute the second limit. In summary, it means that the limit of the term structures, for a maturity tending to infinity, exists and it can be expressed as follows:

$$\lim_{\tau \to \infty} R(r, t, t + \tau) = \frac{2\kappa}{\phi + \psi} \theta.$$

8.1.2. Two-factor Equilibrium Models

In derivation of a general two-factor model of the term structure of interest rates we assume that the factors x and y satisfy the following stochastic differential equations:

$$\begin{aligned} dx &= \mu_x(x,y)dt + \sigma_x(x,y)dw_1, \\ dy &= \mu_y(x,y)dt + \sigma_y(x,y)dw_2, \end{aligned}$$

where the correlation of dw_1 and dw_2 is $E(dw_1 dw_2) = \rho dt$ where $\rho \in [-1, 1]$ is a constant. As it was already mentioned in Chapter 7, we assume that the short rate is a function of these two factors, $r = r(x, y)$. In what follows, we derive a partial differential equation for the price of bond as a derivative of the short rate. We will use the same approach as in the one-factor case. We construct a risk-less portfolio consisting of bonds with different maturities.

Let us denote by $P = P(x, y, t)$ the price of a bond depending on the factors x, y and time t. Using the multidimensional Itō's lemma (see Chapter 2) we obtain

$$dP = \mu dt + \sigma_1 dw_1 + \sigma_2 dw_2, \tag{8.14}$$

where $\mu = \mu(x, y, t)$ and $\sigma_i = \sigma_i(x, y, t)$ are given by

$$\mu = \frac{\partial P}{\partial t} + \mu_x \frac{\partial P}{\partial x} + \mu_y \frac{\partial P}{\partial y} + \frac{\sigma_x^2}{2} \frac{\partial^2 P}{\partial x^2} + \frac{\sigma_y^2}{2} \frac{\partial^2 P}{\partial y^2} + \rho \sigma_x \sigma_y \frac{\partial^2 P}{\partial x \partial y},$$

$$\sigma_1 = \sigma_x \frac{\partial P}{\partial x}, \quad \sigma_2 = \sigma_y \frac{\partial P}{\partial y}.$$

Next we construct a portfolio π consisting of bonds with maturities T_1, T_2 and T_3. We denote the amounts of bonds in the portfolio by V_1, V_2 and V_3. Hence $\pi = P(T_1)V_1 + P(T_2)V_2 + P(T_3)V_3$, where by $P(T_i)$ we have denoted the price of a bond with maturity T_i, $i = 1, 2, 3$. Therefore, change in the portfolio value π can be expressed as follows:

$$\begin{aligned} d\pi &= V_1 dP(T_1) + V_2 dP(T_2) + V_3 dP(T_3) \\ &= (V_1 \mu(T_1) + V_2 \mu(T_2) + V_3 \mu(T_3)) dt \\ &\quad + (V_1 \sigma_1(T_1) + V_2 \sigma_1(T_2) + V_3 \sigma_1(T_3)) dw_1 \\ &\quad + (V_1 \sigma_2(T_1) + V_2 \sigma_2(T_2) + V_3 \sigma_2(T_3)) dw_2. \end{aligned} \tag{8.15}$$

The basic principle of derivation of the model for pricing a bond is again a construction of a risk-less portfolio. It means that we want to eliminate the randomness in evolution of the price of a portfolio. Therefore we choose the amounts of bonds V_1, V_2, V_3 in such a way that

$$V_1\sigma_1(T_1) + V_2\sigma_1(T_2) + V_3\sigma_1(T_3) = 0,$$
$$V_1\sigma_2(T_1) + V_2\sigma_2(T_2) + V_3\sigma_2(T_3) = 0.$$

The second principle is nonexistence of arbitrage opportunities. This principle says that the yield of the risk-less portfolio has to be equal to the risk-less interest rate r, i.e.

$$d\pi = r\pi dt.$$

This way we have deduced the third identity

$$V_1\mu(T_1) + V_2\mu(T_2) + V_3\mu(T_3) = \pi r$$

that has to be satisfied for amounts of bonds V_1, V_2, V_3. The last equality can be also rewritten in the form

$$V_1(\mu(T_1) - rP(T_1)) + V_2(\mu(T_2) - rP(T_2)) + V_3(\mu(T_3) - rP(T_3)) = 0.$$

In summary, we have derived the following system of equations for quantities V_1, V_2, V_3:

$$\begin{pmatrix} \sigma_1(T_1) & \sigma_1(T_2) & \sigma_1(T_3) \\ \sigma_2(T_1) & \sigma_2(T_2) & \sigma_2(T_3) \\ \mu(T_1) - rP(T_1) & \mu(T_2) - rP(T_2) & \mu(T_3) - rP(T_3) \end{pmatrix} \begin{bmatrix} V_1 \\ V_2 \\ V_3 \end{bmatrix} = \begin{bmatrix} 0 \\ 0 \\ 0 \end{bmatrix}.$$

The above system of linear equation has a nonzero solution (V_1, V_2, V_3) if and only if the rows of the matrix are linearly dependent. If the second row is a multiple of the first one (or vice versa), then there is only one source of randomness in the model. This would lead us to a previously studied one-factor interest rate model. Therefore, the third row has to be a linear combination of the previous ones. It means that there are real numbers λ_1, λ_2 such that

$$\mu(T_i) - rP(T_i) = \lambda_1\sigma_1(T_i) + \lambda_2\sigma_2(T_i), \quad \text{for } i = 1, 2, 3.$$

Since the maturities T_i were arbitrary, the functions λ_1, λ_2 cannot depend on maturities of the bonds. Hence

$$\lambda_1 = \lambda_1(x, y, t), \quad \lambda_2 = \lambda_2(x, y, t).$$

Substituting μ, σ_1 and σ_2 we finally obtain the partial differential equation for the bond price:

$$\frac{\partial P}{\partial t} + (\mu_x - \lambda_1\sigma_x)\frac{\partial P}{\partial x} + (\mu_y - \lambda_2\sigma_y)\frac{\partial P}{\partial y}$$
$$+ \frac{\sigma_x^2}{2}\frac{\partial^2 P}{\partial x^2} + \frac{\sigma_y^2}{2}\frac{\partial^2 P}{\partial y^2} + \rho\sigma_x\sigma_y\frac{\partial^2 P}{\partial x \partial y} - r(x,y)P = 0. \qquad (8.16)$$

We remind ourselves that the functions λ_1, λ_2 are called the market prices of risk of corresponding factors.

Two-Factor Vasicek and CIR Models

In this part we consider two-factor models of the term structure of interest rates that are based on the assumption on the evolution of the short rate r. In the Vasicek two-factor model we assume that the short rate is a sum of two factors, i.e., $r = r_1 + r_2$, where the factors r_1, r_2 are solutions to the following stochastic differential equations

$$dr_1 = \kappa_1(\theta_1 - r_1)dt + \sigma_1 dw_1,$$
$$dr_2 = \kappa_2(\theta_2 - r_2)dt + \sigma_2 dw_2.$$

The correlation ρ between the differentials dw_1, dw_2 is assumed to be zero. Now we derive the bond price $P(r_1, r_2, \tau)$ in the two-factor Vasicek model under the assumption that the market prices of risk λ_1, λ_2 are constants. With regard to the previous section, the function P is a solution to the following partial differential equation:

$$-\frac{\partial P}{\partial \tau} + (\kappa_1(\theta_1 - r_1) - \lambda_1 \sigma_1)\frac{\partial P}{\partial r_1} + (\kappa_2(\theta_2 - r_2) - \lambda_2 \sigma_2)\frac{\partial P}{\partial r_2}$$
$$+ \frac{\sigma_1^2}{2}\frac{\partial^2 P}{\partial r_1^2} + \frac{\sigma_2^2}{2}\frac{\partial^2 P}{\partial r_2^2} - (r_1 + r_2)P = 0,$$

where $\tau = T - t$ represents the time to maturity of a bond. We look for a solution P in the separated form

$$P(r_1, r_2, \tau) = P_1(r_1, \tau)P_2(r_2, \tau).$$

From the initial condition $P(r_1, r_2, 0) = 1$, for all $r_1, r_2 > 0$ we can derive the initial condition for the functions P_1 and P_2. We obtain $P_1(r_1, 0) = 1, P_2(r_2, 0) = 1$. Substituting the expression $P(r_1, r_2, \tau) = P_1(r_1, \tau)P_2(r_2, \tau)$ into the equation we obtain, after straightforward algebraic calculation,

$$P_1 \left[-\frac{\partial P_2}{\partial \tau} + (\kappa_2(\theta_2 - r_2) - \lambda_2 \sigma_2)\frac{\partial P_2}{\partial r_2} + \frac{\sigma_2^2}{2}\frac{\partial^2 P_2}{\partial r_2^2} - rP_2 \right]$$
$$+ P_2 \left[-\frac{\partial P_1}{\partial \tau} + (\kappa_1(\theta_1 - r_1) - \lambda_1 \sigma_1)\frac{\partial P_1}{\partial r_1} + \frac{\sigma_1^2}{2}\frac{\partial^2 P_1}{\partial r_1^2} - rP_1 \right] = 0.$$

Since P_1 depends only on r_1, τ and P_2 depends only on r_2, τ, the last identity holds if and only if both the parentheses are zero, i.e.

$$\frac{\partial P_1}{\partial \tau} + (\kappa_1(\theta_1 - r_1) - \lambda_1 \sigma_1)\frac{\partial P_1}{\partial r_1} + \frac{\sigma_1^2}{2}\frac{\partial^2 P_1}{\partial r_1^2} - rP_1 = 0,$$
$$\frac{\partial P_2}{\partial \tau} + (\kappa_2(\theta_2 - r_2) - \lambda_2 \sigma_2)\frac{\partial P_2}{\partial r_2} + \frac{\sigma_2^2}{2}\frac{\partial^2 P_2}{\partial r_2^2} - rP_2 = 0.$$

It means that $P_i, i = 1, 2$, are solutions to the one-factor Vasicek model with the parameters $\kappa_i, \theta_i, \sigma_i$ and market price of risk λ_i, for $i = 1, 2$. Hence the price P can be expressed in a closed form:

$$P(r_1, r_2, \tau) = A_1(\tau)A_2(\tau)e^{-B_1(\tau)r_1 - B_2(\tau)r_2}, \tag{8.17}$$

where the functions $A_i(\tau), B_i(\tau)$ are given by the formulae (8.7) and (8.6).

An analogous result can be obtained for the two-factor CIR model. In this model, the factors r_1, r_2 satisfy the stochastic differential equation (7.17) with zero correlation of the increments $\rho = 0$. In this case the bond price can be again constructed as a product of solutions from the one-factor CIR model in the form (8.17), where the functions $A_i(\tau), B_i(\tau)$ are given by the formulae (8.12) and (8.11).

Term structure of interest rates is then in both models sum of the interest rates from corresponding one-factor models:

$$R(r_1, r_2, t, t+\tau) = -\frac{\ln P(r_1, r_2, \tau)}{\tau} = -\frac{\ln P_1(r_1, \tau)}{\tau} - \frac{\ln P_2(r_2, \tau)}{\tau}.$$

In particular, it means that the interest rate with maturity τ is a linear function of the short rate components r_1 and r_2. Functions P_1 and P_2 have the form $P_i = A_i e^{-B_i r_i}$ and hence

$$R(r_1, r_2, t, t+\tau) = -\frac{\ln A_1 A_2}{\tau} + \frac{B_1}{\tau} r_1 + \frac{B_2}{\tau} r_2.$$

Convergence Models

Recall that in the convergence model due to Corzo and Schwartz [27], the assumption made on the domestic interest rate r_d is, that its evolution is stochastic and it satisfies the stochastic differential equation

$$dr_d = (\alpha + \beta(r_e - r_d))dt + \sigma_d dw_2,$$

where r_e represents the European interest rate[1] It itself is assumed to follow a stochastic process given by

$$dr_e = \gamma(\delta - r_e)dt + \sigma_e dw_e.$$

We show that, under the assumption of constant market prices of risk λ_d, λ_e, the bond prices can be computed in closed form. Prices of European bonds are then obtained from the one-factor Vasicek model in a closed form. Prices of domestic bonds are solutions to the partial differential equation

$$-\frac{\partial P}{\partial \tau} + (\alpha + \beta(r_2 - r_d) - \lambda_d \sigma_d)\frac{\partial P}{\partial r_d} + (\gamma(\delta - r_e) - \lambda_e \sigma_e)\frac{\partial P}{\partial r_e}$$
$$+ \frac{\sigma_d^2}{2}\frac{\partial^2 P}{\partial r_d^2} + \frac{\sigma_e^2}{2}\frac{\partial^2 P}{\partial r_e^2} + \rho \sigma_d \sigma_e \frac{\partial^2 P}{\partial r_d \partial r_e} - r_d P = 0.$$

Its solution can be written in the form

$$P_d(r_d, r_e, \tau) = e^{A(\tau) - r_d B(\tau) - r_e C(\tau)},$$

where $\tau = T - t$ and functions A, B, C satisfy the initial conditions $A(0) = 0, B(0) = 0, C(0) = 0$. Substituting this ansatz form of the solution into the partial differential equation we obtain the following identity

$$r_d \left(B\beta + \dot{B} - 1\right) + r_e \left(C\gamma - B\beta + \dot{C}\right) - \alpha B + \lambda_d \sigma_d B - \gamma \delta C$$
$$+ \lambda_e \sigma_e C + \frac{\sigma_d^2}{2}B^2 + \frac{\sigma_e^2}{2}C^2 + \rho \sigma_d \sigma_e BC - \dot{A} = 0.$$

---
[1] As an example, one can consider the EURIBOR term structure - EURopean Interbank Offered Rate.

This identity has to be satisfied for all r_d, r_e. This is possible if and only if

$$B\beta + \dot{B} - 1 = 0, \quad C\gamma - B\beta + \dot{C} = 0,$$

$$-\alpha B + \lambda_d \sigma_d B - \gamma\delta C + \lambda_e\sigma_e C + \frac{\sigma_d^2}{2}B^2 + \frac{\sigma_e^2}{2}C^2 + \rho\sigma_d\sigma_e BC - \dot{A} = 0.$$

A solution to the above system of ordinary differential equations is given by

$$B(\tau) = \frac{1 - \exp(-b\tau)}{b},$$

$$C(\tau) = \begin{cases} \frac{b}{c-b}\left(\frac{1-\exp(-b\tau)}{b} - \frac{1-\exp(-c\tau)}{c}\right), & \text{if } b \neq c \\ \frac{1-\exp(-c\tau)}{c} - \tau\exp(-c\tau), & \text{if } b = c \end{cases}$$

and the function $A(\tau)$ can be easily obtained in the closed form by evaluating the integral

$$A(\tau) = \int_0^\tau \left[(-a + \lambda_d\sigma_d)B(s) + (-cd + \lambda_e\sigma_e)C(s) + \frac{\sigma_d^2}{2}B(s)^2 \right.$$
$$\left. + \frac{\sigma_e^2}{2}C(s)^2 + \rho\sigma_d\sigma_e B(s)C(s)\right] ds.$$

Domestic interest rates are linear functions of domestic and European short rates:

$$R(r_d, r_e, t, t+\tau) = -\frac{\ln P(\tau, r_e, r_d)}{\tau} = \frac{A(\tau)}{\tau} + \frac{B(\tau)}{\tau}r_d + \frac{C(\tau)}{\tau}r_e.$$

Two-Factor Equilibrium Models. The Fong-Vasicek Model

In this section we recall the Fong–Vasicek model with stochastic volatility derived in the paper by Fong and Vasicek [48]. The short rate r is described by a stochastic differential equation in which the volatility itself is a solution to another stochastic differential equation. More precisely, we have:

$$\begin{aligned} dr &= \kappa_1(\theta_1 - r)dt + \sqrt{y}dw_1, \\ dy &= \kappa_2(\theta_2 - y)dt + v\sqrt{y}dw_2 \end{aligned}$$

(see Chapter 7). Increments dw_1, dw_2 of the Wiener processes w_1 and w_2 can be correlated, $E(dw_1 dw_2) = \rho dt$. Furthemore we assume that the market prices of risk are given by $\lambda_1\sqrt{y}$, respectively, $\lambda_2\sqrt{y}$, where λ_1, λ_2 are some constants. Then the bond price $P = P(\tau, r, y)$, where $\tau = T - t$ is a solution to the following partial differential equation:

$$-\frac{\partial P}{\partial \tau} + (\kappa_1(\theta_1 - r) - \lambda_1 y)\frac{\partial P}{\partial r} + (\kappa_2(\theta_2 - y) - \lambda_2 vy)\frac{\partial P}{\partial y}$$
$$+ \frac{y}{2}\frac{\partial^2 P}{\partial r^2} + \frac{v^2 y}{2}\frac{\partial^2 P}{\partial y^2} + \rho vy\frac{\partial^2 P}{\partial r \partial y} - rP = 0, \qquad (8.18)$$

satisfying the initial condition $P(0, r, y) = 1$. Again we look for a solution in the separated form

$$P(\tau, r, y) = A(\tau)e^{-B(\tau)r - C(\tau)y}.$$

Inserting this form into the partial differential equation we obtain that the functions A, B, C are solutions to the following system of ordinary differential equations:

$$\begin{aligned}
\dot{A} &= -A(\kappa_1\theta_1 B + \kappa_2\theta_2 C), \\
\dot{B} &= -\kappa_1 B + 1, \\
\dot{C} &= -\lambda_1 B - \kappa_2 C - \lambda_2 vC - \frac{B^2}{2} - \frac{v^2 C^2}{2} - v\rho BC
\end{aligned} \qquad (8.19)$$

and satisfy the initial conditions $A(0) = 1, B(0) = 0, C(0) = 0$. Notice that the differential equation for B can be explicitly solved. Knowing the solution B, the differential equation for C can be solved numerically by means of the Runge–Kutta method. Resulting functions B, C are then inserted into the equation for the function A. In summary, we obtain

$$B = \frac{1}{\kappa_1}\left(1 - e^{-\kappa_1 \tau}\right),$$

$$\dot{C} = -\lambda_1 B - \frac{B^2}{2} - (\kappa_2 + \lambda_2 v + v\rho B)C - \frac{v^2}{2}C^2, \; C(0) = 0,$$

$$A = \exp\left(-\theta_1 \tau + \theta_1 B - \kappa_2 \theta_2 \int_0^\tau C(s)ds\right).$$

8.1.3. Non-arbitrage Models, the Ho-Lee and Hull–White Models

One of principal disadvantages of equilibrium models is the fact that they are not able to exactly fit the present term structure of interest rates. The reason is, that there are just few parameters and so they are able to provide an approximation of the present term structure only. This was a motivation for construction of the so-called nonarbitrage models. These models are constructed in such a way that they give the exact present term structure in the market. The price paid for this property is the necessity to introduce time dependent parameters into the process of the short rate.

Ho–Lee Model

Historically the first nonarbitrage model was proposed by Ho and Lee in [62]. A price of bond is again given by compounded interest rate, i.e.

$$P(t, T) = e^{-R(t,T)(T-t)}.$$

In this model the short rate is assumed to follow the stochastic process with time depending drift:

$$dr = \theta(t)dt + \sigma dw. \qquad (8.20)$$

It means that the process is described by a constant volatility σ and time dependent drift $\theta(t)$. The drift is chosen in such a way that the resulting term structure of interest rates $R(t,T), T \geq 0$ at the present date (i.e., $t = 0$) is equal to the present term structure $\{R(0,T), T \geq 0\}$ known from the market data at $t = 0$. Similarly, as in the case of equilibrium models, we obtain the partial differential equation for the bond price $P = P(r,t,T)$

$$\frac{\partial P}{\partial t} + (\theta(t) - \lambda(t)\sigma)\frac{\partial P}{\partial r} + \frac{\sigma^2}{2}\frac{\partial^2 P}{\partial r^2} - rP = 0$$

with the time dependent parameters. Here $\lambda(t)$ denotes the market price of risk. The terminal condition is again given by
$$P(r,T,T) = 1.$$
We will search the solution in the closed form
$$P(r,t,T) = A(T-t)e^{-B(T-t)r},$$
where $A(0) = 1, B(0) = 0$. Substituting it into the partial differential equation for the price P and comparing the coefficients of the orders of r and 1 we finally obtain that A, B are solutions to the system of ordinary differential equations
$$\dot{B}(\tau) = 1, \qquad \dot{A}(\tau) = A(\tau)\left(\frac{\sigma^2}{2}B(\tau)^2 - \phi(T-\tau)B(\tau)\right),$$
where $\phi(t) = \theta(t) - \lambda(t)\sigma$ and $\tau = T - t$ is time to maturity of a bond. Then $B(\tau) = \tau$. For the function A we then obtain
$$\frac{d\ln A}{d\tau}(\tau) = \frac{\sigma^2}{2}\tau^2 - \phi(T-\tau)\tau.$$
Our aim is to find such a function ϕ (and consequently θ), for which the known present term structure of interest rates $\{R(0,T), T \geq 0\}$ is identical to the one corresponding to $P(r,0,T)$. Hence the following equality has to be satisfied:
$$e^{-R(0,T)(T-0)} = P(r,0,T) = A(T-0)e^{-B(T-0)r} = A(T)e^{-R(0,0)T},$$
since $r = R(0,0)$. By comparing the terms in this identity we conclude
$$\ln A(T) = (R(0,0) - R(0,T))T,$$
for all $T > 0$. Hence
$$\frac{d\ln A}{d\tau}(\tau) = R(0,0) - R(0,\tau) - \tau\frac{\partial R}{\partial T}(0,\tau).$$
It means that the drift function $\phi(t)$ can be written as
$$\phi(t) = \frac{\sigma^2}{2}(T-t) + \frac{\partial R}{\partial T}(0, T-t) + \frac{R(0, T-t) - R(0,0)}{T-t}. \tag{8.21}$$

Hull and White Model

Another nonarbitrage model for interest rates was proposed by Hull and White (cf. [65]). Unlike the Ho–Lee model, here we assume the following stochastic differential equation for the short rate:
$$dr = \theta(t)(a(t) - r)dt + \sigma(t)r^\gamma dw, \tag{8.22}$$
where $\gamma = 0$ or $\gamma = \frac{1}{2}$. In this model, it is also possible to find a closed form solution to the corresponding partial differential equation such that the resulting term structure of interest rates coincides with the real one, observed on the market.

8.2. Other Interest Rate Derivatives

Interest Rate Swaps

An interest rate swap is an agreement between two traders. A party A commits to paying a party B the fixed interest rate r^* from the specified amount which can be assumed to be unity. The party B is committed to pay to the party A a floating interest rate r. Assume that the short rate r follows a stochastic process of the form (7.1) and that payments are realized continuously. We can reformulate the interest rate swap agreement as follows: the party A holds a bond paying a coupon $r - r^*$. At the maturity T, i.e., the time of the end of the swap agreement, the bond has the zero value. Derivation of a governing equation for the price $P(t,r)$ of such a bond is very similar to the one of a zero-coupon bond. Using Itō's lemma we obtain a stochastic differential equation for the bond price

$$dP = \mu_P(t,r)dt + \sigma_P(t,r)dw,$$

where

$$\mu_P = \frac{\partial P}{\partial t} + \mu \frac{\partial P}{\partial r} + \frac{1}{2}\sigma^2 \frac{\partial^2 P}{\partial r^2}, \quad \sigma_P = \sigma \frac{\partial P}{\partial r}.$$

We can construct a portfolio consisting of one bond with maturity T_1 and Δ bonds with maturity T_2. Its value is therefore equal to: $\pi = P(T_1) + \Delta P(T_2)$. A change in the value due to the change of bond prices equals $dP(T_1) + \Delta dP(T_1)$. In addition to this, there is a change $(1+\Delta)(r - r^*)dt$ due to coupon payments. In summary, we have

$$\begin{aligned} d\pi &= dP(T_1) + \Delta dP(T_1) + (1+\Delta)(r - r^*)dt \qquad (8.23) \\ &= (\mu_P(T_1) + \Delta \mu_P(T_2) + (1+\Delta)(r - r^*))\, dt + (\sigma_P(T_1) + \Delta \sigma_P(T_2))\, dw. \end{aligned}$$

We make our portfolio risk-neutral by choosing Δ such that $-\frac{\sigma_P(T_1)}{\sigma_P(T_2)}$. Then

$$d\pi = \left(\mu_P(T_1) + -\frac{\sigma_P(T_1)}{\sigma_P(T_2)}\mu_P(T_2) + (1+\Delta)(r - r^*)\right) dt.$$

In order to avoid possibility of an arbitrage opportunity, the yield of the portfolio should be equal to the riskless interest rate r, i.e.

$$d\pi = r\pi dt = r\left(P(T_1) - \frac{\sigma_P(T_1)}{\sigma_P(T_2)}P(T_2)\right) dt.$$

Hence

$$\mu_P(T_1) + -\frac{\sigma_P(T_1)}{\sigma_P(T_2)}\mu_P(T_2) + (1+\Delta)(r - r^*) = r\left(P(T_1) - \frac{\sigma_P(T_1)}{\sigma_P(T_2)}P(T_2)\right)$$

from which we obtain

$$\frac{\mu_P(T_1) + (r - r^*) - rP(T_1)}{\sigma_P(T_1)} = \frac{\mu_P(T_2) + (r - r^*) - rP(T_2)}{\sigma_P(T_2)}.$$

It means that the above expressions are independ of the maturity T. Hence there is a function $\lambda = \lambda(r,t)$ such that

$$\frac{\mu_P(T) + (r - r^*) - rP(T)}{\sigma_P(T)} = \lambda$$

for all maturities T. Similarly as in the case of pricing zero-coupon bonds, this function is called a market price of risk. Finally, substituting μ_P and σ_P yields the partial differential equation for the bond price (and hence for the swap price) $P(t,r)$

$$\frac{\partial P}{\partial t} + (\mu - \lambda \sigma)\frac{\partial P}{\partial r} + \frac{1}{2}\sigma^2 \frac{\partial^2 P}{\partial r^2} - rP + r - r^* = 0. \tag{8.24}$$

The terminal condition $P(T,r) = 0$ follows from the fact that the bond pays only the coupon and there is no such a payment at maturity (see e.g., Kwok [75]).

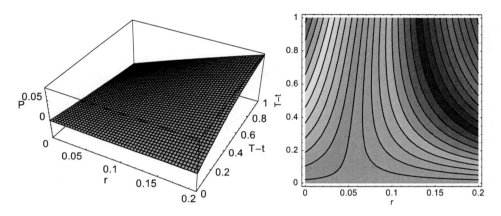

Figure 8.3. A 3D and contour plots of the function $P(r,t)$ describing a price of the interest rate swap.

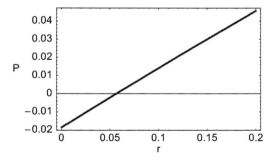

Figure 8.4. A graph of dependence of the interest rate swap price $P(r,0)$ on the floating interest rate $r > 0$.

Equation (8.24) can be solved numerically. For example, let us consider the short rate follows the stochastic differential equation for the CIR process, i.e.

$$dr = \kappa(\theta - r)dt + \sigma\sqrt{r}dw.$$

It means that $\mu(r,t) = \kappa(\theta - r)$ and $\sigma(r,t) = \sigma\sqrt{r}$. If we consider the market price of risk

Table 8.2. The Mathematica source code for pricing the interest rate swap.

```
sigma = 0.4; rstar = 0.05;  kappa = 1;
theta = 0.02; lambda = 0; T = 1;
rmax = 2;

PayOff[r_] := 0;

solution = NDSolve[{
   D[P[r, tau], tau] == (sigma^2/2)r D[P[r, tau], r, r]
   + (kappa*(theta - r) - lambda*sigma*r)* D[P[r, tau], r]
   - r *P[r, tau] + r - rstar,

   P[r, 0] == PayOff[r],
   (D[P[r, tau],r]/.r -> rmax) == 0 },
   P, {tau, 0, T}, {r, 0, rmax}
];

P[r_, tau_] := Evaluate[P[r, tau] /. solution[[1]]  ];
Plot3D[P[r, tau], {r, 0., 0.2}, {tau, 0, T}];
```

$\lambda(r,t) = \lambda\sqrt{r}$ then the governing equation for the price $P(t,r)$ reads as follows:

$$\frac{\partial P}{\partial t} + (\kappa(\theta - r) - \lambda\sigma r)\frac{\partial P}{\partial r} + \frac{1}{2}\sigma^2 r\frac{\partial^2 P}{\partial r^2} - rP + r - r^* = 0, \quad r > 0, t \in (0,T),$$

$$P(r,T) = 0, \quad r > 0.$$

Its numerical solution for model parameters $\sigma = 0.4, r^* = 0.05, \kappa = 1, \theta = 0.02, \lambda = 0, T = 1$ is shown in Fig. 8.3. In Fig. 8.4 we show dependence of the solution $P(r,0)$ on the floating interest rate $r > 0$ at $t = 0$. The Mathematica source code for pricing interest rate swaps is shown in Table 8.2.

The boundary condition for large values of r can be deduced from the limit of the PDE (8.24) when $r \to \infty$. By dividing (8.24) by r, assuming μ, σ growth at most linearly in r and supposing that $P(r,t) \to P(\infty,t)$ as $r \to \infty$ we can postulate the Neumann boundary condition $\frac{\partial P}{\partial r}(r_M,t) = 0$ for right end-point $r_M \gg 1$ of the finite computational domain $r \in (0, r_M)$. Interestingly enough, in the case when the structural condition

$$\frac{2\kappa\theta}{\sigma^2} \geq 1$$

is satisfied, we do not need to prescribe the boundary condition for our numerical solution at $r = 0$. This is a consequence of the well-known Fichera condition. Recall that in the paper [46] Fichera introduced the so-called Fichera condition in order to determine wheather it is necessary or not to specify the boundary conditions (b.c.) on the boundary of the domain. In particular, for a parabolic PDE of the form

$$-\frac{\partial P}{\partial \tau} + \frac{1}{2}\sigma^2(r,\tau)\frac{\partial^2 P}{\partial r^2} + \beta(r,\tau)\frac{\partial P}{\partial r} + c(r,\tau)P = d(r,\tau), \quad r > 0, \tau > 0,$$

where $\sigma(0,\tau) = 0$, the Fichera condition at $r = 0$ reads as follows: if

$$\lim_{r \to 0^+} \left[\beta(r,\tau) - \frac{1}{2} \frac{\partial}{\partial r} \sigma^2(r,\tau) \right] \begin{cases} \geq 0 & \text{then no b.c. at } r = 0 \text{ is needed,} \\ < 0 & \text{then a b.c. at } r = 0 \text{ must be prescribed,} \end{cases} \quad (8.25)$$

in order to guarantee uniqueness of a solution P. In our case when $\beta(r,\tau) \equiv \kappa(\theta - r) - \lambda \sigma r$ and $\sigma(r,t) \equiv \sigma \sqrt{r}$ the Fichera condition reduces to the inequality $\kappa \theta - \frac{1}{2} \sigma^2 \geq 0$.

Swaptions

A swaption is a right, but not an obligation, to enter the swap contract at time $T < T_s$ for the exercise price X. It can be shown (see Kwok [75]) that the price of this right $V(r,t)$ is a solution to the same partial differential equation as in the case of the one-factor bond pricing equation. The only difference is the terminal condition. In the case of a call swaption it is given by

$$V(r,T) = (W(r,T) - X)^+$$

and for a put swaption,

$$V(r,T) = (X - W(r,T))^+.$$

The term $W(r,t)$ is the value of the swap contract at the time $t \in [0,T)$. In fact, it is an underlying asset for the swaption.

Problem Section and Exercises

1. Consider the Vasicek model with parameters $\theta = 0.0364, \kappa = 41.9624, \sigma = 0.0888$ (estimated from BRIBOR overnight rates from the year 2007). Construct the yield curves for different values of market prices of risk λ. How do interest rates depend on this parameter?

2. Consider the CIR model with parameters $\theta = 0.0364, \kappa = 44.9889, \sigma = 0.4917$ (estimated from BRIBOR overnight rates from the year 2007). Construct the yield curves for different values of parameter λ from market prices of risk. How do interest rates depend on this parameter?

3. Derive a closed form expression for the bond price computed by means of the two-factor CIR model if the market prices of risk are given by $\lambda_1 \sqrt{r_1}$ and $\lambda_2 \sqrt{r_2}$.

4. Compute the limit of the domestic interest rates as time to maturity approaches zero in the convergence model due to Corzo and Schwartz, i.e. $\lim_{\tau \to \infty} R(r_d, r_e, t, t+\tau)$. What is the relation of this limit and the corresponding limit of the European interest rates?

5. In the Hull and White model (8.22) construct a solution P in the closed form $P(r,t) = A(T-t)e^{-B(T-t)r}$ and derive the system of ordinary differential equations for the functions $A(\tau)$ and $B(\tau)$ where $\tau = T - t$.

6. For the Chan, Karolyi, Longstaff and Sanders short rate interest rate model $dr = \kappa(\theta - r)dt + \sigma r^\gamma dw$ (see (7.5) derive a partial differential equation for pricing a zero coupon bond $P(r,t)$. Using the Fichera condition (8.25) show that no boundary conditions at $r = 0$ are needed provided that $\gamma > \frac{1}{2}$. In the case $0 < \gamma < \frac{1}{2}$ (or $\gamma = \frac{1}{2}$ and $2\kappa\theta/\sigma^2 < 1$) prove that the boundary condition at $r = 0$ has to be prescribed.

Chapter 9

American Style of Derivative Securities

In this chapter, we are interested in mathematical modeling of American style of derivative contracts. Unlike European style of derivative contracts, the American style of derivatives is characterized by the possibility of early exercising of an option contract at some time $t^* \in [0, T)$ prior the obligatory expiration time T. For instance, the American call or put option can be exercised anytime before the obligatory expiration time T. It should be noted that most of derivative contracts traded nowadays are of the American style.

An American call (put) option is an agreement (contract) between the writer and the holder of an option. It represents the right but not the obligation to purchase (sell) the underlying asset at the prescribed exercise price E anytime in the forecoming interval $[0, T]$ with the specified time of obligatory expiration at $t = T$.

Similarly as in the case of European style of options we ask the question: what is the price of such an option (the option premium) at the time $t = 0$ of contracting. In other words, how much should the holder of the option pay the writer for such a derivative security. For the American call (put) option the problem is to price the contract, i.e., to find a price $V^{ac}(S,t)$ of the American call option ($V^{ap}(S,t)$ for the put option), at the time $t \in [0, T]$.

Since American options give the holder more rights when compared to corresponding European options their price should be higher, i.e.

$$V^{ac}(S,t) \geq V^{ec}(S,t), \quad V^{ap}(S,t) \geq V^{ep}(S,t),$$

for any time $t \in [0, T]$ and underlying asset price $S \geq 0$. Moreover, a price of the American call and put options should be greater or equal than their price at expiry given by the pay-off diagram:

$$V^{ac}(S,t) \geq V^{ec}(S,T) = (S-E)^+,$$
$$V^{ap}(S,t) \geq V^{ep}(S,T) = (E-S)^+,$$

for any time $t \in [0, T]$ and $S \geq 0$. Indeed, if the price $V^{ac}(S,t)$ of an American call option at the time $t < T$ before the expiry T is less by one dollar than its terminal pay–off diagram $(S-E)^+$ then by buying such an option and its immediate exercising (which is allowed for American style of options) we can receive from the writer the underlying asset for the

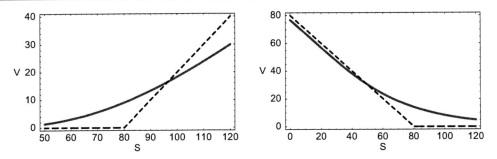

Figure 9.1. Graphs of solutions corresponding to the European call option (left) on asset paying nontrivial dividends and the European put option on asset paying no dividends (right).

exercise price E. If we sell it on the market we receive its spot price S. This way the holder can earn one dollar without bearing any risk. Such a situation obviously would lead to an arbitrage opportunity. Since there is a demand for such an option the market will increase its price to the level that is greater or equal to the pay-off diagram.

In Fig. 9.1 (left) we can observe that the price of the European call option on the underlying asset paying continuous dividends with a rate $q > 0$ always intersects the pay-off diagram. It is easy to justify such a behavior since it follows from the explicit formula (3.8) for pricing a European call option that

$$\lim_{S \to \infty} \frac{V^{ec}(S,t)}{S} = e^{-q(T-t)} < 1.$$

Hence $V^{ec}(S,t) < S - E$ for a sufficiently large $S \gg E$ and $0 \leq t < T$. Similarly, for the European put option on the asset with a dividend rate $q \geq 0$) we can deduce from (3.14) the following inequality: $V^{ep}(0,t) = Ee^{-r(T-t)} < E$. Therefore the solution $V^{ep}(S,t)$ always intersects the pay–off diagram of the put option. In Fig. 9.1 (right) we can see a graph of a solution representing the European put option and its comparison with the pay-off diagram.

In the case of an American call option on the underlying asset paying no dividends ($q = 0$), its pricing is simple and the price coincides with the European one, i.e.

$$\text{if } q = 0 \text{ then } V^{ac}(S,t) = V^{ec}(S,t), \text{ for each } S \geq 0, t \in [0,T]. \tag{9.1}$$

The reason is that it is not worth to exercise the American call option before the expiry T. Indeed, if we early exercise the option at the time $t \in [0,T)$ then its value falls down to the value given by the pay–off diagram $(S-E)^+$. It means that its value is strictly less than the value of the European call option because $V^{ec}(S,t) > (S-E)^+$ for the case $q = 0$.

The situation is more complicated in the case of an American call option on the underlying asset paying nontrivial dividends ($q > 0$). In such a case, the solution $V^{ec}(S,t)$ intersects the pay–off diagram $(S-E)^+$. Hence we are unable to follow the same argument as in the case of vanishing dividends $q = 0$. Furthermore, holding an American call option until the expiry $t = T$ would mean that its value is identical with the European style of a call option. But this is not possible because $V^{ec}(S,t) < (S-E)^+$ for large values of the underlying asset

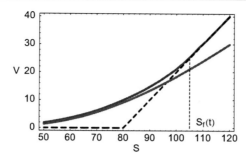

Figure 9.2. A comparison of solutions of the European and American call options at some time $0 \leq t < T$. The position of the early exercise boundary is labeled by $S_f(t)$.

price $S \gg E$. Therefore the price of the American call option is strictly higher than that of the European call option, i.e.

$$\text{if } q > 0, r > 0, \text{ then } V^{ac}(S,t) > V^{ec}(S,t), \text{ for each } S > 0, t \in [0,T]. \tag{9.2}$$

Since the graph of a solution of the European put option always intersects the pay-off diagram of a put option for $q \geq 0$ we obtain the strict inequality:

$$\text{if } q \geq 0, r > 0, \text{ then } V^{ap}(S,t) > V^{ep}(S,t), \text{ for each } S \geq 0, t \in [0,T]. \tag{9.3}$$

9.1. Pricing of American Options by Solutions to Free Boundary Problems

In the previous section we presented an analysis of basic properties of the American call and put options. We showed that pricing of American style options leads to a free boundary problem for pricing the American call option on the underlying asset paying continuous dividends $q > 0$. In addition to a function $V(S,t) = V^{ac}(S,t)$ we have to find the free boundary position, i.e., the function $S_f(t)$ depending on time $t \in [0,T]$. This function forms the so-called early exercise boundary having the following properties:

1. If $S < S_f(t)$ for $t \in [0,T]$ then $V^{ac}(S,t) > (S-E)^+$ and we hold the call option because its value is strictly higher than the pay-off diagram of a call option. In order to hedge the portfolio we make use of the Black–Scholes model, i.e., for $0 < t < T$ and $S < S_f(t)$ the Black–Scholes equation holds true.

2. If $S \geq S_f(t)$ for $t \in [0,T]$ then $V^{ac}(S,t) = (S-E)^+$ and we have to exercise the call option because its value coincides with the terminal pay-off diagram.

In Fig. 9.2 we can see the behavior of a price of the American call option $S \mapsto V^{ac}(S,t)$ at some time $t \in [0,T)$. We also plot a comparison with the pay-off diagram $S \mapsto (S-E)^+$

and the lower price of the European call option on asset paying continuous dividends. The value $S_f(t)$ splits the interval of underlying asset prices into two subintervals: $0 < S < S_f(t)$ in which we hold the call option and $S \geq S_f(t)$ in which the call option has to be exercised at the time t.

Now we are in a position to state a mathematical formulation of the problem of pricing American style of a call option. The problem is to find a function $V = V^{ac}(S,t)$ and the function $S_f : [0,T] \to \mathbb{R}$ determining the early exercise boundary with the following properties:

(the free boundary problem for pricing the American call option)

1. The function $V(S,t)$ is a solution to the Black–Scholes partial differential equation :

$$\frac{\partial V}{\partial t} + \frac{\sigma^2}{2}S^2\frac{\partial^2 V}{\partial S^2} + (r-q)S\frac{\partial V}{\partial S} - rV = 0 \qquad (9.4)$$

defined on a time dependent domain $0 < t < T$, $0 < S < S_f(t)$.

2. It satisfies the terminal pay-off diagram:

$$V(S,T) = (S-E)^+ \qquad (9.5)$$

3. and the boundary conditions:

$$V(0,t) = 0, \quad V(S_f(t),t) = S_f(t) - E, \quad \frac{\partial V}{\partial S}(S_f(t),t) = 1, \qquad (9.6)$$

at $S = 0$ and $S = S_f(t)$.

So far, we did not explain a financial meaning of the boundary condition $\frac{\partial V}{\partial S}(S_f(t),t) = 1$ imposed on a solution at the point $S = S_f(t)$ of the early exercise of a call option. Notice that this condition together with the continuity condition $V(S_f(t),t) = S_f(t) - E$ guarantee the C^1 continuity in the S variable of the function $V^{ac}(S,t)$ at the point $S = S_f(t)$ for each $0 < t < T$. It should be obvious that prescribing Dirichlet boundary conditions $V(0,t) = 0$ at $S = 0$ and $V(S_f(t),t) = S_f(t) - E$ at $S = S_f(t)$ is not sufficient for unique solvability of the free boundary problem. Indeed, it follows from the basic properties of solutions to parabolic equations (see e.g., Ševčovič [105]), that for any function $t \mapsto S_f(t)$ we can find a (unique) solution to the Black–Scholes equation satisfying the above mentioned Dirichlet conditions at $S = 0$ and $S = S_f(t)$. Hence we would have no other condition determining the free boundary profile $t \mapsto S_f(t)$. Therefore an additional condition that connects the free boundary position $S_f(t)$ and the solution (S,t) is needed.

In what follows, we will show that the condition $\frac{\partial V}{\partial S}(S_f(t),t) = 1$ guaranteeing C^1 continuity of the contact of a solution $V(S,t)$ and its pay-off diagram $(S-E)^+$ is indeed the boundary condition fullfiled by an American call option. We will follow the idea of derivation of the boundary condition due to Merton (cf. Kwok [75]). It is based on the financial argument stating that the price $V^{ac}(S,t)$ of an American call option should be given as the maximal value among all call option prices whose early exercise boundary is prescribed by a continuous function of time. More precisely,

$$V^{ac}(S,t) = \max_{\eta} V(S,t;\eta),$$

where the maximum is taken over all positive continuous functions $\eta : [0,T] \to \mathbb{R}^+$. Here $V(S,t;\eta)$ denotes the price of a call option given by a solution to the Black–Scholes equation on a time dependent domain $0 < t < T, 0 < S < \eta(t)$ and satisfying the Dirichlet boundary conditions $V(0,t;\eta) = 0, V(\eta(t),t;\eta) = \eta(t) - E$, for $t \in [0,T]$. The early exercise boundary function S_f is then the argument of maximum of the above variational problem. This is indeed a variational problem since our aim is to find a maximum of the functional $\eta \mapsto V(S,t;\eta)$ defined on the infinite dimensional Banach space of all continuous functions. The first order necessary condition read as follows:

$$D_\eta V(S,t;S_f)\xi = 0, \quad \text{for any function } \xi \in C([0,T]),$$

where $D_\eta V(S,t;\eta) : C([0,T]) \to \mathbb{R}$ is a linear operator representing the Fréchet derivative of V with respect to η. Here $C([0,T])$ is a Banach space of all continuous functions defined on the interval $[0,T]$ endowed with the maximum norm. Let $t \in [0,T)$ be a fixed time. Since for any function $\eta \in C([0,T])$ we have $V(\eta(t),t;\eta) - \eta(t) + E = 0$ then, by differentiating this equality with respect to the function η in the direction $\xi \in C([0,T])$, we conclude, for any $t \in (0,T)$, the following identity:

$$\frac{\partial V}{\partial S}(\eta(t),t;\eta)\xi(t) + D_\eta V(\eta(t),t;\eta)\xi - 1.\xi(t) = 0.$$

With regard to the first order necessary condition $D_\eta V(S,t;S_f)\xi = 0$ we obtain $\frac{\partial V}{\partial S}(S_f(t),t;S_f)\xi(t) = \xi(t)$ at the maximizer $\eta = S_f$. As the function $\xi \in C([0,T])$ was arbitrary, we end up with the desired boundary condition that has to be fullfiled by the solution $V(S,t)$ and the early exercise boundary function S_f:

$$\frac{\partial V}{\partial S}(S_f(t),t;S_f) = 1.$$

In the case of the American put option one can argue similarly. The free boundary problem for pricing the American put option consists in construction of a function $V = V^{ap}(S,t)$ together with the early exercise boundary profile $S_f : [0,T] \to \mathbb{R}$ satisfying the following conditions:

(the free boundary problem for pricing the American put option)

1. The function $V(S,t)$ is a solution to the Black–Scholes partial differential equation:

$$\frac{\partial V}{\partial t} + \frac{\sigma^2}{2}S^2\frac{\partial^2 V}{\partial S^2} + (r-q)S\frac{\partial V}{\partial S} - rV = 0 \tag{9.7}$$

defined on the time dependent domain $S > S_f(t)$ where $0 < t < T$.

2. It satisfies the terminal pay-off diagram:

$$V(S,T) = (E-S)^+ \tag{9.8}$$

3. and boundary conditions for the put option:

$$V(+\infty,t) = 0, \quad V(S_f(t),t) = E - S_f(t), \quad \frac{\partial V}{\partial S}(S_f(t),t) = -1, \tag{9.9}$$

for $S = S_f(t)$ and $S = \infty$.

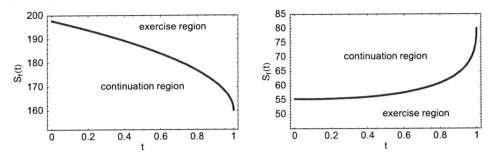

Figure 9.3. A plot of the early exercise boundary function $t \mapsto S_f(t)$ for an American call (left) and put (right) options.

Next, we shall present several useful facts concerning the early exercise boundary position for American call and put options. First, we will consider the case of a call option. Notice that the early exercise boundary position should be greater or equal than the exercise price E. Indeed, it is not reasonable to exercise a call option with the expiration price E when the spot price S of the underlying asset is less than E.

Since the function $S \mapsto V(S,t)$ is continuously differentiable with respect to the S variable at $S = S_f(t)$ we obtain, by differentiating the identity $V(S_f(t),t) = S_f(t) - E$ with respect to time t, the identity: $\frac{\partial V}{\partial S}(S_f(t),t)\dot{S}_f(t) + \frac{\partial V}{\partial t}(S_f(t),t) = \dot{S}_f(t)$. Taking into account the boundary conditions $\frac{\partial V}{\partial S}(S,t) = 1$ for $S = S_f(t)$ we conclude[1]

$$\frac{\partial V}{\partial t}(S_f(t),t) = 0, \quad \text{for each } t \in (0,T).$$

Using the above expression and the fact that the Black–Scholes equation is valid within the interval $0 < S < S_f(t)$ we obtain, by passing to the limit $S \to S_f(t)$:

$$\begin{aligned} qS_f(t) - rE &= -(r-q)S_f(t)\frac{\partial V}{\partial S}(S_f(t),t) + rV(S_f(t),t) \\ &= \frac{\sigma^2}{2}S_f(t)^2 \frac{\partial^2 V}{\partial S^2}(S_f(t),t) \geq 0, \end{aligned} \quad (9.10)$$

because the function $S \mapsto V(S,t)$ has nonnegative second derivative at $S = S_f(t)$. Indeed, if $\frac{\partial^2 V}{\partial S^2}(S,t) < 0$ at $S = S_f(t)$ then, with regard to the boundary condition $\frac{\partial V}{\partial S}(S_f(t),t) = 1$, we would obtain $V(S,t) < (S-E)^+$ for all $S < S_f(t)$, where S is close to $S_f(t)$, a contradiction. Now, it follows from (9.10) that

$$S_f(t) \geq E \max\left(\frac{r}{q}, 1\right), \quad \text{for each } t \in [0,T]. \quad (9.11)$$

It remains to determine the terminal value $S_f(T)$ at the expiration T. Either $S_f(T) = E$ or $S_f(T) > E$. If $S_f(T) > E$ then, with regard to the limit $V(S,t) \to S - E$ for $t \to T$, we can deduce that the second derivative $\frac{\partial^2 V}{\partial S^2}$ converges to zero for $S = S_f(t) > E$ as $t \to T$. Taking

---

[1] A proof of the identity for the put option is similar.

into account identity (9.10) we obtain, in the limit $t \to T$, $E < S_f(T) = rE/q$. But this is possible only if $r > q > 0$. In both cases we conclude

$$S_f(T) = E \max\left(\frac{r}{q}, 1\right). \tag{9.12}$$

In the case of an American put option one can argue similarly as before. It can be shown that the early exercise boundary position $S_f(t)$ has the following properties:

$$S_f(T) = E, \qquad S_f(t) \leq E, \text{ for each } t \in [0,T]. \tag{9.13}$$

One of important problems in the field of mathematical finance is the analysis of the early exercise boundary $S_f(t)$ and the optimal stopping time (an inverse function to $S_f(t)$) for American call (or put) options on assets paying a continuous dividend yield with a rate $q > 0$ (or $q \geq 0$). However, an exact analytical expression for the free boundary profile is not known yet. Many authors have investigated various approximation models leading to approximate expressions for valuing American call and put options: analytic approximations (Barone–Adesi and Whaley [12], Kuske and Keller [74], Dewynne et al. [36], Geske et al. [53, 54], MacMillan [79], Mynemi [85]); methods of reduction to a nonlinear integral equation (Alobaidi [5], Kwok [75], Mallier et al. [80, 81], Ševčovič [103], Stamicar et al. [109]). In recent papers [125, 126], Zhu derived a closed form of the analytic approximation of the free boundary position in terms of an infinite parametric integral. We also refer to a survey paper by Chadam [22] focusing on free boundary problems in mathematical finance. For example, in the case of an American call option, it follows from the detailed analysis of the early exercise boundary behavior close to expiry T that its asymptotic expression has the form

$$S_f(t) \approx K\left(1 + 0.638\sigma\sqrt{T-t}\right), \quad K = E\max(r/q, 1) \tag{9.14}$$

for $t \to T$. The above asymptotic formula has been derived by Dewynne et al. in [36] and by Ševčovič in [103]. In the latter reference one can furthermore find a nonlinear integral equation for the early exercise boundary position $S_f(t)$ for the entire interval $t \in [0,T]$.

In the case of the American put option the profile of the early exercise boundary has a different asymptotics. The function $S_f(t)$ can be expressed as:

$$S_f(t) = E e^{-(r-\frac{\sigma^2}{2})(T-t)} e^{\sigma\sqrt{2(T-t)}\eta(t)}.$$

In the above formula the auxiliary function η can be approximated for $t \to T$ by the following expression:

$$\eta(t) \approx -\sqrt{-\ln\left[\frac{2r}{\sigma}\sqrt{2\pi(T-t)}e^{r(T-t)}\right]}. \tag{9.15}$$

The asymptotic formula (9.15) has been derived by Stamicar, Chadam and Ševčovič in [109]. In Fig. 9.3 we present a graph of the early exercise boundary function for the call option (left) and put option (right) computed by means of the approximative formulae (9.14) and (9.15) for the model parameters $T = 1, E = 80, r = 0.04, \sigma = 0.37$ and $q = 0.02$ for the call option and $q = 0$ for the put option. Notice that we present a survey of the transformation methods for calculating the early exercise boundary for American style of options in the forecoming Chapter 12.

9.2. Pricing American Style of Options by Solutions to a Linear Complementarity Problem

The purpose of this section is to analyze the Black–Scholes partial differential equation for the entire range of values $0 < S < \infty$ of the underlying asset price. In contrast to the case of European style of options, we will show that, for American options the Black–Scholes inequality holds true. More precisely, we will show that, for the American call (put) option the following partial differential inequality is satisfied:

$$\frac{\partial V}{\partial t} + \frac{\sigma^2}{2} S^2 \frac{\partial^2 V}{\partial S^2} + (r-q) S \frac{\partial V}{\partial S} - rV \leq 0, \tag{9.16}$$

for each $0 < S < \infty, 0 < t < T$.

First, let us consider the case of an American call option. According to the results from the previous section, we know that the Black–Scholes equation is satisfied on the time dependent spatial interval $0 < S < S_f(t)$ in which we hold the option, i.e., we have the equality sign in (9.16). At the same time, for such values of the underlying asset S we have the strict inequality $V(S,t) > (S-E)^+$. On the other hand, if $S \geq S_f(t)$ then $V(S,t) = (S-E)^+ = S-E$ because $S_f(t) \geq E$. Now, if we insert the linear function $S-E$ into the Black–Scholes equation we obtain

$$\begin{aligned}\frac{\partial V}{\partial t} &+ \frac{\sigma^2}{2} S^2 \frac{\partial^2 V}{\partial S^2} + (r-q) S \frac{\partial V}{\partial S} - rV \\ &= (r-q)S - r(S-E) = rE - qS \leq rE - qS_f(t) \leq 0,\end{aligned}$$

because $S_f(t) \geq E \max(\frac{r}{q}, 1)$.

In the case of an American put option on the underlying asset paying no dividends ($q=0$) we can argue similarly. In the continuation interval $S > S_f(t)$ where we hold the put option the Black–Scholes equation is satisfied. Hence the equality is fulfilled in equality (9.16). At the same time we have the strict inequality $V(S,t) > (E-S)^+$. Now, if $0 < S \leq S_f(t)$ then $V(S,t) = (E-S)^+ = E-S$ because $S_f(t) \leq E$. If we insert the linear function $E-S$ into the Black–Scholes equation we obtain

$$\begin{aligned}\frac{\partial V}{\partial t} &+ \frac{\sigma^2}{2} S^2 \frac{\partial^2 V}{\partial S^2} + rS \frac{\partial V}{\partial S} - rV \\ &= -rS - r(E-S) = -rE < 0.\end{aligned}$$

In summary, we have shown the following property of a solution to the problem of pricing the American style of call and put options:

Linear complementarity formulation for American options (9.17)

$$\frac{\partial V}{\partial t} + \frac{\sigma^2}{2} S^2 \frac{\partial^2 V}{\partial S^2} + (r-q) S \frac{\partial V}{\partial S} - rV \leq 0, \quad V(S,t) \geq \bar{V}(S),$$

$$\left(\frac{\partial V}{\partial t} + \frac{\sigma^2}{2} S^2 \frac{\partial^2 V}{\partial S^2} + (r-q) S \frac{\partial V}{\partial S} - rV\right)(V(S,t) - \bar{V}(S)) = 0,$$

for any $0 < S < \infty, 0 < t < T$, where $\bar{V}$ denotes the terminal pay–off diagram, i.e.

$$\bar{V}(S) = \begin{cases} (S-E)^+, & \text{for the call option,} \\ (E-S)^+, & \text{for the put option.} \end{cases} \quad (9.18)$$

Pricing an American call (put) option by means of a solution to the linear complementarity problem can be mathematically stated as follows: find a continuously differentiable function $V(S,t)$ such that it is a solution to (9.17) and it satisfies the terminal condition (9.18) and corresponding boundary conditions.

At the end of this chapter, we will show how to transform the linear complementarity problem for pricing American call or put options in terms of a solution to a parabolic variational inequality. Similarly as in Chapter 3, we can transform the Black–Scholes equation by using the following change of independent variables (see (3.4)):

$$S = Ee^x, \quad t = T - \tau,$$

where $x \in (0, \infty), \tau \in (0, T)$ and transformed function

$$V(S,t) = Ee^{-\alpha x - \beta \tau} u(x, \tau),$$

where

$$\alpha = \frac{r-q}{\sigma^2} - \frac{1}{2}, \quad \beta = \frac{r+q}{2} + \frac{\sigma^2}{8} + \frac{(r-q)^2}{2\sigma^2}. \quad (9.19)$$

After straightforward algebraic calculations we conclude that the Black–Scholes equation can be rewritten in the form:

$$\frac{\partial u}{\partial \tau} = \frac{\sigma^2}{2} \frac{\partial^2 u}{\partial x^2}, \quad (9.20)$$

for each $x \in \mathbb{R}, \tau \in (0, T)$. Since the American call (put) option should satisfy the condition $V(S,t) \geq V(S,T) \equiv \bar{V}(S)$ we end up with the following condition for the transformed function:

$$u(x, \tau) \geq g(x, \tau), \quad (9.21)$$

for each $x \in \mathbb{R}, \tau \in (0, T)$. The function g corresponds to the transformed pay-off diagram of the call (put) option, i.e.

$$\begin{aligned} g(x, \tau) &= e^{\alpha x + \beta \tau} \max(e^x - 1, 0), & \text{for a call option,} \\ g(x, \tau) &= e^{\alpha x + \beta \tau} \max(1 - e^x, 0), & \text{for a put option.} \end{aligned} \quad (9.22)$$

The initial condition for the function u reads as follows:

$$u(x, 0) = g(x, 0) \quad (9.23)$$

for each $x \in \mathbb{R}$. As for the boundary conditions for a call option we obtain

$$u(-\infty, \tau) = g(-\infty, \tau) = 0, \ \lim_{x \to \infty} u(x, \tau)/g(x, \tau) = 1, \quad (9.24)$$

for each $\tau \in (0, T)$. In the case of a put option we have

$$\lim_{x \to -\infty} u(x, \tau)/g(x, \tau) = 1, \quad u(+\infty, \tau) = g(+\infty, \tau) = 0, \quad (9.25)$$

for each $\tau \in (0,T)$.

Summarizing, the linear complementarity problem for pricing the American call (put) option can be written in the form of a parabolic variational inequality:

$$\frac{\partial u}{\partial \tau} - \frac{\sigma^2}{2}\frac{\partial^2 u}{\partial x^2} \geq 0, \quad u(x,\tau) - g(x,\tau) \geq 0, \qquad (9.26)$$
$$\left(\frac{\partial u}{\partial \tau} - \frac{\sigma^2}{2}\frac{\partial^2 u}{\partial x^2}\right)(u(x,\tau) - g(x,\tau)) = 0,$$

for each $x \in \mathbb{R}, 0 < \tau < T$. The problem is to find a function $u : \mathbb{R} \times (0,T) \to \mathbb{R}$ such that u is a continuously differentiable function satisfying the transformed linear complementarity (9.26) and corresponding initial and boundary conditions.

9.3. Pricing Perpetual Options

In this concluding section we discuss the problem of pricing the so-called perpetual call and put options. Perpetual options are options with a very large maturity $T \to \infty$. In what follows, we will show that perpetual American options can be priced by an explicit formula. Indeed, suppose there exists a limit $\bar{V}(S)$ of a solution $V(S,t)$ for the maturity $T \to \infty$. Then, setting $\tau = T - t$, we obtain $\lim_{\tau \to \infty} V = \lim_{T \to \infty} V = \bar{V}$. Hence

$$\lim_{T \to \infty} \frac{\partial V}{\partial t} = \lim_{\tau \to \infty} \frac{\partial V}{\partial \tau} = 0.$$

Furthermore, we suppose that the early exercise boundary S_f tends to its limiting value $S_f(\infty)$ as $T \to \infty$. It means that for the function $\rho(\tau) = S_f(T-t)$ we have $\lim_{\tau \to \infty} \rho(\tau) = \rho_\infty = S_f(\infty)$.

In the case of an American put option on the underlying asset paying no dividends, the price $\bar{V}(S)$ and the limiting early exercise boundary position ρ^∞ of a perpetual option is a solution to the stationary Black–Scholes partial differential equation:

$$\frac{\sigma^2}{2}S^2\frac{\partial^2 \bar{V}}{\partial S^2} + rS\frac{\partial \bar{V}}{\partial S} - r\bar{V} = 0, \quad S > \rho^\infty, \qquad (9.27)$$

and

$$\bar{V}(+\infty) = 0, \quad \bar{V}(\rho_\infty) = E - \rho_\infty, \quad \frac{\partial \bar{V}}{\partial S}(\rho_\infty) = -1. \qquad (9.28)$$

Fortunately, the above boundary value problem for the function $\bar{V}$ and the limiting early exercise boundary position ρ_∞ has a simple explicit solution discovered by Merton. Its explicit form reads as follows:

$$\bar{V}(S) = CS^{-\gamma}, \quad S > \rho_\infty.$$

where $C, \gamma > 0$ are positive constants. If we insert this ansatz into (9.27) we obtain a simple relationship for the constant γ:

$$\frac{\sigma^2}{2}\gamma(\gamma+1) - r\gamma - r = 0.$$

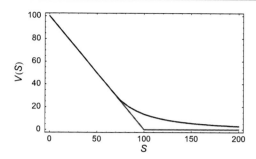

Figure 9.4. A plot of a perpetual American put option $S \mapsto \bar{V}(S)$ for the parameters: $E = 100, \sigma = 0.3, r = 0.1$.

This is a quadratic equation for γ. Its positive solution $\gamma > 0$ is given by

$$\gamma = \frac{2r}{\sigma^2}.$$

The constant $C > 0$ and the limiting value ρ_∞ can be deduced from the smooth pasting boundary conditions (9.28). It leads to a system of two equations:

$$-\gamma C (\rho^\infty)^{-\gamma-1} = -1, \qquad C\rho_\infty^{-\gamma} = E - \rho_\infty,$$

from which we deduce

$$\rho^\infty = E \frac{\gamma}{1+\gamma}, \qquad C = \frac{1}{\gamma}\rho_\infty^{\gamma+1}.$$

A graph of a perpetual American put option is shown in Fig. 9.4.

In the case of a perpetual American call option on the underlying asset paying nontrivial dividends $q > 0$ one can proceed similarly as in the case of the put option (see Exercise 7).

Problem Section and Exercises

1. Show that the early exercise boundary function $S_f(t)$ for the American call option is a decreasing function of the time t. On the other hand, the function $S_f(t)$ is an increasing function for the American put option.

2. Show that the put–call parity (see Chapter 3) need not to be satisfied for the American style of call and put options.

3. How the early exercise boundary $S_f(t)$ for the American call option depends on the volatility σ of the underlying asset price? Is it an increasing or decreasing function?

4. Denote by $V^{ac}(S,t;E,r,q,\sigma)$ and $V^{ap}(S,t;E,r,q,\sigma)$ the price of American call and put options for the underlying asset price S, time t, expiration price E, interest rate r and dividend yield q. Derive the so-called call–put symmetry relation for prices of the American call and put options, i.e., prove the equality:

$$V^{ap}(S,t;E,r,q,\sigma) = V^{ac}(E,t;S,q,r,\sigma).$$

5. In the case of European call and put options prove the call–put symmetry by using the explicit formulae for their prices derived in Chapter 3.

6. Using the approximative formulae for the early exercise boundary position for the American call and put options show that, in the case of a call option, we obtain the following limit of the time derivative $\lim_{t\downarrow T} \dot{S}_f(t) = -\infty$ whereas $\lim_{t\downarrow T} \dot{S}_f(t) = +\infty$ for a put option. Based on these limits discuss option holder's behavior for times t close to expiry T.

7. Derive an explicit formula for pricing a perpetual American call option on the underlying asset paying nontrivial dividends $q > 0$. Find a solution $\bar{V}(S)$ in the form $V(S) = CS^\gamma$, where $C, \gamma > 0$ are positive constants.

8. Derive an explicit formula for pricing a perpetual American put option on the underlying asset paying nontrivial dividends $q > 0$.

Chapter 10

Numerical Methods for Pricing of Simple Derivatives

The aim of this chapter is to propose and discuss effective numerical methods for the valuation of selected types of derivatives. We focus our attention to the valuation of European and American call and put options. However, numerical techniques and method discussed in this chapter can be easily extended to other types of financial derivatives. At the beginning of this chapter, we analyze explicit and implicit numerical schemes for pricing European types of derivatives. Although there exist explicit formulae for solutions to the Black–Scholes equation (see Chapter 3), the reason for presentation of numerical methods for pricing European style options consists in the fact that it provides a possibility of qualitative and quantitative comparison of a numerical approximate solution and the exact one. Such an information can be useful in the case when we try to extend numerical methods for derivatives for which an analytical solution is not known. For example, it is the case of American style of derivatives. In the second part of this chapter we present efficient numerical methods for pricing American types of call and put options in a more detail. In the process of numerical discretization of the Black-Scholes equation it will turn out that efficient and fast numerical methods and techniques for solving systems of linear equations are needed. This is why we also discuss efficient numerical algorithms for solving linear systems of equations and linear complementarity problems.

10.1. Explicit Numerical Finite Difference Method for Solving the Black–Scholes Equation

In this section, we present a numerical scheme for construction of an approximation of a solution to the Black–Scholes partial differential equation for pricing European types of derivatives. The numerical scheme is based on finite difference approximation of all partial derivatives entering the Black–Scholes equation. Knowledge of an explicit solution to the Black–Scholes equation (see Chapter 3) enables us to compare a numerical approximate solution with the exact one. Such a comparison will enable us to extend the numerical approximation scheme also to the case of other financial derivatives for which an explicit solution is not know like, e.g., American style options or some exotic types of derivatives.

At the beginning of this section, we recall the transformation of the Black–Scholes equation to a parabolic heat equation $\partial_t u = \partial_x^2 u$. The transformation will be subsequently used as a basis for construction of explicit and implicit in time numerical schemes based on finite differences approximation of partial derivatives entering the Black–Scholes equation.

Recall that by introducing new independent variables x, τ and the transformed function u (see (3.4) or (9.19)):

$$x = \ln(S/E) \in (-\infty, \infty), \ \tau = T - t \in (0, T), \ V(S, t) = E e^{-\alpha x - \beta \tau} u(x, \tau),$$

where $\alpha = \frac{r-q}{\sigma^2} - \frac{1}{2}$, $\beta = \frac{r+q}{2} + \frac{\sigma^2}{8} + \frac{(r-q)^2}{2\sigma^2}$, the Black–Scholes partial differential equation

$$\frac{\partial V}{\partial t} + \frac{\sigma^2}{2} S^2 \frac{\partial^2 V}{\partial S^2} + (r - q) S \frac{\partial V}{\partial S} - rV = 0$$

can be transformed into a form of the standard parabolic equation

$$\frac{\partial u}{\partial \tau} = \frac{\sigma^2}{2} \frac{\partial^2 u}{\partial x^2}. \tag{10.1}$$

The initial condition for the transformed function u depends on the terminal pay–off diagram of a chosen financial derivative. For plain vanilla call and put options, the initial conditions have the form:

$$u(x, 0) = \begin{cases} e^{\alpha x} \max(e^x - 1, 0), & \text{for a call option,} \\ e^{\alpha x} \max(1 - e^x, 0), & \text{for a put option.} \end{cases} \tag{10.2}$$

In what follows, we shall present key ideas of a finite difference approximation of the partial differential equation (10.1). The proposed numerical approximation method consists in considering a discrete mesh of points in the domain of independent variables $(x, \tau) \in \mathbb{R} \times (0, T)$ and replacement of a solution u and its partial derivatives by finite differences in each nodal point of the discretization mesh.

Choose a spatial step $h > 0$ and a time step $k > 0$ so that $k = T/m$, where $m \in \mathbb{N}$ is a number of time discretization steps in the interval $[0, T]$. In the domain of independent variables $(x, \tau) \in \mathbb{R} \times (0, T)$ we consider a mesh of grid points

$$x_i = ih, \ i = \ldots, -2, -1, 0, 1, 2, \ldots, \quad \tau_j, \ j = 0, 1, \ldots, m.$$

By u_i^j we denote an approximation of a solution u in the grid point (x_i, τ_j), i.e.

$$u_i^j \approx u(x_i, \tau_j).$$

Derivation of a finite difference numerical scheme for approximation of equation (10.1) is based on replacement of all partial derivatives by finite differences, which can be easily derived by expanding of a function into Taylor series. In a grid point (x_i, τ_j), let us consider the Taylor series expansion of the function u up to the third order. Since $x_{i+1} - x_i = h$ and $x_i - x_{i-1} = h$ we have

$$u(x_{i+1}, \tau_j) \approx u(x_i, \tau_j) + \frac{\partial u}{\partial x} h + \frac{1}{2!} \frac{\partial^2 u}{\partial x^2} h^2 + \frac{1}{3!} \frac{\partial^3 u}{\partial x^3} h^3, \tag{10.3}$$

$$u(x_{i-1}, \tau_j) \approx u(x_i, \tau_j) - \frac{\partial u}{\partial x} h + \frac{1}{2!} \frac{\partial^2 u}{\partial x^2} h^2 - \frac{1}{3!} \frac{\partial^3 u}{\partial x^3} h^3, \tag{10.4}$$

where the approximation error is of the order $O(h^4)$ for $h \to 0$. Subtracting equation (10.4) from (10.3), dividing by $2h$, we obtain the so-called central finite difference approximation of the first partial derivatives u with respect to x:

$$\frac{\partial u}{\partial x}(x_i, \tau_j) \approx \frac{u_{i+1}^j - u_{i-1}^j}{2h}, \tag{10.5}$$

with an approximation error of the order $O(h^2)$ for small values of h. By summing equations (10.3) and (10.4) we end up with the approximation of the second partial derivative of u with respect to the variable x:

$$\frac{\partial^2 u}{\partial x^2}(x_i, \tau_j) \approx \frac{u_{i+1}^j - 2u_i^j + u_{i-1}^j}{h^2} \tag{10.6}$$

with an approximation error of the second derivative of the order $O(h^2)$ as $h \to 0$. Analogously, for the time derivative $\frac{\partial u}{\partial \tau}$, we can deduce, from Taylor series expansion at the point (x_i, τ_j), the following approximation:

$$u(x_i, \tau_{j+1}) \approx u(x_i, \tau_j) + k \frac{\partial u}{\partial \tau}(x_i, \tau_j)$$

for which the order of the approximation is equal to $O(k^2)$ for $k \to 0$. Taking into account the above expansion, we obtain the forward in time approximation of the time derivative:

$$\frac{\partial u}{\partial \tau}(x_i, \tau_j) \approx \frac{u_i^{j+1} - u_i^j}{k}, \tag{10.7}$$

with an approximation error in the τ variable of the order $O(k)$. Now, if we insert approximations of partial derivatives at the nodal point (x_i, τ_j) into the parabolic partial differential equation (10.1) then we conclude that the approximate solution u_i^j at (x_i, τ_j) satisfies the equation

$$\frac{u_i^{j+1} - u_i^j}{k} = \frac{\sigma^2}{2} \frac{u_{i+1}^j - 2u_i^j + u_{i-1}^j}{h^2}, \tag{10.8}$$

where an approximation error of the equation is of the order $O(k + h^2)$ for $k, h \to 0$. It means that the value of u_i^{j+1} for a new time level $j+1$ can be explicitly expressed by using the values of the solution from the previous time level j as follows:

$$u_i^{j+1} = \gamma u_{i-1}^j + (1 - 2\gamma) u_i^j + \gamma u_{i+1}^j, \quad \text{where } \gamma = \frac{\sigma^2 k}{2h^2}, \tag{10.9}$$

for $i = \ldots, -2, -1, 0, 1, 2, \ldots$, and $j = 0, 1, \ldots, m-1$.

Let us choose $N \in \mathbb{N}$ such that the interval of the spatial discretization $(-L, L) = (x_{-N+1}, x_{N-1})$ is sufficiently large and the values u_{-N}^j and u_N^j can be approximated by the Dirichlet boundary conditions. From the practical point of view, it is sufficient to choose $L \approx 1.2$. Indeed, it means that the original financial variable S then belongs to a sufficiently wide interval $(Ee^{-L}, Ee^L) = (0.3E, 3.32E)$.

For a European call option we have $V(0, t) = 0$ and $V(S, t)/S \to e^{-q(T-t)}$ for $S \to \infty$. On the other hand, for a European put option it holds: $V(0, t) = Ee^{-r(T-t)}$ and $V(S, t) \to 0$

as $S \to \infty$. It means that, for a large value of N, the boundary values u^j_{-N} and u^j_N can be approximated by taking the limiting values of the transformed solution, i.e.

$$u^j_{-N} = \phi^j := \begin{cases} 0, & \text{for a European call option,} \\ e^{-\alpha Nh + (\beta-r)jk}, & \text{for a European put option,} \end{cases} \quad (10.10)$$

$$u^j_N = \psi^j := \begin{cases} e^{(\alpha+1)Nh + (\beta-q)jk}, & \text{for a European call option,} \\ 0, & \text{for a European put option.} \end{cases}$$

If we denote by u^j approximation of a solution at the time level τ_j, i.e.

$$u^j = (u^j_{-N+1}, \ldots, u^j_{-1}, u^j_0, u^j_1, \ldots, u^j_{N-1}) \in \mathbb{R}^n,$$

where $n = 2N - 1$, then we can rewrite the explicit numerical approximation scheme (10.9) in the vector form as follows:

$$u^{j+1} = \mathbf{A}u^j + b^j, \quad \text{for } j = 0, 1, \ldots, m-1, \quad (10.11)$$

where $\mathbf{A}$ is a tridiagonal matrix given by

$$\mathbf{A} = \begin{pmatrix} 1-2\gamma & \gamma & 0 & \cdots & 0 \\ \gamma & 1-2\gamma & \gamma & & \vdots \\ 0 & \ddots & \ddots & \ddots & 0 \\ \vdots & & \gamma & 1-2\gamma & \gamma \\ 0 & \cdots & 0 & \gamma & 1-2\gamma \end{pmatrix}, \quad b^j = \begin{pmatrix} \gamma\phi^j \\ 0 \\ \vdots \\ 0 \\ \gamma\psi^j \end{pmatrix}.$$

The advantage of the vector notation consists in simplification of the stability and convergence analysis of the explicit numerical scheme (10.11) by means of the qualitative properties of the matrix $\mathbf{A}$. Under the so-called Courant–Fridrichs–Lewy (CFL) stability condition:

$$0 < \gamma \leq \frac{1}{2}, \quad \text{i.e.} \quad \frac{\sigma^2 k}{h^2} \leq 1, \quad (10.12)$$

the numerical discretization scheme (10.11) is stable. It means that

$$\lim_{\substack{k \to 0 \\ h \to 0 \\ \sigma^2 k \leq h^2}} \tilde{u}_{k,h}(x, \tau) = u(x, \tau), \quad (10.13)$$

where $u(x, \tau)$ is a solution to the parabolic partial differential equation (10.1) and $\tilde{u}_{k,h}$ is a piece-wise linear function with values $\tilde{u}_{k,h}(x_i, \tau_j) = u^j_i$ in nodal points (x_i, τ_j). The above limit is considered for values of parameters h, k satisfying the CFL condition (10.12).

The iteration matrix $\mathbf{A}$ entering recurrent relation (10.11) has an important property for the parameter value γ satisfying the CFL condition (10.12). Its maximum L_∞-norm is at most one. It is worth noting that the coefficients in the matrix $\mathbf{A}$ are non-negative, i.e., $\gamma > 0, 1 - 2\gamma \geq 0$. If $u \in \mathbb{R}^n$, then for the i-th component of the vector $(\mathbf{A}u)_i$ we have $(\mathbf{A}u)_i = \gamma u_{i-1} + (1-2\gamma)u_i + \gamma u_{i+1}$ and so $|(\mathbf{A}u)_i| \leq \gamma|u_{i-1}| + (1-2\gamma)|u_i| + \gamma|u_{i+1}| \leq (\gamma +$

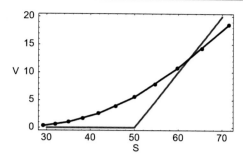 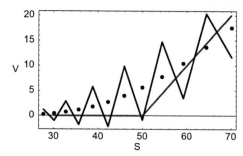

Figure 10.1. A solution $S \mapsto V(S,t)$ for the price of a European call option obtained by means of the binomial tree method with $\gamma = 1/2$ (left) and comparison with the exact solution (dots). The oscillating solution $S \mapsto V(S,t)$ which does not converge to the exact solution for the parameter value $\gamma = 0.56 > 1/2$, where $\gamma > 1/2$, does not fulfill the CFL condition.

$(1 - 2\gamma) + \gamma) \|u\|_\infty = \|u\|_\infty$, where $\|u\|_\infty = \max_i |u_i|$ is the maximum L_∞-norm of the vector u. Hence,

$$\|\mathbf{A}u\|_\infty \leq \|u\|_\infty, \qquad \text{for each } u \in \mathbb{R}^n. \tag{10.14}$$

If we denote $m_j = \min_i u_i^j$, $M_j = \max_i u_i^j$ the minimum and maximum of the vector u^j, then, for $0 < \gamma \leq 1/2$, we obtain

$$M^{j+1} \leq \max(M^j, \phi^j, \psi^j), \quad m^{j+1} \geq \min(m^j, \phi^j, \psi^j), \tag{10.15}$$

for $j = 0, 1, \ldots, m - 1$. Indeed, for inner indices $i = -N+2, \ldots, N-2$ we have $u_i^{j+1} = \gamma u_{i-1}^j + (1 - 2\gamma) u_i^j + \gamma u_{i+1}^j \leq M^j$. For outer indices $i = -N+1$ and $N-1$ we moreover have to take into account the boundary conditions ϕ_j and ψ_j (see (10.10)). The above system of inequalities (10.15) is referred to as the discrete maximum (minimum) principle, which is a discrete counterpart of the maximum principle for solutions to parabolic partial differential equations.

A numerical solution for pricing European call options by means of the explicit numerical scheme is depicted in Fig. 10.1. We chose the following financial and numerical parameters: $\sigma = 0.4, r = 0.04, q = 0.12, T = 1, E = 50$ and $N = 100, m = 20$ (see the source code shown in Table 10.1). It is important to emphasize that the explicit scheme works fine only in the case when the CFL condition $\gamma \leq 1/2$ is satisfied (see Fig. 10.1 (left)). On the other hand, if $\gamma > 1/2$ then the numerical solution need not to converge to the exact analytical solution. It may oscillate as it can be obvious from Fig. 10.1 (right), where we computed a solution for the parameter value $\gamma = 0.56 > 1/2$. Notice that in this case the discrete maximum principle is violated.

10.1.1. Discrete Methods Based on Binomial and Trinomial Trees

In this part, we focus our attention to a special case of the explicit numerical scheme (10.9). If we choose the ratio between the spatial and time discretization steps such that

$$h = \sigma \sqrt{k}, \tag{10.16}$$

Table 10.1. The Mathematica source code for an explicit numerical scheme for pricing the European call option.

```
sigma = 0.4; r = 0.04; q = 0.12;
T = 1; X = 50;

alfa = (r - q)/sigma^2 - 1/2;
beta = (r + q)/2 + sigma^2/8 + (r - q)^2/(2sigma^2);

NN = 100; n = 2 NN - 1;
m = 20;

k = T/m;
gama = 0.5;
h = sigma Sqrt[k/(2 gama)];

A=Table[Table[If[i==j, 1 - 2gama,
        If[i==j-1, gama, If[i==j+1, gama, 0 ]]],
{j, 1, n}], {i, 1, n}];


u0 = Table[Exp[alfa i h] Max[Exp[i h] - 1, 0],
{i, -NN + 1, NN - 1}];
phi[j_] := 0.;
psi[j_] := Exp[(alfa + 1)NN h + (beta - q)j k];

uold = u0;
For[j = 0, j <= m - 1, 1,
{
b = Table[If[i == -NN + 1, gama phi[j],
        If[i == NN - 1, gama psi[j], 0]],
        {i, -NN + 1, NN - 1}];
unew = A.uold + b;
uold = unew;
Vnew = Table[
{X Exp[i h],
 X Exp[-alfa i h - beta j k] unew[[i+NN]]
},
{i,-NN+1, NN-1}];
j++;
}];

ListPlot[Vnew];
```

i.e., $\gamma = 1/2$, then the term $1 - 2\gamma$ vanishes in (10.9). The explicit scheme has a simpler form:

$$u_i^{j+1} = \frac{1}{2}u_{i-1}^j + \frac{1}{2}u_{i+1}^j. \tag{10.17}$$

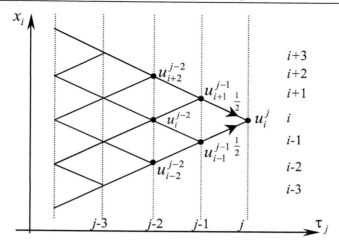

Figure 10.2. A binomial tree as an illustration of the algorithm for solving a parabolic equation by an explicit method with $2\gamma = \sigma^2 k/h^2 = 1$.

It means that the solution u_i^{j+1} at the time τ_{j+1} is the arithmetic average between values u_{i-1}^j and u_{i+1}^j calculated at the previous time τ_j. A graphical illustration of the algorithm is depicted in Fig. 10.2. Since there is a similarity with a binomial tree we will henceforth refer to the explicit method with $\gamma = 1/2$ to as the binomial tree method. In general, if $0 < \gamma < \frac{1}{2}$, the explicit method is referred to as the trinomial tree method.

Risk Neutral Probabilities and the Binomial Tree Method

We summarize the discussion on the binomial tree method by pointing out its relation to the discrete binomial model proposed by Cox, Ross and Rubinstein in 1979 (see e.g., Melicherčík et al. [83]). The key idea of the binomial model consists in construction of the so-called risk-neutral option price V^{j+1} at the time $t_{j+1} = T - \tau_{j+1}$ by means of the option and underlying stock price at the time $t_j = T - \tau_j$. Suppose that the underlying stock price at the time t_{j+1} has a price S and, with a probability $p \in (0,1)$, it attains a higher value $S_+ > S$ and, with a complementary probability $1 - p \in (0,1)$, it attains a lower value $S_- < S$ at the time t_j. Denote by V_+ and V_-, respectively, the option prices corresponding to the upward and downward movement of underlying prices. Let us construct a portfolio consisting of one option in a long position and δ underlying stocks in a short position. The principle of nonexistence of long lasting arbitrage opportunities enables us to conclude that the value of the portfolio at the time $t_{j+1} < t_j$, (when discounted by the risk-free interest rate r of a zero coupon bond over the time interval of the length k) should be equal to the value of the portfolio at the time t_j, i.e.

$$e^{rk}(V - \delta S) = V_- - \delta S_- = V_+ - \delta S_+.$$

Hence,

$$\delta = \frac{V_+ - V_-}{S_+ - S_-},$$

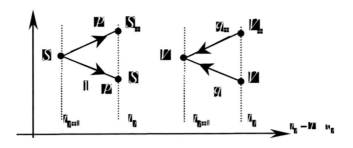

Figure 10.3. A binomial tree illustrating a solution to a parabolic equation by the explicit scheme with the parameter $2\gamma = \sigma^2 k/h^2 = 1$.

and the price V can be expressed as follows:

$$V = e^{-rk}(q_+V_+ + q_-V_-), \quad \text{where } q_+ = \frac{Se^{rk} - S_-}{S_+ - S_-}, \; q_- = 1 - q_+ \qquad (10.18)$$

(see [75, Chapter 2.1]). Again, using the principle of nonexistence of long lasting arbitrage opportunities, we can conclude that $S_- < Se^{rk} < S_+$. It means that $q_+ > 0$ and so the values q_+, q_- can be interpreted as the so-called risk-neutral probabilities. Notice that the pricing formula (10.18) does not contain any information regarding the real probabilities p and $1 - p$ of upward and downward change of the underlying stock price S to the value S_+ or S_-, respectively. It is in accordance with the fact that the option price is independent of the deterministic drift of the underlying stock price (see Chapter 3). In Fig. 10.3, we present an illustration of the option price computation by means of a binomial model. The one stage binomial model can be recursively applied for the entire term structure $t_0 = T, \ldots, t_m = 0$ in order to compute the option price at the time option contract $t_m = 0$. Recall that the price of the option at the expiration time $t_0 = T$ is given by its pay–off diagram.

The binomial model can be also derived from the explicit numerical scheme (10.17). Indeed, let us denote

$$V_i^j \approx V(S_i, T - \tau_j), \quad \text{where } S_i = Ee^{x_i} = Ee^{ih}.$$

Taking into account the transformation $V(S,t) = Ee^{-\alpha x - \beta t}u(x,t)$, we obtain $V_i^j = Ee^{-\alpha ih - \beta jk}u_i^j$. In terms of the original variable V_i^j, the numerical scheme (10.17) can be expressed as follows:

$$V_i^{j+1} = e^{-rk}\left(q_-V_{i-1}^j + q_+V_{i+1}^j\right), \quad \text{where } q_\pm = \frac{1}{2}e^{\pm\alpha h - (\beta - r)k}. \qquad (10.19)$$

Since $\sigma^2 k/(2h^2) = \gamma = 1/2$, then, with regard to relations determining the constants α, β, we obtain

$$\frac{\alpha^2}{2}h^2 - (\beta - r)k = \left(\sigma^2\frac{\alpha^2}{2} - (\beta - r)\right)k = 0.$$

It means that, for sufficiently small values of the time step k, it holds that:

$$e^{\pm\alpha h - (\beta - r)k} \approx 1 \pm \alpha h - (\beta - r)k + \frac{\alpha^2}{2}h^2 + O(k^{\frac{3}{2}}) = 1 \pm \alpha h + O(k^{\frac{3}{2}}),$$

for $k \to 0$ and $h = \sigma\sqrt{k} \to 0$. Thus,

$$q_+ = \frac{1+\alpha h}{2}, \quad q_- = \frac{1-\alpha h}{2}, \quad q_- + q_+ = 1. \tag{10.20}$$

Since $q_- + q_+ = 1, q_\pm > 0$, these constants are again referred to as risk-neutral probabilities.

10.2. Implicit Numerical Method for Solving the Black–Scholes Equation

In the previous section, we have derived and analyzed the explicit numerical scheme for solving the Black–Scholes equation. The method was based on approximation of partial derivatives by means of finite differences. In this part, we focus our attention to the time implicit finite difference approximation of the transformed Black–Scholes equation. The idea behind the construction of the time implicit method is based on approximation of the partial derivative $\partial u/\partial \tau$ in the nodal point x_i^j by means of the backward time difference, i.e.

$$\frac{\partial u}{\partial \tau}(x_i, \tau_j) \approx \frac{u_i^j - u_i^{j-1}}{k}. \tag{10.21}$$

Hence we can conclude that the approximation u_i^j of a solution at (x_i, τ_j) satisfies the equation

$$\frac{u_i^j - u_i^{j-1}}{k} = \frac{\sigma^2}{2}\frac{u_{i+1}^j - 2u_i^j + u_{i-1}^j}{h^2}. \tag{10.22}$$

Therefore, the values $u_{i-1}^j, u_i^j, u_{i+1}^j$ at the new time level j can be implicitly expressed in terms of the value of the solution at the previous time level $j-1$,

$$-\gamma u_{i-1}^j + (1+2\gamma)u_i^j - \gamma u_{i+1}^j = u_i^{j-1}, \quad \text{where } \gamma = \frac{\sigma^2 k}{2h^2}, \tag{10.23}$$

for $i = \ldots, -2, -1, 0, 1, 2, \ldots$, and $j = 1, \ldots, m$. If we restrict ourselves to a finite number of spatial nodal points $x_i, i = -N+1, \ldots, -1, 0, 1, \ldots, N-1$, we can rewrite the implicit scheme (10.23) in the matrix form as follows:

$$\mathbf{A}u^j = u^{j-1} + b^{j-1}, \quad \text{for } j = 1, 2, \ldots, m, \tag{10.24}$$

where $\mathbf{A}$ is a tridiagonal $n \times n$ square matrix, $n = 2N-1$,

$$\mathbf{A} = \begin{pmatrix} 1+2\gamma & -\gamma & 0 & \cdots & 0 \\ -\gamma & 1+2\gamma & -\gamma & & \vdots \\ 0 & \cdot & \cdot & \cdot & 0 \\ \vdots & & -\gamma & 1+2\gamma & -\gamma \\ 0 & \cdots & 0 & -\gamma & 1+2\gamma \end{pmatrix}, \quad b^j = \begin{pmatrix} \gamma\phi^{j+1} \\ 0 \\ \vdots \\ 0 \\ \gamma\psi^{j+1} \end{pmatrix}.$$

Here $\gamma = \sigma^2 k/(2h^2)$. In Fig. 10.4 we present the shape of the solution $S \mapsto V(S,t)$ representing the price of the European call option for model parameters $\sigma = 0.4, r = 0.04, q = $

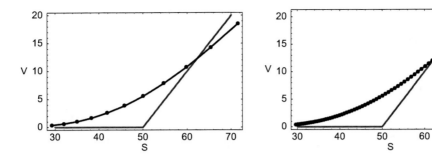

Figure 10.4. A solution $S \mapsto V(S,t)$ for pricing a European call option obtained by means of the implicit finite difference method with $\gamma = 1/2$ (left) and comparison with the exact analytic solution (dots). The numerical scheme is also stable for a large value of the parameter $\gamma = 20 > 1/2$ not satisfying the CFL condition (right).

$0.12, T-t = 1, E = 50$. In the left figure we plot a solution corresponding to the parameter $\gamma = 1/2$, whereas the right figure depicts a solution for larger value of the parameter, $\gamma = 20$. Both numerical examples almost coincide with the exact analytical solution, which is available for the European style of plain vanilla options. The Mathematica source code for a numerical implicit scheme for pricing a European call option is presented in Table 10.2.

It is worth noting that the advantage of the implicit numerical scheme consists in removing the restrictive assumption $\sigma^2 k/(2h^2) = \gamma \leq 1/2$, which is necessary for the stability of the explicit scheme and constitutes a too restrictive relationship between the time and spatial discretization steps k and h, respectively. Furthermore, it can be shown (see Vitásek [121] or Faddeev, Faddeeva [45]) that the implicit numerical approximation scheme (10.24) is unconditionally stable. It means that the following limit holds true:

$$\lim_{(k,h)\to(0,0)} \tilde{u}_{k,h}(x,\tau) = u(x,\tau), \qquad (10.25)$$

where the piece-wise linear function $\tilde{u}_{k,h}(x,\tau)$ has the same meaning as in the case of the explicit scheme (10.11). Therefore, by using the implicit approximation scheme, we can numerically solve the parabolic equation with a larger time discretization step k keeping the discretization step h sufficiently small enough in order to capture a fine spatial resolution of the underlying asset price. This advantage is however negatively compensated by the requirement of solving systems of linear equations. As it will be shown in the subsequent section, there are efficient and fast numerical methods for solving sparse systems of linear equations. In particular, we will discuss the LU decomposition method and Gauss–Seidel iterative method. In contrast to the classical Gauss elimination method, for sparse linear systems, these methods have lower memory requirements and they are faster than the elimination method.

Similarly as in the case of the explicit finite difference scheme (10.11), we are yet able to derive several useful properties of a solution obtained by the implicit numerical scheme (10.24) from which we can derive unconditional stability of the scheme. Using the inverse matrix $\mathbf{A}^{-1}$, the implicit scheme can be rewritten in the form:

$$u^{j+1} = \mathbf{A}^{-1} u^j + \mathbf{A}^{-1} b^j.$$

Table 10.2. The Mathematica source code for an implicit method for pricing the European call option.

```
sigma = 0.4; r = 0.04; q = 0,12; T = 1; X = 50;
alfa = (r - q)/sigma^2 - 1/2;
beta = (r + q)/2 + sigma^2/8 + (r - q)^2/(2sigma^2);

NN = 100; n = 2 NN - 1;
m = 20;

k = T/m;
gama = 20;
h = sigma Sqrt[k/(2 gama)];

A = Table[Table[ If[i == j, 1 + 2gama,
         If[i == j - 1, -gama, If[i == j + 1, -gama, 0 ]]],
         {j, 1, n}], {i, 1, n}];

u0 = Table[Exp[alfa i h] Max[Exp[i h] - 1, 0],
{i,-NN+1, NN-1}];

phi[j_]:=0.; psi[j_]:=Exp[(alfa + 1)NN h + (beta-q) j k];

uold = u0;
For[j = 0, j <= m-1, 1,
{
b = Table[If[i == -NN + 1, gama phi[j+1],
         If[i==NN-1, gama psi[j+1], 0]],{i,-NN+1, NN-1}];
unew = LinearSolve[A, uold + b];
uold = unew;
Vnew = Table[{ X Exp[i h],
         X Exp[-alfa i h -beta j k] unew[[i+NN]]},
         {i, -NN+1, NN-1}];
j++;
}];

ListPlot[Vnew];
```

First, we will show that the inverse matrix $\mathbf{A}^{-1}$ has the maximum L_∞-norm $\|\mathbf{A}^{-1}\|_\infty$ bounded by one, independently of the parameter $\gamma > 0$. Indeed, let $\mathbf{A}u = b$, i.e., $u = \mathbf{A}^{-1}b$. Let us denote $M = \max_i |u_i|$. As

$$-\gamma u_{i-1} + (1+2\gamma)u_i - \gamma u_{i+1} = b_i,$$

we obtain

$$(1+2\gamma)|u_i| = |b_i + \gamma u_{i-1} + \gamma u_{i+1}| \leq |b_i| + 2\gamma M.$$

Hence $(1+2\gamma)M = (1+2\gamma)\max_i |u_i| \leq \max_i |b_i| + 2\gamma M \leq \|b\|_\infty + 2\gamma M$ from which we can easily deduce the inequality: $M \leq \|b\|_\infty$. It means that

$$\|\mathbf{A}^{-1}b\|_\infty \leq \|b\|_\infty, \qquad \text{for each } b \in \mathbb{R}^n. \tag{10.26}$$

If we again denote by $m_j = \min_i u_i^j$, $M_j = \max_i u_i^j$ the minimum and maximum of the vector u^j, respectively, then, for arbitrary value of the parameter $\gamma > 0$, we obtain

$$M^{j+1} \leq \max(M^j, \phi^j, \psi^j), \quad m^{j+1} \geq \min(m^j, \phi^j, \psi^j), \tag{10.27}$$

for $j = 0, 1, \ldots, m-1$. Indeed, suppose that the maximum M^{j+1} is attained for the index i_o, i.e., $M^{j+1} = u_{i_o}^{j+1}$. Then, for the case $= -N + 2 \leq i_o \leq N - 2$ (i_o is an inner index), we obtain $(1+2\gamma)M^{j+1} = (1+2\gamma)u_{i_o}^{j+1} = u_{i_o}^j + \gamma u_{i_o-1}^{j+1} + \gamma u_{i_o+1}^{j+1} \leq M^j + 2\gamma M^{j+1}$, and so $M^{j+1} \leq M^j$. For boundary indices $i = -N + 1$ and $N - 1$ we have to take into account the boundary conditions ϕ_j and ψ_j (see (10.10)). The system of inequalities (10.15) is again referred to as the discrete maximum (minimum) principle for the implicit numerical method for solving parabolic partial differential equations.

10.3. Compendium of Numerical Methods for Solving Systems of Linear Equations

The purpose of this section is to present to the reader a survey of basic numerical methods for solving systems of linear equations. We focus our attention to the problem of numerical computation of the system of linear equations:

$$\mathbf{A}u = b, \tag{10.28}$$

where $\mathbf{A}$ is a square $n \times n$ matrix of real numbers, $b \in \mathbb{R}^n$ is a given vector and $u \in \mathbb{R}^n$ is a solution of the system (10.28). For a detailed overview of numerical methods of linear algebra we refer the reader to books by Faddeev and Faddeeva [45], Vitásek [121] or Fiedler [47].

10.3.1. LU Decomposition Method

A key idea of the method of the so-called LU decomposition is rather simple and it consists in multiplicative decomposition of the matrix $\mathbf{A}$ into a product of two matrices, i.e., $\mathbf{A} = \mathbf{L}.\mathbf{U}$. Here, $\mathbf{L}$ is a lower triangle matrix. It means that $L_{ij} = 0$ for $i < j$. The matrix $\mathbf{U}$ is an upper triangular matrix, i.e., $U_{ij} = 0$ for $i > j$. Now, problem (10.28) can be easily rewritten into the form $\mathbf{L}\mathbf{U}u = b$. Therefore, the solution vector u can be constructed by means of solutions to lower and upper diagonal systems of linear equations:

$$\mathbf{L}y = b \quad \text{and} \quad \mathbf{U}u = y.$$

Notice that a linear problem with an upper or lower diagonal matrix can be easily solved by backward or forward substitution of the solution vector components. The method of LU decomposition is suitable for linear problems having tridiagonal matrix **A**, i.e.

$$\mathbf{A} = \begin{pmatrix} \alpha_1 & \gamma_1 & 0 & \cdots & 0 \\ \beta_2 & \alpha_2 & \gamma_2 & & \vdots \\ 0 & \cdot & \cdot & \cdot & 0 \\ \vdots & \cdots & \beta_{n-1} & \alpha_{n-1} & \gamma_{n-1} \\ 0 & \cdots & 0 & \beta_n & \alpha_n \end{pmatrix}. \tag{10.29}$$

Suppose that the tridiagonal matrix **A** is diagonally dominant. It means that

$$\alpha_i > |\beta_i| + |\gamma_i|, \quad \text{for each } i = 1, 2, \ldots, n. \tag{10.30}$$

For such a matrix, there exists a unique LU decomposition with matrices **L** and **U** of the form

$$\mathbf{L} = \begin{pmatrix} 1 & 0 & 0 & \cdots & 0 \\ l_2 & 1 & \cdot & & \vdots \\ 0 & \cdot & \cdot & \cdot & 0 \\ \vdots & & \cdot & \cdot & 0 \\ 0 & \cdots & 0 & l_n & 1 \end{pmatrix}, \quad \mathbf{U} = \begin{pmatrix} d_1 & \gamma_1 & 0 & \cdots & 0 \\ 0 & d_2 & \gamma_2 & & \vdots \\ 0 & \cdot & \cdot & \cdot & 0 \\ \vdots & & \cdot & d_{n-1} & \gamma_{n-1} \\ 0 & \cdots & 0 & 0 & d_n \end{pmatrix},$$

where the entries of matrices **L** and **U** can be expressed as follows:

$$d_1 = \alpha_1, \quad d_i = \alpha_i - \frac{\gamma_{i-1}\beta_i}{d_{i-1}}, \quad l_i = \frac{\beta_i}{d_{i-1}}, \quad \text{for } 2 \leq i \leq n.$$

It can be shown by a recursive argument that the assumption of diagonal dominance (10.30) enables us to conclude that the diagonal elements $d_1, d_2, \ldots, d_n$ are nonzero. A solution y of the system of equations $\mathbf{L}y = b$ with a lower diagonal matrix **L** can be constructed as follows:

$$y_1 = b_1, \quad y_i = b_i - l_i y_{i-1}, \quad \text{for } i = 2, \ldots, n.$$

Finally, a solution to the system of equations $\mathbf{U}u = y$ with the upper diagonal matrix **U** can be constructed as follows:

$$u_n = \frac{y_n}{d_n}, \quad u_i = \frac{y_i - \gamma_i u_{i+1}}{d_i}, \quad \text{for } i = n-1, \ldots, i.$$

10.3.2. Gauss–Seidel Successive Over-relaxation Method

The method of the LU decomposition described in the previous section yields the exact solution to a system of linear equations. Nevertheless, in many practical situations, we do not need the exact solution and an approximate solution is sufficient. This is often the case when the linear solver is just a part of other numerical approximations. In such situations, we can take an approximate solution to a system of linear equations because its approximation error can be lower when compared to errors of other numerical approximation used in the algorithm.

Figure 10.5. Carl Friedrich Gauss (1777-1855).

The Gauss–Seidel successive over-relaxation method[1] (SOR method) is a popular iterative method for solving linear systems of equations. Its idea consists in construction of an approximative solution by means of a recursively constructed sequence of approximations. In what follows, we shall explain the SOR method in more details.

A matrix $\mathbf{A}$ can be additively decomposed into a sum of a lower-diagonal, upper-diagonal and diagonal matrices, i.e.

$$\mathbf{A} = \mathbf{L} + \mathbf{D} + \mathbf{U},$$

where

$$\mathbf{L}_{ij} = \mathbf{A}_{ij}, \quad \text{for } j < i, \quad \text{otherwise } \mathbf{L}_{ij} = 0,$$
$$\mathbf{D}_{ij} = \mathbf{A}_{ij}, \quad \text{for } j = i, \quad \text{otherwise } \mathbf{D}_{ij} = 0,$$
$$\mathbf{U}_{ij} = \mathbf{A}_{ij}, \quad \text{for } j > i, \quad \text{otherwise } \mathbf{U}_{ij} = 0.$$

Concerning the diagonal matrix $\mathbf{D}$, we shall assume its invertibility. It means that $A_{ii} \neq 0$ for $i = 1, \ldots, n$. Let $\omega > 0$ be a given relaxation parameter. A solution to the system of linear equations $\mathbf{A}u = b$ is equivalent with a solution to the problem

$$\mathbf{D}u = \mathbf{D}u + \omega(b - \mathbf{A}u).$$

With regard to the decomposition $\mathbf{A} = \mathbf{L} + \mathbf{D} + \mathbf{U}$, we obtain, after straightforward calculations, that u is a solution to the linear problem

$$(\mathbf{D} + \omega \mathbf{L})u = (1 - \omega)\mathbf{D}u + \omega(b - \mathbf{U}u). \tag{10.31}$$

The matrix $\mathbf{D} + \omega \mathbf{L}$ is invertible because $A_{ii} \neq 0$. Therefore, u is a solution to the following problem:

$$u = \mathbf{T}_\omega u + c_\omega, \quad \text{where } \mathbf{T}_\omega = (\mathbf{D} + \omega \mathbf{L})^{-1}\left((1 - \omega)\mathbf{D} - \omega \mathbf{U}\right), \tag{10.32}$$

---

[1]Carl Friedrich Gauss, 1777-1855, mathematician, physicist, geophysicists and astronomer. He worked in the field of mathematical analysis, differential geometry, probability theory (Gauss normal distribution, least square regression method). He developed one of the first theories of electromagnetism.

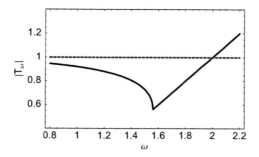

Figure 10.6. A graph of the spectral norm of the operator $\|\mathbf{T}_\omega\|$ as a function of the relaxation parameter ω.

and $c_\omega = \omega(\mathbf{D} + \omega\mathbf{L})^{-1}b$. Using the linear iteration operator $\mathbf{T}_\omega$, we can define a recursive sequence of approximate solutions to the problem $\mathbf{A}u = b$:

$$u^0 = 0, \quad u^{p+1} = \mathbf{T}_\omega u^p + c_\omega \quad \text{for } p = 1, 2, \ldots. \tag{10.33}$$

Notice that the initial condition u^0 can be also chosen differently respecting the character of the underlying problem. Clearly, if the sequence of approximate solution vectors u^p converges (in the Euclidean space $\mathbb{R}^n$) to a vector u for $p \to \infty$, then, with regard to the continuity of the linear operator $\mathbf{T}_\omega$, we obtain $u = \mathbf{T}_\omega u + c_\omega$. Hence, the vector u is a solution to the original system of linear equations $\mathbf{A}u = b$.

It is worthwhile noting, that the condition

$$\|\mathbf{T}_\omega\| < 1 \tag{10.34}$$

imposed on the norm of the linear operator $\mathbf{T}_\omega$ enables us to conclude contractivity of the mapping $\mathbb{R}^n \ni u \mapsto \mathbf{T}_\omega u + c_\omega \in \mathbb{R}^n$. According to the Banach fixed point theorem (see e.g., [26]), a sequence of vectors u^p indeed converges to the vector u as $p \to \infty$.

With regard to (10.31), we can compute the sequence of approximate solutions as follows:

$$u_i^{p+1} = \frac{\omega}{A_{ii}} \left(b_i - \sum_{j<i} A_{ij} u_j^{p+1} - \sum_{j>i} A_{ij} u_j^p \right) + (1-\omega) u_i^p, \tag{10.35}$$

for $p = 1, 2, \ldots$, and $i = 1, 2, \ldots, n$. At the end of this part, we discuss the problem of optimal choice of the relaxation parameter $\omega \neq 0$. Taking into account the above condition guaranteeing the convergence of a recurrently defined sequence (10.33), it is important to choose optimal relaxation parameter ω in such way that the norm $\|\mathbf{T}_\omega\|$ of the linear operator $\mathbf{T}_\omega$ is minimal.

As a norm of the linear operator $\mathbf{B} = \mathbf{T}_\omega$ we can consider, for instance, the so-called Frobenius matrix norm defined as: $\|\mathbf{B}\| = (\sum_{i,j=1}^n \mathbf{B}_{ij}^2)^{\frac{1}{2}}$ or the spectral norm $\|\mathbf{B}\| = \max_{\lambda \in \sigma(\mathbf{B})} |\lambda|$, where $\sigma(\mathbf{B})$ is the set of eigenvalues of the matrix $\mathbf{B}$. In Fig. 10.6, we show the dependence of the spectral norm of the operator $\mathbf{T}_\omega$ as a function of the relaxation parameter ω. As a matrix $\mathbf{A}$ we considered the tridiagonal matrix having the form of (10.29), where $\alpha_i = 2, \beta_i = \gamma_i = -1$ and $n = 10$. It should be obvious from Fig. 10.6 that there is an optimal choice of the relaxation parameter $\omega \approx 1.5$ for our matrix $\mathbf{A}$ such that the spectral norm of the operator $\mathbf{T}_\omega$ is minimal. In general, for tridiagonal and diagonally dominant matrices, there exists an optimal choice of the relaxation parameter

$$\omega \in (1,2),$$

for which the norm $\|\mathbf{T}_\omega\|$ is minimal and such that $\|\mathbf{T}_\omega\| < 1$ (see e.g., Vitásek [121]).

10.4. Methods for Solving Linear Complementarity Problems

The aim of this part is to recall the basic facts concerning solution of the so-called linear complementarity problem. As we have already mentioned in Chapter 9, solving the linear complementarity problem is a key tool for construction of a solution to the problem of pricing American style of derivatives. In the first part, we will present a simple modification of the SOR method yielding a solution to the linear complementarity problem. In the second part, of this section we will concentrate on practical examples of application of the numerical method for solving linear complementarity problem.

Our aim is to solve numerically the following linear complementarity problem:

$$\begin{aligned} \mathbf{A}u \geq b, \quad u &\geq g, \\ (\mathbf{A}u - b)_i (u_i - g_i) &= 0, \quad \text{for } i = 1, \ldots, n. \end{aligned} \tag{10.36}$$

Notations $\mathbf{A}u \geq b$ and $u \geq g$, respectively, mean that the inequalities between vectors are fulfilled component-wise. We shall suppose that the matrix $\mathbf{A}$ is diagonally dominant and satisfying the condition (10.30).

10.4.1. Projected Successive Over-relaxation Method

A modification of the Gauss–Seidel SOR method is the basis for construction of an efficient numerical method for solving the linear complementarity problem (10.36). In each iteration step p we shall modify the approximate solution u^{p+1} in such a way that it satisfies the constraint condition $u \geq g$. Finally, we shall prove that the limit of such iterative approximation is indeed a solution of the linear complementarity problem (10.36).

Let us define a recurrent sequence of approximate solutions of the linear complementarity problem as follows:

$$u^0 = 0, \quad u^{p+1} = \max\left(\mathbf{T}_\omega u^p + c_\omega, g\right), \quad \text{for } p = 1, 2, \ldots, \tag{10.37}$$

where the maximum is again taken component-wise, i.e.

$$u_i^{p+1} = \max\left((\mathbf{T}_\omega u^p + c_\omega)_i, g_i\right), \quad \text{for each } i = 1, 2, \ldots, n.$$

For a moment, let us suppose that convergence of the sequence of vectors u^p towards a vector $u \in \mathbb{R}^n$ as $p \to \infty$ is guaranteed. Since $u^p \geq g$, so does the limit $u = \lim_{p \to \infty} u^p \geq g$. Taking into account the relation (10.35), we obtain

$$u_i^{p+1} = \max\left(\frac{\omega}{A_{ii}}\left(b_i - \sum_{j<i} A_{ij} u_j^{p+1} - \sum_{j>i} A_{ij} u_j^p\right) + (1-\omega) u_i^p, g_i\right), \tag{10.38}$$

for each $i = 1, 2, \ldots, n$. It means that, in the limit $p \to \infty$, the following inequality holds true:
$$u_i \geq \frac{\omega}{A_{ii}}\left(b_i - \sum_{j<i} A_{ij} u_j - \sum_{j>i} A_{ij} u_j\right) + (1-\omega) u_i.$$

With regard to the diagonal dominance of the matrix $\mathbf{A}$ we have $A_{ii} > 0$. Hence
$$A_{ii} u_i \geq \omega\left(b_i - \sum_{j<i} A_{ij} u_j - \sum_{j>i} A_{ij} u_j\right) + (1-\omega) A_{ii} u_i.$$

Since $\omega > 0$, then, after a short manipulation, we conclude the inequality
$$(\mathbf{A}u)_i \geq b_i, \quad \text{for each } i = 1, 2, \ldots, n.$$

Finally, if for some index i the strict inequality $u_i > g_i$ holds then for all sufficiently large values of the iteration index p we also have $u_i^{p+1} > g_i$. Now, it follows from the definition of the sequence of vectors u^p that
$$u_i^{p+1} = \frac{\omega}{A_{ii}}\left(b_i - \sum_{j<i} A_{ij} u_j^{p+1} - \sum_{j>i} A_{ij} u_j^p\right) + (1-\omega) u_i^p$$

for all sufficiently large iteration indices p. Passing to the limit $p \to \infty$ we conclude the equality $(\mathbf{A}u)_i = b_i$ for the index i. It means that $(\mathbf{A}u - b)_i (u_i - g_i) = 0$ and so the vector u is indeed a solution to the linear complementarity problem (10.36).

At the end of this part, we discuss the question regarding the convergence of the sequence recurrently defined as in (10.37). Let us denote
$$F(u) = \max(\mathbf{T}_\omega u^p + c_\omega, g).$$

Clearly, $F : \mathbb{R}^n \to \mathbb{R}^n$ is a nonlinear mapping. In what follows, we will prove that it is a contractive mapping. Consequently, by using the Banach fixed point theorem, the sequence u^p has a limit in $\mathbb{R}^n$. For arbitrary two vectors $u, v \in \mathbb{R}^n$ it holds:

$$F(u)_i - F(v)_i = \begin{cases} \phi_i - \psi_i, & \text{if } \phi_i, \psi_i \geq g_i, \\ 0, & \text{if } \phi_i, \psi_i \leq g_i, \\ \phi_i - g_i, & \text{if } \phi_i \geq g_i, \psi_i < g_i, \\ g_i - \psi_i, & \text{if } \phi_i < g_i, \psi_i \geq g_i, \end{cases}$$

where $\phi = \mathbf{T}_\omega u + c_\omega, \psi = \mathbf{T}_\omega v + c_\omega$. In the case $\phi_i \geq g_i$ and $\psi_i < g_i$ we have $0 \leq \phi_i - g_i < \phi_i - \psi_i$. Analogously, for $\psi_i \geq g_i$ and $\phi_i < g_i$ we obtain $0 \leq \psi_i - g_i < \psi_i - \phi_i$. In both cases, we may conclude that, for each $i = 1, \ldots, n$, it holds
$$|F(u)_i - F(v)_i| \leq |\phi_i - \psi_i| \leq |(\mathbf{T}_\omega u)_i - (\mathbf{T}_\omega v)_i|.$$

Therefore $\|F(u) - F(v)\| \leq \|\mathbf{T}_\omega u - \mathbf{T}_\omega v\| = \|\mathbf{T}_\omega(u-v)\| \leq \|\mathbf{T}_\omega\| \|u-v\|$. This way we have shown the following result:

$$\begin{aligned}&\text{if } \|\mathbf{T}_\omega\| < 1 \quad \text{then } F \text{ is contractive in } \mathbb{R}^n, \\ &\text{the sequence } u^p \text{ converges to } u \text{ for } p \to \infty, \\ &\text{the vector } u \in \mathbb{R}^n \text{ is a solution to the problem (10.36)}.\end{aligned} \quad (10.39)$$

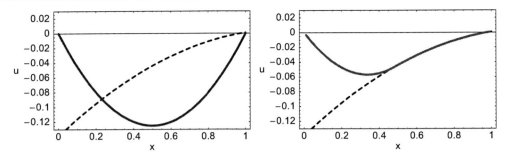

Figure 10.7. Left: the bold curve represents a solution $\tilde{u}$ of the unconstrained problem $-\tilde{u}''(x) = b(x), \tilde{u}(0) = \tilde{u}(1) = 0$. Right: the bold curve represents a solution u of the constrained problem with a constraint given by a function g (dashed curve).

Iterative method for solving the problem (10.36) described by the recurrent relation (10.37) and (10.38), respectively, is called the Projected successive over-relaxation method (PSOR for short). It has been proposed and analyzed by Elliott and Ockendon [42] in the context of solutions to variational inequalities.

10.4.2. Numerical Solutions of the Obstacle Problem

In this part, we present a basic idea how to solve free boundary problems. Recall that the pricing of American style of options can be also transformed into a free boundary problem (see Chapter 9). Notice that free boundary problems often arise in other applied fields including, in particular, physics or mechanics. One of the simplest examples is the obstacle problem, which is directly connected to the problem of pricing American options with early exercise opportunities.

We consider an elastic spring (or a thin beam) extended over the interval $[0,1]$. There is an external force with the size b pushing the spring in the downward direction. If we denote by function $u : [0,1] \to \mathbb{R}$ the vertical displacement of the spring clamped in the boundary points $x = 0, 1$, then it follows from the theory of elasticity that the equation describing the displacement has the form of the boundary value problem:

$$-u''(x) = b(x), \quad \text{for } x \in [0,1], \quad u(0) = u(1) = 0.$$

Recall that the above equation can be derived as the solution to the so-called variational problem in which the goal is to minimize the total energy composed from the energy of deformation and potential energy: $\Phi(u) = \frac{1}{2}\int_0^1 |u_x|^2 dx - \int_0^1 bu\, dx$. A numerical solution to the above boundary value problem can be easily constructed by a finite difference approximation. Let us denote $x_i = i/n, i = 0, 1, \ldots, n$, points of a partition of the interval $[0,1]$ into n equidistant subintervals. The displacement $u(x_i)$ at the point x_i will be denoted by u_i. Then the numerical solution can be constructed by solving the system of linear equations:

$$-\frac{u_{i-1} - 2u_i + u_{i+1}}{h^2} = b_i, \quad \text{for } i = 1, \ldots, n-1,$$

where $h=1/n, b_i = b(x_i)$ and the second derivative $u''(x)$ at the point x_i has been approximated by means of finite differences. Notice that $u_0 = 0 = u_n$. The system of equation can be rewritten in the matrix form:

$$\mathbf{A} u = b,$$

where the matrix $\mathbf{A}$ is a $(n-1) \times (n-1)$ tridiagonal matrix having the form of (10.29) with entries $\alpha_i = 2/h^2, \beta_i = \gamma_i = -1/h^2$. A graph of a parabolic-like solution for the constant gravitational-like force $b = -1$ is depicted in Fig. 10.7 (left).

Now, let us consider a situation where we have given a constraint described by a function g defined on $[0,1]$. We require that the solution of our elastic spring problem should be above the prescribed constraint, i.e.

$$u(x) \geq g(x), \quad \text{for } x \in (0,1).$$

If the spring is strictly above the constraint $u(x) > g(x)$ at x then we require that the differential equation is satisfied at x, i.e., $-u''(x) = b(x)$. In any case (including the equality $u(x) = g(x)$) we require that the solution solves a differential inequality $-u''(x) \geq b(x)$. Such a linear complementarity formulation for the constrained elastic spring can be rigorously derived by inspecting the first order conditions for a minimizer of the total energy functional Φ restricted to the set $\{u \in C^1(0,1), u(x) \geq g(x), \text{ for } x \in (0,1)\}$. If we denote $g_i = g(x_i)$ then the discrete approximation of the solution u can be represented by a solution to the linear complementarity problem of the form (10.36). More precisely,

$$\mathbf{A} u \geq b, \quad u \geq g, \quad (\mathbf{A} u - b)_i (u_i - g_i) = 0, \quad \text{for } i = 1, \ldots, n.$$

An illustrative description of a solution obtained by the Projected SOR method is shown in Fig. 10.7 (right). Notice that the solution of constrained problem is continuously differentiable at the point of pasting of the spring and constraint. This feature is in analogy with a C^1 smooth pasting principle valid for a solution to the American style of vanilla options (see Chapter 9).

10.5. Numerical Methods for Pricing of American Style Options

In what follows, we shall present a numerical scheme for solving the linear complementarity problem arising in the problem of pricing the American style of options. We will make use of the PSOR algorithm discussed and analyzed in the previous section. We remind ourselves (see Chapter 9) that the problem of pricing American call and put options can be transformed into the form of a linear complementarity problem (9.26). More precisely,

$$\left(\frac{\partial u}{\partial \tau} - \frac{\sigma^2}{2} \frac{\partial^2 u}{\partial x^2} \right) (u(x,\tau) - g(x,\tau)) = 0,$$

$$\frac{\partial u}{\partial \tau} - \frac{\sigma^2}{2} \frac{\partial^2 u}{\partial x^2} \geq 0, \quad u(x,\tau) - g(x,\tau) \geq 0,$$

for each $x \in \mathbb{R}, 0 < \tau < T$. Our goal is to find a continuously differentiable function $u: \mathbb{R} \times (0,T) \to \mathbb{R}$ solving the above linear complementarity problem. Recall that the function

$g(x,\tau)$ represents the transformed pay–off diagram of the option, i.e.

$$g(x,\tau) = e^{\alpha x + \beta \tau} \max(e^x - 1, 0), \quad \text{for a call option,}$$
$$g(x,\tau) = e^{\alpha x + \beta \tau} \max(1 - e^x, 0), \quad \text{for a put option,}$$

where $\alpha = \frac{r-q}{\sigma^2} - \frac{1}{2}$, $\beta = \frac{r+q}{2} + \frac{\sigma^2}{8} + \frac{(r-q)^2}{2\sigma^2}$. The initial condition for the function u is given by

$$u(x,0) = g(x,0), \quad \text{for each } x \in \mathbb{R}.$$

Next we shall discretize the linear complementarity problem by means of finite differences. By symbols u^j and g^j we shall denote approximation of a solution u and the transformed pay–off diagram at the time level τ_j. It means that

$$u^j = (u^j_{-N+1}, \ldots, u^j_{-1}, u^j_0, u^j_1, \ldots, u^j_{N-1}) \in \mathbb{R}^n,$$
$$g^j = (g^j_{-N+1}, \ldots, g^j_{-1}, g^j_0, g^j_1, \ldots, g^j_{N-1}) \in \mathbb{R}^n,$$

where $n = 2N - 1$. Choose $N \in \mathbb{N}$ sufficiently large and such that the interval of spatial discretization (x_{-N+1}, x_{N-1}) is large enough in order to approximate the boundary values u^j_{-N} and u^j_N by Dirichlet boundary conditions. With regard to the boundary conditions (9.24) and (9.25) occurring in the continuous formulation of the problem of pricing American call and put options, we can postulate the following boundary conditions for the discrete solution:

$$u^j_{-N} = \phi^j := g(x_{-N}, \tau_j), \qquad u^j_N = \psi^j := g(x_N, \tau_j). \tag{10.40}$$

Now the problem of linear complementarity can be restated in its discrete form as follows:

$$\mathbf{A}u^{j+1} \geq u^j + b^j, \quad u^{j+1} \geq g^{j+1}, \quad \text{for } j = 0, 1, \ldots, m-1,$$
$$(\mathbf{A}u^{j+1} - u^j - b^j)_i (u^{j+1} - g^{j+1})_i = 0, \quad \text{for each } i, \tag{10.41}$$

where $u^0 = g^0$. The matrix $\mathbf{A}$ is the same $(n-1) \times (n-1)$ tridiagonal matrix as the one appearing in the implicit numerical scheme for pricing the European style of vanilla options, i.e.

$$\mathbf{A} = \begin{pmatrix} 1+2\gamma & -\gamma & 0 & \cdots & 0 \\ -\gamma & 1+2\gamma & -\gamma & & \vdots \\ 0 & \cdot & \cdot & \cdot & 0 \\ \vdots & & -\gamma & 1+2\gamma & -\gamma \\ 0 & \cdots & 0 & -\gamma & 1+2\gamma \end{pmatrix}, \quad b^j = \begin{pmatrix} \gamma \phi^{j+1} \\ 0 \\ \vdots \\ 0 \\ \gamma \psi^{j+1} \end{pmatrix},$$

where $\gamma = \sigma^2 k/(2h^2)$. The discrete form of the linear complementarity problem can be solved by using the PSOR (Projected SOR) method discussed in section 10.4.1. In Fig. 10.8 we depict a graph of a solution $S \mapsto V(S,t)$ describing the price of American call and put options. We chose the following financial parameters: $\sigma = 0.4, r = 0.04, q = 0.12, T - t = 1, E = 50$ in the case of the call option and $\sigma = 0.6, r = 0.08, q = 0, T - t = 1, E = 50$ in the case of the put option. In both cases, we chose numerical parameters $N = 100, m = 20$ whereas $\sigma^2 k/(2h^2) = \gamma = 1$. At each time level, we used 20 iterations of the PSOR algorithm with a relaxation parameter $\omega = 1.7$. A simplified Mathematica source code for solving the American call option problem can be found in Table 10.3.

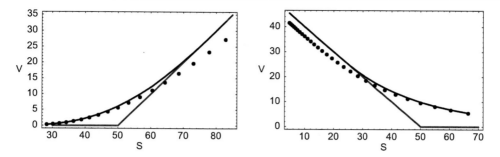

Figure 10.8. A solution $S \mapsto V(S,t)$ describing the price of an American call option (left) and put option (right) obtained by means of the Projected SOR algorithm. Corresponding solutions to the problem of pricing European call and put options are plotted by dots.

10.5.1. Identification of the Early Exercise Boundary for American Options

The main purpose of this section is to identify a position of the early exercise boundary for American call and put options. Notice that the Projected SOR algorithm for solving the problem of pricing American options does not provide explicit information concerning the early exercise boundary position function $S_f(t)$. On the other hand, it should be obvious that the value of $S_f(t)$ can be *a posteriori* calculated from the solution $V(S,t)$. Suppose that we already know the solution $V(S,t)$ at the time $t \in [0,T]$. Then the early exercise position $S_f(t)$ is uniquely determined through the relation:

$$S_f(t) = \begin{cases} \min\left(S > 0,\; V(S,t) = S - E\right), & \text{for a call option,} \\ \max\left(S > 0,\; V(S,t) = E - S\right), & \text{for a put option.} \end{cases} \quad (10.42)$$

Using a numerical approximation, we obtain values of the approximate solution evaluated in the nodal points $(S_i, T - \tau_j)$, where $S_i = E e^{ih}$ and $\tau_j = jk$. In order to construct the position of the early exercise boundary, we can make use of the following algorithm:

$$S_f(T - \tau_j) = \begin{cases} \min\left(S_i > 0,\; |V(S_i, T - \tau_j) - (S_i - E)| < \varepsilon\right) \\ \qquad \text{for a call option,} \\ \max\left(S_i > 0,\; |V(S_i, T - \tau_j) - (E - S_i)| < \varepsilon\right) \\ \qquad \text{for a put option,} \end{cases} \quad (10.43)$$

where $0 < \varepsilon \ll 1$ is a prescribed tolerance level. In practical examples in which the exercise price $E \approx 10$, it is usually sufficient to take $\varepsilon \approx 10^{-5}$.

In Fig. 10.9, we plot the early exercise boundary for an American call option and we compare it with an analytic approximation introduced and analyzed in Chapter 9. The computation was performed by using financial parameters $\sigma = 0.2, r = 0.1, q = 0.05, E = 10, T = 1$. The value $S_f(0) = 22.3893$ of the early exercise boundary position for the call option at $t = 0$ is almost identical (the relative error is less than 0.1%) to the one computed using a different method in the paper [103] by Ševčovič. In Fig. 10.10 we present a 3D graph of a solution $(S,t) \mapsto V(S,t)$ for the price of an American call option. We also depict

Table 10.3. The Mathematica source code for the PSOR method for pricing American call options.

```
sigma = 0.4; r = 0.04; q = 0.12; T = 1; X = 50;
alfa = (r - q)/sigma^2 - 1/2;
beta = (r + q)/2 + sigma^2/8 + (r - q)^2/(2sigma^2);

NN=100; n=2 NN-1; m=20; MaxSORiter = 20; omega = 1.7;
gama = 1; k = T/m; h = sigma Sqrt[k/(2 gama)];

A = Table[Table[ If[i == j, 1 + 2gama,
          If[i == j-1, -gama, If[i == j+1, -gama, 0 ]]],
       {j, 1, n}], {i, 1, n}];

g[x_, tau_] := Exp[alfa x + beta tau] Max[Exp[x]-1, 0];
uold = Table[g[i h, 0], {i, -NN + 1, NN - 1}];
phi[j_] := g[-NN h, j k];
psi[j_] := g[NN h, j k];

For[j = 1, j <= m, 1,
{
b = Table[If[i == -NN + 1, gama phi[j],
          If[i == NN - 1, gama psi[j], 0]],
        {i, -NN + 1, NN - 1}];
gvec = Table[g[i h, j k], {i, -NN + 1, NN - 1}];
unew = uold;
  For[p = 1, p <= MaxSORiter, 1,
    {
    For[ii = 1, ii <= n, 1,
      {
      upom = (1 - omega) unew[[ii]]
        + (omega/A[[ii, ii]])*(b[[ii]] + uold[[ii]]
        - If[ii > 1, A[[ii, ii - 1]]*unew[[ii - 1]], 0]
        - If[ii < n, A[[ii, ii + 1]]*unew[[ii + 1]], 0] );
      unew=ReplacePart[unew, Max[upom, gvec[[ii]] ], ii];
      ii++;}];
    p++;}];
uold = unew;
Vnew = Table[
{X Exp[i h], X Exp[-alfa i h -beta j k] unew[[i+NN]]},
{i, -NN+1, NN-1}];
j++;}];

ListPlot[Vnew];
```

the early exercise boundary position. The computation was realized for the parameters: $\sigma = 0.6, r = 0.1, q = 0.09, E = 10, T = 1$.

At the end of this section, we present an example of computation of price of the Amer-

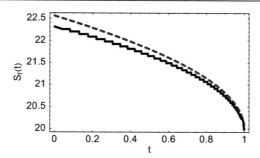

Figure 10.9. A comparison of the early exercise boundary position computed by the PSOR algorithm when using (10.43) and the analytic approximation discussed in Chapter 9 (dashed line).

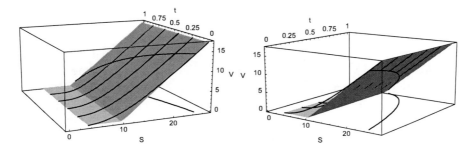

Figure 10.10. Two different view points on the 3D graph of the solution $(S,t) \mapsto V(S,t)$ describing the price of an American call option. Five selected solution profiles are compared with the pay-off diagram surface.

ican binary option. Such type of an option was discussed in Chapter 6.4 for the case of a European style of option. Let us recall that the pay-off diagram of a binary option is given by:

$$V^{bin}(S,T) = \begin{cases} 1, & \text{if } S \in [E_1, E_2], \\ 0, & \text{otherwise}, \end{cases}$$

where $0 < E_1 < E_2$. The formula for pricing the European style of a binary option at the time $t \in [0,T]$ has the following form: $V^{bin}(S,t) = e^{-r(T-t)}(N(d_1^{E_1}) - N(d_1^{E_2}))$. If we consider the transformed pay–off diagram

$$g(x,\tau) = e^{\alpha x + \beta \tau} V^{bin}(S,T),$$

where $S = e^x$, then, by using the PSOR algorithm for pricing the American style of a binary option, we obtain numerical results shown in Fig. 10.11. Notice that, in comparison to the European style of a binary option, the value of the American binary option is always above the pay–off diagram.

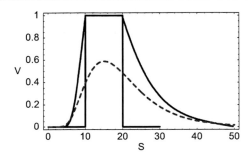

Figure 10.11. A comparison of solutions corresponding to American and European style (dashed line) of a binary option for the parameters: $E_1 = 10, E_2 = 20, r = 0.04, q = 0.02, T = 1$.

10.5.2. Implied Volatility for American Options

Similarly as in section 4.2., for the American style of an option we can also find the value of the volatility parameter $\sigma > 0$ such that the calculated price of the American style of an option coincides with the market one. The implied volatility $\sigma_{impl} > 0$ is therefore such a value of the volatility parameter for which the theoretical price of the American call (put) option $V(S,t;\sigma)$ at the time t and the underlying asset price $S = S_{real}(t)$ is identical with the real market value of the American call (put) option $V_{real}(t)$. In other words, in order to find the implied volatility σ_{impl} of a given option we have to solve the implicit equation

$$V_{real}(t) = V(S_{real}(t), t; \sigma_{impl}). \tag{10.44}$$

To calculate the theoretical price of the American option $V(S_{real}(t), t; \sigma)$ for known values of the underlying asset price $S_{real}(t)$ at the time $t \in [0,T)$, we make use of the PSOR algorithm. The option price for $S_i < S_{real}(t) < S_{i+1}$ can be calculated by a piece-wise linear interpolation of the option prices at points S_i, S_{i+1} of a numerical discretization mesh. Since the price of the American option is again an increasing function of the volatility, we can repeat the calculation of the option price for a discrete set of volatilities $0 < \sigma_1 < \sigma_2 < \cdots < \sigma_P$. The discrete set of volatilities should represent a sufficiently fine mesh of values. The value σ_p, for which the matching error $|V_{real}(t) - V(S_{real}(t), t; \sigma_p)|$ is minimal, can be identified with the implied volatility of an American option.

10.5.3. Source codes for Numerical Algorithms

Source codes of all numerical algorithms presented in this and other chapters of this book can be found and freely downloaded from the address:

www.iam.fmph.uniba.sk/institute/sevcovic/derivaty.

As the login name use: derivaty and as the password: opcie. In the archive repository you can find samples of numerical codes written in Mathematica and Matlab programming environments.

Problem Section and Exercises

1. Calculate the value of the American call option on a dividend paying asset with a dividend yield $q = 5\%$ p.a. ($q = 0.05$). It is known that the historical volatility of the underlying asset has been estimated to $\sigma = 20\%$ p.a. and the riskless bond interest rate $r = 10\%$. The call option is written on the exercise price $E = 10$ with the expiration time $T = 1$ (one year). Calculate the option price at $t = 0$ for the asset prices $S = 24, S = 23, S = 21, S = 20, S = 18$. Based on these calculations, estimate the position of the early exercise boundary at $t = 0$.

2. Calculate the value of the bought straddle strategy consisting of purchasing one call and one put option with the same exercise price E for a) European type of call and put options, b) American type of call and put options. It is known that the price of the underlying asset $S = 55$ paying continuous dividends with $q = 0.03$, its historical volatility is $\sigma = 0.4$, interest rate of a risk-less bond $r = 0.05$. Expiration time is three months, i.e., $T = 0.25$ and $E = 60$.

3. The present market price of IBM shares is USD 118.86. The value of the European put option written for the exercise price $E = 120$ with exercise time two months is equal to USD 5.5. The interest rate of riskless bond is 5% p.a., the dividend yield $q = 2\%$ p.a. Calculate the implied volatility for the European style put option. Compare the value of the implied volatility to the one calculated with an assumption that the put option is of the American style.

Chapter 11

Nonlinear Extensions of the Black–Scholes Pricing Model

In this chapter, we deal with various generalizations of the classical linear Black–Scholes equation for pricing derivative securities. We will analyze generalized option pricing models which are capable of capturing several important phenomena like e.g., transaction costs, investor's risk from unprotected portfolio, investor's expected utility maximization, illiquid markets, large traders feedback influence, etc. We will show that these generalizations can be mathematically stated in the form of a nonlinear Black–Scholes equation in which the volatility is adjusted to be a function of the option price itself.

11.1. Overview of Nonlinear Extensions to the Black–Scholes Option Pricing Model

According to the classical theory due to Black, Scholes and Merton discussed in Chapter 2, the price of an option in an idealized financial market can be computed from a solution to the well-known Black–Scholes linear parabolic equation derived by Black and Scholes in [14], and, independently by Merton (cf. Kwok [75], Dewynne et al. [36], Hull [65], Wilmott et al. [122]). Recall that a European call (put) option is the right but not obligation to purchase (sell) an underlying asset at the expiration price E at the expiration time T. Assuming that the underlying asset S follows a geometric Brownian motion

$$dS = (\rho - q)S dt + \sigma S dW, \qquad (11.1)$$

where ρ is a drift, q is the asset dividend yield rate, σ is the volatility of the asset and W is the standard Wiener process (cf. [75]), one can derive a governing partial differential equation for the price of an option. With regard to Chapter 2, the mathematical model for the price $V(S,t)$ of an option is the following parabolic PDE:

$$\frac{\partial V}{\partial t} + (r-q)S\frac{\partial V}{\partial S} + \frac{\sigma^2}{2}S^2\frac{\partial^2 V}{\partial S^2} - rV = 0, \qquad (11.2)$$

where σ is the volatility of the underlying asset price process, $r > 0$ is the interest rate of a zero-coupon bond, $q \geq 0$ is the dividend yield rate. A solution $V = V(S,t)$ represents the price of an option if the price of an underlying asset is $S > 0$ at time $t \in [0, T]$.

The case when the diffusion coefficient $\sigma > 0$ in (11.2) is constant represents a classical Black–Scholes equation originally derived by Black and Scholes in [14]. On the other hand, if we assume the volatility coefficient $\sigma > 0$ to be a function of the solution V itself then (11.2) with such a diffusion coefficient represents a nonlinear generalization of the Black–Scholes equation. It is a purpose of this chapter to focus our attention to the case when the diffusion coefficient σ^2 may depend on the time $T - t$ to expiry, the asset price S and the second derivative $\partial_S^2 V$ of the option price (hereafter referred to as Γ), i.e.

$$\sigma = \sigma(S^2 \partial_S^2 V, S, T - t). \tag{11.3}$$

A motivation for studying the nonlinear Black–Scholes equation (11.2) with a volatility σ having a general form of (11.3) arises from option pricing models taking into account nontrivial transaction costs, market feedbacks and/or risk from a volatile (unprotected) portfolio. Recall that the linear Black–Scholes equation with constant σ has been derived under several restrictive assumptions like e.g., frictionless, liquid and complete markets, etc. We also recall that the linear Black–Scholes equation provides a perfectly replicated hedging portfolio. In the last decades some of these assumptions have been relaxed in order to model, for instance, the presence of transaction costs (see e.g., Leland [78], Hoggard et al. [64], Avellaneda and Paras [10]), feedback and illiquid market effects due to large traders choosing given stock-trading strategies (Frey [50], Frey and Patie [51], Frey and Stremme [52], During et al.[39], Schönbucher and Wilmott [99]), imperfect replication and investor's preferences (Barles and Soner [11]), risk from unprotected portfolio (Kratka [73], Jandačka and Ševčovič [66] or [104]). One of the first nonlinear models is the so-called Leland model (cf. [78]) for pricing call and put options under the presence of transaction costs. It has been generalized for more complex option strategies by Hoggard, Whaley and Wilmott in [64]. In this model the volatility σ is given by

$$\sigma^2(S^2 \partial_S^2 V, S, \tau) = \hat{\sigma}^2(1 + \text{Le} \, \text{sgn}(\partial_S^2 V)), \tag{11.4}$$

where $\hat{\sigma} > 0$ is a constant historical volatility of the underlying asset price process and Le > 0 is the so-called Leland constant given by Le $= \sqrt{2/\pi} C/(\hat{\sigma}\sqrt{\Delta t})$ where $C > 0$ is a constant round trip transaction cost per unit dollar of transaction in the assets market and $\Delta t > 0$ is the time-lag between portfolio adjustments. We refer the reader to discussion in Chapter 5 for details of derivation of the Leland model.

Notice that dependence of volatility adjustment on the second derivative of the price is quite natural. Indeed, in the idealized Black–Scholes theory, the delta hedging strategy yields $\delta = \pm \partial_S V$ depending on the type of an option. Therefore one may expect more frequent transaction in regions with the high second derivative $\partial_S^2 V$ (cf. [14]).

Another nonlinear generalization of the Black–Scholes equation has been proposed by Avellaneda, Levy and Paras [9] for description of incomplete markets and uncertain but bounded volatility. In their model we have

$$\sigma^2(S^2 \partial_S^2 V, S, \tau) = \begin{cases} \hat{\sigma}_1^2, & \text{if } \partial_S^2 V < 0, \\ \hat{\sigma}_2^2, & \text{if } \partial_S^2 V > 0, \end{cases} \tag{11.5}$$

where σ_1 and σ_2 represent a lower and upper a-priori bound on the otherwise unspecified asset price volatility.

If transaction costs are taken into account perfect replication of the contingent claim is no longer possible and further restrictions are needed in the model. By assuming that investor's preferences are characterized by an exponential utility function Barles and Soner derived a nonlinear Black–Scholes equation with the volatility σ given by

$$\sigma^2(S^2\partial_S^2 V, S, \tau) = \hat{\sigma}^2 \left(1 + \Psi(a^2 e^{r\tau} S^2 \partial_S^2 V)\right) \tag{11.6}$$

where Ψ is a solution to the ODE: $\Psi'(x) = (\Psi(x)+1)/(2\sqrt{x\Psi(x)}-x), \Psi(0) = 0$ and $a > 0$ is a given constant representing investor's risk aversion (see [11]). Notice that $\Psi(x) = O(x^{\frac{1}{3}})$ for $x \to 0$ and $\Psi(x) = O(x)$ for $x \to \infty$.

A generalized nonlinear Black–Scholes model has been also derived for the case when the asset dynamics takes into account the presence of feedback and illiquid market effects. Frey and Stremme (cf. [52, 51]) introduced directly the asset price dynamics in the case when a large trader chooses a given stock-trading strategy (see also [99]). The diffusion coefficient σ is again nonconstant and it can be expressed as:

$$\sigma^2(S^2\partial_S^2 V, S, \tau) = \hat{\sigma}^2 \left(1 - \rho\lambda(S)S\partial_S^2 V\right)^{-2}, \tag{11.7}$$

where $\hat{\sigma}^2, \rho > 0$ are constants and $\lambda(S)$ is a strictly convex function, $\lambda(S) \geq 1$. Interestingly enough, explicit solutions to the Black–Scholes equation with varying volatility as in (11.7) have been derived by Bordag and Chankova[16] and Bordag and Frey [17]. The nonlinear model with $\lambda \equiv 1$ has been derived by Frey in [50]. It describes the option price in a stylized market where a large investor can influence the underlying stock price by his/her stock-holding strategy.

The last example of the Black–Scholes equation with a nonlinearly depending volatility is the so-called Risk Adjusted Pricing Methodology model proposed by Kratka in [73] and revisited by Jandačka and Ševčovič in [66]. In order to maintain (imperfect) replication of a portfolio by the delta hedge one has to make frequent portfolio adjustments leading to a substantial increase in transaction costs. On the other hand, rare portfolio adjustments may lead to an increase of the risk arising from a volatile (unprotected) portfolio. In the Risk adjusted pricing methodology (RAPM) model our purpose is to optimize the time-lag Δt between consecutive portfolio adjustments. By choosing $\Delta t > 0$ in such way that the sum of the rate of transaction costs and the rate of a risk from unprotected portfolio is minimal one can find the optimal time lag $\Delta t > 0$. In the RAPM model, it turns out that the volatility is again nonconstant and it has the following form:

$$\sigma^2(S^2\partial_S^2 V, S, \tau) = \hat{\sigma}^2 \left(1 + \mu(S\partial_S^2 V)^{\frac{1}{3}}\right). \tag{11.8}$$

Here $\hat{\sigma}^2 > 0$ is a constant historical volatility of the asset price returns and $\mu = 3(C^2 R/2\pi)^{\frac{1}{3}}$, where $C, R \geq 0$ are nonnegative constants representing the transaction cost measure and the risk premium measure, respectively. (see [66] for details).

Notice that all the above mentioned nonlinear models are consistent with the original Black–Scholes equation in the case the additional model parameters (e.g., Le, a, ρ, μ) are vanishing. If plain call or put vanilla options are concerned then the function $V(S,t)$ is convex in S variable and therefore each of the above mentioned models has a diffusion coefficient strictly larger than $\hat{\sigma}^2$ leading to a larger values of computed option prices. They can be therefore identified with higher ask option prices, i.e., offers to sell an option.

11.2. Risk Adjusted Pricing methodology Model

The aim of this section is to present, in a more detail, one of nonlinear generalizations of the classical Black–Scholes equation with a volatility σ of the form (11.3) . We focus on the so-called Risk adjusted pricing methodology model due to Kratka [73] and its generalization by Jandačka and Ševčovič [66] (see also [104]). In this model both the risk arising from nontrivial transaction costs as well as the risk from unprotected volatile portfolio are taken into account. Their sum representing the total risk is subject of minimization. The original model was proposed by [73]. In [66] we modified Kratka's approach by considering a different measure for risk arising from unprotected portfolio in order to construct a model, which is scale invariant and the resulting partial differential equation is tractable and mathematically well posed. These two important features were missing in the original model of Kratka. The model is based on the Black–Scholes parabolic PDE in which transaction costs are described by the Hoggard, Whalley and Wilmott extension of the Leland model (cf. [64, 75, 65]) whereas the risk from a volatile portfolio is described by the average value of the variance of the synthesized portfolio. Transaction costs as well as the volatile portfolio risk depend on the time-lag between two consecutive transactions. We define the total risk premium as a sum of transaction costs and the risk cost from the unprotected volatile portfolio. By minimizing the total risk premium functional we obtain the optimal length of the hedge interval.

Concerning the dynamics of an underlying asset we will assume that the asset price $S = S(t), t \geq 0$, follows a geometric Brownian motion (11.1) with a drift ρ, standard deviation $\hat{\sigma} > 0$ and it may pay continuous dividends, i.e., $dS = (\rho - q)Sdt + \hat{\sigma}SdW$ where dW denotes the differential of the standard Wiener process and $q \geq 0$ is a continuous dividend yield rate. This assumption is made when deriving the classical Black–Scholes equation (see e.g., [65, 75]). Similarly as in the derivation of the classical Black–Scholes equation we construct a synthesized portfolio Π consisting of a one option with a price V and δ assets with a price S per one asset:

$$\Pi = V + \delta S. \tag{11.9}$$

We recall that the key idea in the Black–Scholes theory is to examine the differential $\Delta \Pi$ of equation (11.9). The right-hand side of (11.9) can be differentiated by using Itō's formula whereas portfolio's increment $\Delta \Pi(t) = \Pi(t + \Delta t) - \Pi(t)$ of the left-hand side can be expressed as follows:

$$\Delta \Pi = r\Pi \Delta t - \delta q S \Delta t, \tag{11.10}$$

where $r > 0$ is a risk-free interest rate of a zero-coupon bond. In the real world, such a simplified assumption is not satisfied and a new term measuring the total risk should be added to (11.10). More precisely, the change of the portfolio Π is composed of two parts: 1) the risk-free interest rate part $r\Pi \Delta t$ and the dividend yield term $\delta q S \Delta t$ and 2) the total risk premium: $r_R S \Delta t$ where r_R is a risk premium per unit asset price. We consider a short positioned call option. Therefore the writer of an option is exposed to this total risk. Hence we are going to price the higher ask option price – an offer to sell an option. It means that $\Delta\Pi = r\Pi\Delta t - \delta q S\Delta t - r_R S\Delta t$. The total risk premium r_R consists of the transaction risk premium r_{TC} and the portfolio volatility risk premium r_{VP}, i.e. $r_R = r_{TC} + r_{VP}$. Hence

$$\Delta\Pi = r\Pi\Delta t - \delta q S\Delta t - (r_{TC} + r_{VP})S\Delta t. \qquad (11.11)$$

Our next goal is to show how these risk premium measures r_{TC}, r_{VP} depend on the time lag and other quantities, like e.g., $\hat{\sigma}, S, V$ and derivatives of V. The problem can be decomposed in two parts: modeling the transaction costs measure r_{TC} and volatile portfolio risk measure r_{VP}.

We begin with modeling transaction costs. In practice, we have to adjust our portfolio by frequent buying and selling of assets. In the presence of nontrivial transaction costs, continuous portfolio adjustments may lead to infinite total transaction costs. A natural way how to consider transaction costs within the frame of the Black–Scholes theory is to follow the well known Leland approach extended by Hoggard, Whalley and Wilmott (cf. [64, 75]). Next we recall an idea how to incorporate the effect of transaction costs into the governing equation. More precisely, we will derive the coefficient of transaction costs r_{TC} occurring in (11.11). Let us denote by C the round trip transaction cost per unit dollar of transaction. Then

$$C = (S_{ask} - S_{bid})/S, \qquad (11.12)$$

where S_{ask} and S_{bid} are the so-called ask and bid prices of the asset, i.e., the market price offers for selling and buying assets, respectively. Here $S = (S_{ask} + S_{bid})/2$ denotes the mid value of the underlying asset price.

In order to derive the term r_{TC} in (11.11) measuring transaction costs we will assume, for a moment, that there is no risk from volatile portfolio, i.e., $r_{VP} = 0$. Then $\Delta V + \delta\Delta S = \Delta\Pi = r\Pi\Delta t + \delta q S\Delta t + r_{TC}S\Delta t$. Following Leland's approach (cf. [64]), using Itō's formula and assuming δ-hedging of a synthetised portfolio Π one can derive that the coefficient r_{TC} of transaction costs is given by the formula:

$$r_{TC} = \frac{C\hat{\sigma}S}{\sqrt{2\pi}}|\partial_S^2 V|\frac{1}{\sqrt{\Delta t}} \qquad (11.13)$$

(see [64, Eq. (3)]). It leads to the well known Leland generalization of the Black–Scholes equation (11.2) in which the diffusion coefficient is given by (11.4) (see Chapter 5 for details).

Next we focus our attention to the problem how to incorporate a risk from a volatile portfolio into the model. In the case when a portfolio, consisting of options and assets, is highly volatile, an investor usually asks for a price compensation. Notice that exposure to risk is higher when the time-lag between portfolio adjustments is higher. We shall propose a measure of such a risk based on the volatility of a fluctuating portfolio. It can be measured by the variance of relative increments of the replicating portfolio $\Pi = V + \delta S$, i.e., by the term $Var((\Delta\Pi)/S)$. Hence it is reasonable to define the measure r_{VP} of the portfolio volatility risk as follows:

$$r_{VP} = R\frac{var\left(\frac{\Delta\Pi}{S}\right)}{\Delta t}. \qquad (11.14)$$

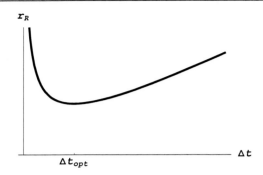

Figure 11.1. The total risk premium $r_R = r_{TC} + r_{VP}$ as a function of the time-lag Δt between two consecutive portfolio adjustments.

In other words, r_{VP} is proportional to the variance of a relative change of a portfolio per time interval Δt. The constant R represents the so-called *risk premium coefficient*. It can be interpreted as the marginal value of investor's exposure to risk. If we apply Itō's formula to the differential $\Delta \Pi = \Delta V + \delta \Delta S$ we obtain $\Delta \Pi = (\partial_S V + \delta) \hat{\sigma} S \Delta W + \frac{1}{2} \hat{\sigma}^2 S^2 \Gamma (\Delta W)^2 + \mathcal{G}$, where $\Gamma = \partial_S^2 V$ and $\mathcal{G} = (\partial_S V + \delta) \rho S \Delta t + \partial_t V \Delta t$ is a deterministic term, i.e., its expected value $E(\mathcal{G}) = \mathcal{G}$ in the lowest order Δt - term approximation. Thus

$$\Delta \Pi - E(\Delta \Pi) = (\partial_S V + \delta) \hat{\sigma} S \phi \sqrt{\Delta t} + \frac{1}{2} \hat{\sigma}^2 S^2 (\phi^2 - 1) \Gamma \Delta t,$$

where ϕ is a random variable with the standard normal distribution such that $\Delta W = \phi \sqrt{\Delta t}$. Hence the variance of $\Delta \Pi$ can be computed as follows:

$$var(\Delta \Pi) = E\left([\Delta \Pi - E(\Delta \Pi)]^2\right)$$
$$= E\left([(\partial_S V + \delta) \hat{\sigma} S \phi \sqrt{\Delta t} + \frac{1}{2} \hat{\sigma}^2 S^2 \Gamma (\phi^2 - 1) \Delta t]^2\right).$$

Similarly, as in the derivation of the transaction costs measure r_{TC} we assume the δ-hedging of portfolio adjustments, i.e., we choose $\delta = -\partial_S V$. Since $E((\phi^2 - 1)^2) = 2$ for a normally distributed random variable $\phi \sim N(0,1)$, we obtain an expression for the risk premium r_{VP} in the form:

$$r_{VP} = \frac{1}{2} R \hat{\sigma}^4 S^2 \Gamma^2 \Delta t. \tag{11.15}$$

Notice that in our approach the increase in the time-lag Δt between consecutive transactions leads to a linear increase of the risk from a volatile portfolio where the coefficient of proportionality depends the asset price S, option's Gamma, $\Gamma = \partial_S^2 V$, as well as the constant historical volatility $\hat{\sigma}$ and the risk premium coefficient R via equation (11.15).

11.2.1. Nonlinear Black–Scholes Equation for the Risk Adjusted Pricing Model

The total risk premium $r_R = r_{TC} + r_{VP}$ consists of two parts: transaction costs premium r_{TC} and the risk from a volatile portfolio r_{VP} premium defined as in (11.13) and (11.15), respectively. We assume that an investor is risk aversive and he/she wants to minimize the value of the total risk premium r_R. For this purpose one has to choose the optimal time-lag Δt between two consecutive portfolio adjustments. As both r_{TC} as well as r_{VP} depend on the time-lag Δt so does the total risk premium r_R. In order to find the optimal value of Δt we have to minimize the following function:

$$\Delta t \mapsto r_R = r_{TC} + r_{VP} = \frac{C|\Gamma|\hat{\sigma} S}{\sqrt{2\pi}} \frac{1}{\sqrt{\Delta t}} + \frac{1}{2} R\hat{\sigma}^4 S^2 \Gamma^2 \Delta t.$$

The unique minimum of the function $\Delta t \mapsto r_R(\Delta t)$ depicted in Fig. 11.1 is attained at the time-lag $\Delta t_{opt} = K^2/(\hat{\sigma}^2 |S\Gamma|^{\frac{2}{3}})$ where $K = (C/(R\sqrt{2\pi})^{\frac{1}{3}}$. Therefore, for the minimal value of the function $\Delta t \mapsto r_R(\Delta t)$ we have

$$r_R(\Delta t_{opt}) = \frac{3}{2} \left(\frac{C^2 R}{2\pi} \right)^{\frac{1}{3}} \hat{\sigma}^2 |S\Gamma|^{\frac{4}{3}}. \tag{11.16}$$

Taking into account both transaction costs as well as risk from a volatile portfolio effects we have shown that the equation for the change $\Delta \Pi$ of a portfolio Π read as:

$$\Delta V + \delta \Delta S = \Delta \Pi = r\Pi \Delta t - \delta q S \Delta t - r_R S \Delta t,$$

where r_R represents the total risk premium, $r_R = r_{TC} + r_{VP}$. Applying Itō's lemma to a smooth function $V = V(S, t)$ and assuming the δ-hedging strategy for the portfolio adjustments we finally obtain the following generalization of the Black–Scholes equation for valuing options:

$$\frac{\partial V}{\partial t} + \frac{\hat{\sigma}^2}{2} S^2 \frac{\partial^2 V}{\partial S^2} + (r - q) S \frac{\partial V}{\partial S} - rV - r_R S = 0.$$

By taking the optimal value of the total risk coefficient r_R derived as in (11.16) the option price V is a solution to the following nonlinear parabolic equation:

(*Risk adjusted pricing methodology Black–Scholes equation*)

$$\frac{\partial V}{\partial t} + \frac{\hat{\sigma}^2}{2} S^2 \left(1 + \mu (S\partial_S^2 V)^{\frac{1}{3}} \right) \frac{\partial^2 V}{\partial S^2} + (r - q) S \frac{\partial V}{\partial S} - rV = 0, \tag{11.17}$$

where $\mu = 3 \left(\frac{C^2 R}{2\pi} \right)^{\frac{1}{3}}$ and Γ^p with $\Gamma = S\partial_S^2 V$ and $p = 1/3$ stands for the signed power function, i.e., $\Gamma^p = |\Gamma|^{p-1}\Gamma$. In the case there are neither transaction costs ($C = 0$) nor the risk from a volatile portfolio ($R = 0$) we have $\mu = 0$. Then equation (11.17) reduces to the original Black–Scholes linear parabolic equation (11.2). We note that equation (11.17) is a backward parabolic PDE if and only if the function

$$\beta(H) = \frac{\hat{\sigma}^2}{2}(1 + \mu H^{\frac{1}{3}})H \tag{11.18}$$

is an increasing function in the variable $H := S\Gamma = S\partial_S^2 V$. It is clearly satisfied if $\mu \geq 0$ and $H \geq 0$.

11.2.2. Derivation of the Gamma Equation

Our next goal is to transform the fully nonlinear parabolic equation (11.17) into a quasilinear parabolic equation for which one can construct an effective numerical scheme for approximation of the solution $V(S,t)$. Equation (11.17) can be rewritten in the form

$$\partial_t V + S\beta(S\Gamma) = (r-q)(V - S\partial_S V) + qV, \quad S > 0, t \in (0,T), \tag{11.19}$$

where $\Gamma = \partial_S^2 V$. Using the standard change of independent variables: $x = \ln(S/E), x \in \mathbb{R}, \tau = T - t, \tau \in (0,T)$. Since the above equation contains the term $S\Gamma = S\partial_S^2 V$ it is convenient to introduce the following transformation:

$$H(x,\tau) = S\Gamma = S\partial_S^2 V(S,t). \tag{11.20}$$

Now we are in position to derive an equation for the function H. It turns out that the function $H(x,\tau)$ is a solution to a nonlinear parabolic equation subject to the initial and boundary conditions. More precisely, by taking the second derivative of equation (11.17) with respect to x we obtain, after some calculations, that $H = H(x,\tau)$ is a solution to the quasilinear parabolic equation

$$\frac{\partial H}{\partial \tau} = \frac{\partial^2}{\partial x^2}\beta(H) + \frac{\partial}{\partial x}\beta(H) + (r-q)\frac{\partial H}{\partial x} - qH, \tag{11.21}$$

$\tau \in (0,T), x \in \mathbb{R}$ (see [66]). Henceforth, we will refer to (11.21) as the Γ equation. A solution H to (11.21) is subjected to the initial condition at $\tau = 0$:

$$H(x,0) = \bar{H}(x), \quad x \in \mathbb{R}, \tag{11.22}$$

where $\bar{H}(x)$ is the Dirac δ function $\bar{H}(x) = \delta(x)$. Recall that the Dirac function is a function in distributional sense such that

$$\int_{-\infty}^{\infty} \delta(x - x_0)\phi(x)dx = \phi(x_0), \quad \int_{-\infty}^{\infty} \delta(x)dx = 1,$$

for any smooth function ϕ. For the purpose of construction of a numerical scheme we approximate the initial Dirac delta function by the function $\bar{H}(x) = N'(d)/(\hat{\sigma}\sqrt{\tau^*})$ where $\tau^* > 0$ is sufficiently small, $N(d)$ is the cumulative distribution function of the normal distribution, and $d = \left(x + (r - q - \hat{\sigma}^2/2)\tau^*\right)/\hat{\sigma}\sqrt{\tau^*}$. It corresponds to the value $H = S\partial_S^2 V$ of a call (put) option valued by a linear Black–Scholes equation with a constant volatility $\hat{\sigma} > 0$ at the time $T - \tau^*$ close to expiry T when the time parameter $0 < \tau^* \ll 1$ is sufficiently small. In the case of call or put options the function H is subjected to boundary conditions at $x = \pm\infty$,

$$H(-\infty,\tau) = H(\infty,\tau) = 0, \quad \tau \in (0,T). \tag{11.23}$$

It is important to emphasize that the solution $V(S,t)$ can be directly computed from the function $H = H(x,\tau)$ defined in (11.20). Indeed, in the case of a call option we have

$$\partial_S V(S,t) = \partial_S V(0,t) + \int_0^S \frac{1}{s}H(\ln(s/E), T-t)ds = \int_{-\infty}^{\ln(S/E)} H(x, T-t)dx$$

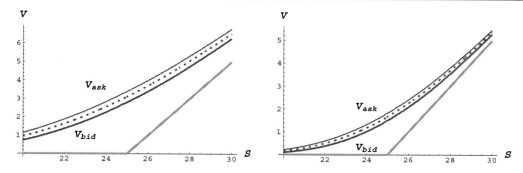

Figure 11.2. A comparison of bid and ask option prices computed by means of the RAPM model. The middle dotted line is the option price computed from the Black–Scholes equation. We chose $\sigma = 0.3, \mu = 0.2, r = 0.011, E = 25$ and $T = 1$ (left) and $T = 0.3$ (right). Source: Jandačka and Ševčovič [66].

from which we deduce, by integration,

(call option) $$V(S,t) = \int_{-\infty}^{\infty} (S - Ee^x)^+ H(x, T-t) dx, \qquad (11.24)$$

because $\partial_S V(0,t) = V(0,t) = 0$ for the call option. Similarly, for the put option we have

(put option) $$V(S,t) = \int_{-\infty}^{\infty} (Ee^x - S)^+ H(x, T-t) dx. \qquad (11.25)$$

11.2.3. Pricing of European Style of Options by the RAPM Model

Let us denote $V(S,t;C,\hat{\sigma},R)$ the value of a solution to (11.17) with parameters $C,\hat{\sigma},R$. Suppose that the coefficient of transaction costs C is known from the analysis of bid-ask spreads of the underlying stock price. The constant $C \geq 0$ is given by (11.12). In real option market data we can observe different bid and ask prices for an option, $V_{bid} < V_{ask}$, respectively. Let us denote by V_{mid} the mid value, i.e., $V_{mid} = \frac{1}{2}(V_{bid} + V_{ask})$. By the RAPM model we are able to explain such a bid-ask spread in option prices. The higher ask price corresponds to a solution of the RAPM model with some nontrivial risk premium $R > 0$ and $C > 0$ whereas the mid value V_{mid} corresponds to a solution $V(S,t)$ for vanishing risk premium $R = 0$, i.e., to a solution of the linear Black–Scholes equation (11.2). An illustrative example of bid-ask spreads captured by the RAPM model is shown in Fig. 11.2.

To calibrate the RAPM model we seek for a couple $(\hat{\sigma}_{RAPM}, R)$ of implied RAPM volatility $\hat{\sigma}_{RAPM}$ and the risk premium R such that $V_{ask} = V(S,t;C,\hat{\sigma}_{RAPM},R)$ and $V_{mid} = V(S,t;C,\hat{\sigma}_{RAPM},0)$. Such a system of two nonlinear equations for unknowns $\hat{\sigma}_{RAPM}$ and R can be easily solved by means of the Newton-Kantorovich iterative method (cf. [66]). Notice that $\hat{\sigma}_{RAPM}$ is, in fact, the implied volatility computed from the mid option price V_{mid} (see Chapter 4).

As an example we considered sample data sets for call options on Microsoft stocks. We considered a flat interest rate $r = 0.02$, a constant transaction cost coefficient $C = 0.01$ estimated from (11.12) and we assumed that the underlying asset pays no dividends, i.e., $q =$

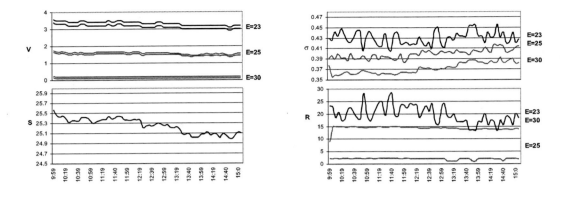

Figure 11.3. Intra-day behavior of Microsoft stocks (April 4, 2003) and shortly expiring call options with expiry date April 19, 2003. Computed implied volatilities $\hat{\sigma}_{RAPM}$ and risk premium coefficients R. Source: Jandačka and Ševčovič [66].

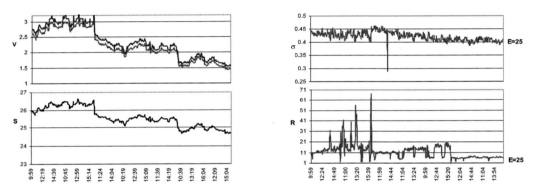

Figure 11.4. One week behavior of Microsoft stocks (March 20 - 27, 2003) and call options with expiration date April 19, 2003. Computed implied volatilities $\hat{\sigma}_{RAPM}$ and risk premiums R. Source: Jandačka and Ševčovič [66].

0. In Fig. 11.3 we present results of calibration of implied couple $(\hat{\sigma}_{RAPM}, R)$. Interestingly enough, two call options with higher strike prices $E = 25, 30$ had almost constant implied risk premium R. On the other hand, the risk premium R of an option with lowest $E = 23$ was highly fluctuating.

Finally, in Fig. 11.4 we present one week behavior of implied volatilities and risk premium coefficients for the Microsoft call option on $E = 25$ expiring at $T =$ April 19, 2003. In the beginning of the investigated period the risk premium coefficient R was rather high and fluctuating. On the other hand, it tends to a flat value of $R \approx 5$ at the end of the week. Interesting feature can be observed at the end of the second day when both stock and option prices went suddenly down. The time series analysis of the implied volatility $\hat{\sigma}_{RAPM}$ from first two days was unable to predict such a behavior. On the other hand, high fluctuation in the implied risk premium R during first two days can send a signal to an investor that

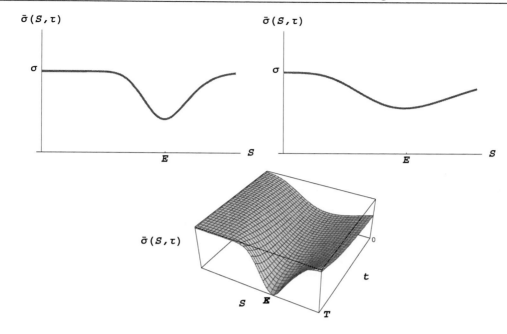

Figure 11.5. Explanation of the volatility smile by the RAPM model. Dependence of $\bar{\sigma}(S,t)$ on S is depicted in (top-left) for t close to T and, in (top-right), for a time $0 < t \ll T$. The mapping $(S,t) \mapsto \bar{\sigma}(S,t)$ is shown in (bottom). Source Jandačka and Ševčovič [66].

sudden changes can be expected in the near future.

11.2.4. Explanation of the Volatility Smile by the RAPM Model

One of the most striking phenomena in the Black-Scholes theory is the so-called *volatility smile* phenomenon. Notice that derivation of the classical Black-Scholes equation relies on the assumption of a constant value of the volatility σ. On the other hand, as it might be documented by many examples observed in market options data sets (see e.g., [10, 118]) such an assumption is often violated. More precisely, the implied volatility σ_{impl} is no longer constant and it can depend on the asset price S, the strike price E as well as the time t.

Following Kratka [73] and Jandačka and Ševčovič [66], we are able to explain the volatility smile analytically by taking into account the RAPM approach. The Risk adjusted Black-Scholes equation (11.17) can be viewed as an equation with a variable volatility coefficient, i.e.

$$\partial_t V + \frac{\bar{\sigma}^2(S,t)}{2} S^2 \Gamma = r(V - S\partial_S V)$$

where $\Gamma = \partial_S^2 V$ and the volatility $\bar{\sigma}^2(S,t)$ depends itself on a solution $V = V(S,t)$ as follows:

$$\bar{\sigma}^2(S,t) = \sigma^2 \left(1 - \mu(S\Gamma)^{1/3}\right). \tag{11.26}$$

In Fig. 11.5 we show the dependence of the function $\bar{\sigma}(S,t)$ on the asset price S and time t. It should be obvious that the function $S \mapsto \bar{\sigma}(S,t)$ has a convex shape near the exercise price E. We have used the RAPM model in order to compute values of $\Gamma = \partial_S^2 V$. We chose $\mu = 0.2, \sigma = 0.3, r = 0.011$, and $T = 0.5$.

With regard to scale invariance property of the RAPM model (see section 2.7) if we express both the asset price S as well as the option price V in terms of units of E (i.e., we introduce scaling $s \leftrightarrow S/E$ and $v \leftrightarrow V/E$) then the volatility $\bar{\sigma}$ defined as in (11.26) is a function of the ratio $s = S/E$ and time t only.

11.3. Modeling Feedback Effects

In this section we present the model for pricing derivative securities in the presence of feedback effects. The model was derived by Frey in [50]. Its mathematical representation is again a nonlinear generalization of the Black–Scholes equation with a volatility depending on the second derivative of the option prices itself. We follow the derivation of the model as it was presented in [50].

The classical Black–Scholes equation was derived under the crucial assumption made on the completeness and perfect liquidity of financial markets (see Chapter 3). The standard perfect liquid market assumption of the Black–Scholes theory claiming that an investor can trade any large amount of assets without influencing the underlying asset price may fail. However, debacles of derivative contracts pointed out the importance of the market liquidity, which is of high concern to large investors and risk managers.

In a stylized market with riskless assets like e.g., bonds and risky stocks we consider a trader whose goal is to replicate a derivative contract with expiration T by using a dynamic trading strategy given by a pair of adapted processes (α_t, β_t). Here α_t (β_t) denotes the number of risky shares (riskless bonds) in the portfolio at time t. We will construct a model for the case when our hedger is a large trader who can influence the price of the stock by her hedging strategy.

We will assume that the stock-holding process α_t is left continuous, i.e., $\alpha_t = \lim_{s \to t^-} \alpha_s$ and the right continuous process $\alpha_t^+ := \lim_{s \to t^+} \alpha_s$ is a semimartingale (cf. [50]). Furthermore, we will suppose that the downward jumps of the stock-holding strategy are bounded from bellow,

$$\alpha_t^+ - \alpha_t > -\frac{1}{\bar{\rho}},$$

for some constant $\bar{\rho} > 0$. Next, suppose that our large trader uses stock-holding strategy α_t and the underlying stock prices process satisfies the following SDEs

$$dS_t = \mu S_t dt + \sigma S_t dW_t + \rho S_t d\alpha_t, \tag{11.27}$$

where μ is a drift parameter, $\sigma > 0$ is the volatility of the process and $0 \leq \rho < \bar{\rho}$ is the so-called market liquidity parameter. It is worth noting that the quantity $1/(\rho S_t)$ measures the size of the change in the stock-holding position of our large trader. Notice that if $\alpha_t \equiv 0$ or $\rho = 0$ then the stock price S_t follows the geometric Brownian motion as described in Chapter 2.

The dynamic hedging strategy α_t of our large trader is assumed to be a smooth function $\phi(S_t,t)$ of the time t and the current stock price S_t,

$$\alpha_t = \phi(S_t,t).$$

In order to guarantee nondegeneracy of the stochastic differential equation for S we will suppose that the following regularity assumption $\rho S \frac{\partial \phi}{\partial S}(S,t) < 1$ holds for any $S > 0, t \in (0,T)$.

Let us emphasize that the differential of the underlying stock price dS_t is given implicitly in equation (11.27) because $d\alpha_t$ depends on dS_t via the hedging strategy $\alpha_t = \phi(S_t,t)$. In what follows, we will show that the stock process for S_t satisfies the SDE having the explicit form:

$$dS_t = b(S_t,t)S_t dt + v(S_t,t)S_t dW_t. \tag{11.28}$$

To derive the form of the volatility function $v(S,t)$ and the drift term $b(S,t)$ we make use of Itō's lemma applied to the function $\alpha_t = \phi(S_t,t)$. We have

$$d\alpha_t = \left(\frac{\partial \phi}{\partial t} + \frac{v^2}{2} S_t^2 \frac{\partial^2 \phi}{\partial S^2}\right) dt + \frac{\partial \phi}{\partial S} dS_t. \tag{11.29}$$

Inserting the differential $d\alpha_t$ into (11.27) we obtain

$$\left[1 - \rho S_t \frac{\partial \phi}{\partial S}\right] dS_t = \mu S_t dt + \sigma S_t dW_t + \rho S_t \left(\frac{\partial \phi}{\partial t} + \frac{v^2}{2} S_t^2 \frac{\partial^2 \phi}{\partial S^2}\right) dt.$$

Therefore

$$v(S,t) = \frac{\sigma}{1 - \rho S \frac{\partial \phi}{\partial S}},$$

$$b(S,t) = \frac{1}{1 - \rho S \frac{\partial \phi}{\partial S}} \left(\mu + \rho \left(\frac{\partial \phi}{\partial t} + \frac{v^2}{2} S^2 \frac{\partial^2 \phi}{\partial S^2}\right)\right). \tag{11.30}$$

We can derive the partial differential equation for the option price $V = V(S,t)$ as a derivative security of the underlying stock price S_t driven by the stochastic process (11.28). It can be readily done by following the derivation of the classical Black–Scholes equation in Chapter 3 and replacing the constant volatility σ by the volatility function $v = v(S,t)$. The resulting equation for the option price $V(S,t)$ has the form

$$\frac{\partial V}{\partial t} + \frac{1}{2} v(S,t)^2 S^2 \frac{\partial^2 V}{\partial S^2} + (r-q)S\frac{\partial V}{\partial S} - rV = 0.$$

In order to simplify further notations, in the rest of this section, we shall assume that the risk-free interest rate and dividend yield are zero, i.e., $r = 0, q = 0$. In such case, the Black–Scholes equation reduces to:

$$\frac{\partial V}{\partial t} + \frac{1}{2} v(S,t)^2 S^2 \frac{\partial^2 V}{\partial S^2} = 0, \quad S > 0, t \in (0,T). \tag{11.31}$$

The solution is subject to the terminal pay-off diagram $V(S,T)$ at expiry T corresponding to either call option ($V(S,T) = (S-E)^+$) or put option ($V(S,T) = (E-S)^+$).

Our purpose is to specify the trading stock-holding strategy α_t of our large trader. Following the idea due to Frey [50], we will construct the hedging strategy $\alpha_t = \phi(S_t,t)$ taking into account the goal of vanishing the so-called tracking error of the strategy. By definition (see [50, (3.1)]), the tracking error e_T^M of the strategy α_t is given by

$$e_T^M = V(S_T,T) - V_T^M := V(S_T,T) - \left(V_0 + \int_0^T \alpha_t dS_t\right). \quad (11.32)$$

By superscript M we have denoted mark-to-market values not accounting for liquidation costs. It should be obvious that a positive (negative) value of the tracking error indicates loss (profit) in hedge done by α_t strategy. Therefore, it measures the profit or loss (P&L) function of a trader who is using the hedging strategy α_t.

11.3.1. The Case of the Standard Black–Scholes Delta Hedging Strategy

First we consider the standard Black–Scholes delta hedging strategy as it was introduced in Chapter 3. Let V^{bs} be the solution of the linear Black–Scholes equation with a constant volatility σ of the underlying stock process σ. It means that the function V^{bs} is a solution to the equation:

$$\frac{\partial V^{bs}}{\partial t} + \frac{\sigma^2}{2}S^2 \frac{\partial^2 V^{bs}}{\partial S^2} = 0, \quad S > 0, t \in (0,T). \quad (11.33)$$

Notice that we have assumed zero risk-free interest rate and dividend yield. The solution V^{bs} is subject to the terminal pay-off diagram $V(S,T)$ representing the chosen derivative security, e.g., a call or put option.

Now suppose that a large trader chooses the trading strategy $\alpha_t = \phi(S_t,t)$ given by the simple delta hedging strategy based on the V^{bs} - a solution to the Black–Scholes equation (11.33), i.e.

$$\phi(S,t) = \frac{\partial V^{bs}}{\partial S}(S,t).$$

Following [50, Proposition 3.2] we will show that the tracking error e_T^M at expiry T for the large trader is always positive. Indeed, the difference between the terminal pay-off value $V_T = V(S_T,T)$ and the initial value $V_0 = V(S_0,0)$ can be expressed by the integral

$$V(S_T,T) - V(S_0,0) = \int_0^T dV^{bs}(S_t,t).$$

Now assuming the process S_t follows the SDEs (11.28) and applying Itō's lemma to the differential $dV^{bs}(S_t,t)$ we obtain

$$V(S_T,T) - V(S_0,0) = \int_0^T \frac{\partial V^{bs}}{\partial S}(S_t,t)dS_t$$
$$+ \int_0^T \left(\frac{\partial V^{bs}}{\partial t}(S_t,t) + \frac{v^2(S_t,t)}{2}S_t^2 \frac{\partial^2 V^{bs}}{\partial S^2}(S_t,t)\right) dt.$$

If we express the derivative $\frac{\partial V^{bs}}{\partial t}$ from the Black–Scholes equation (11.33) and insert the expression (11.30) for $v(S,t)$ we obtain

$$e_T^M = \frac{\sigma^2}{2} \int_0^T \left[\frac{1}{(1-\rho S \frac{\partial^2 V^{bs}}{\partial S^2})^2} - 1\right] S_s^2 \frac{\partial^2 V^{bs}}{\partial S^2}(S_s,s)ds.$$

For $\rho > 0$, it is easy to verify that the above integrand term is always positive for both positive or negative values of $\frac{\partial^2 V^{bs}}{\partial S^2}$. Hence the tracking error e_T^M for the standard Black–Scholes delta hedging strategy is always positive. In other words, in non-perfectly liquid markets the Black–Scholes delta hedging is a costly strategy for a large trader (cf. [50, 52]).

11.3.2. Zero Tracking Error Strategy and the Nonlinear Black–Scholes

Now the idea of derivation of the nonlinear Black–Scholes equation comes very quickly. It is based on construction of the zero tracking error strategy α_t given by the delta of the price V solving a nonlinear generalization of the Black–Scholes equation. More precisely, it is given by function $\alpha_t = \phi(S_t, t)$, where

$$\phi(S,t) = \frac{\partial V}{\partial S}(S,t)$$

and V is a solution of the Black–Scholes equation (11.31) with the volatility function $v(S,t)$ given by

$$v(S,t) = \frac{\sigma}{1 - \rho S \frac{\partial^2 V}{\partial S^2}(S,t)}.$$

Suppose that the option price $V = V(S,t)$ is a solution to the nonlinear parabolic equation

$$\frac{\partial V}{\partial t} + \frac{1}{2} \frac{\sigma^2}{\left[1 - \rho S \frac{\partial^2 V}{\partial S^2}\right]^2} S^2 \frac{\partial^2 V}{\partial S^2} = 0, \quad S > 0, t \in (0,T) \tag{11.34}$$

and it is subjected to the terminal pay-off diagram $V(S,T)$ corresponding to the chosen type of derivative security, e.g., a call or put option. According to [50, Proposition 4.2] the self-financing strategy $\alpha_t = \frac{\partial V}{\partial S}(S_t, t)$ is a perfect replication strategy, i.e., its tracking error e_T^M is equal to zero.

For further details concerning the model and its qualitative and quantitative analysis the reader is referred to the paper by Frey [50]. Notice that a characterization of option replicating strategies for a large trader in terms of a nonlinear parabolic differential equation have been also obtained by other authors including Schönbucher and Wilmott [99], Frey [49] and Sircar and Papanicolaou [108].

11.3.3. The Gamma Equation for Frey's Nonlinear Model

Similarly as in the case of the RAPM model (see section 11.2.2.) we can construct a quasi-linear parabolic equation for the new variable $H(x,\tau) = S\partial_S^2 V(S,t)(=S\Gamma)$.

Equation (11.34) can be rewritten in the form

$$\partial_t V + S\beta(S\Gamma) = 0, \quad S > 0, t \in (0,T), \tag{11.35}$$

where $\Gamma = \partial_S^2 V$ and

$$\beta(H) = \frac{\sigma^2}{2} \frac{H}{[1 - \rho H]^2}. \tag{11.36}$$

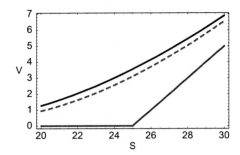 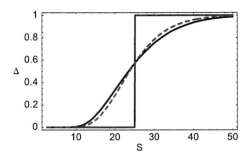

Figure 11.6. A comparison of the call option price $V(S,0)$ computed by the linear Black–Scholes equation (dashed line) and that of the nonlinear Frey's model (solid line). The corresponding profile of the $\Delta = \frac{\partial V}{\partial S}$ of the call option (right).

Again using the change of independent variables: $x = \ln(S/E), x \in \mathbb{R}, \tau = T - t, \tau \in (0,T)$, we end up with the following quasilinear parabolic equation for the transformed variable $H = H(x,\tau)$:

$$\frac{\partial H}{\partial \tau} = \frac{\partial^2}{\partial x^2} \beta(H) + \frac{\partial}{\partial x} \beta(H), \qquad (11.37)$$

$\tau \in (0,T), x \in \mathbb{R}$. In the case of a call or put option the solution H to (11.37) is subjected to the initial condition at $\tau = 0$:

$$H(x,0) = \bar{H}(x), \quad x \in \mathbb{R}, \qquad (11.38)$$

where $\bar{H}(x)$ is the Dirac delta function $\bar{H}(x) = \delta(x)$. The solution H is subjected to boundary conditions at $x = \pm\infty$,

$$H(-\infty,\tau) = H(\infty,\tau) = 0, \quad \tau \in (0,T). \qquad (11.39)$$

Once the solution $H(x,\tau)$ is known then the option price $V = V(S,t)$ is given by the formula (11.24) for the call option case or (11.25) for the put option case. A numerical discretization scheme of the Γ equation (11.37) will be derived and discussed in subsequent section 11.6.

In Fig. 11.6 we present a comparison of the call option price profile $V(S,0)$ and its delta $\Delta = \frac{\partial V}{\partial S}$ for the linear Black–Scholes equation and the nonlinear Frey's model with the parameter $\rho = 0.1$. We chose the model parameters: $E = 25, \sigma = 0.3, T = 1, r = q = 0$.

11.4. Modeling Investor's Preferences

In this section we recall a nonlinear model derived by Barles and Soner in [11]. It again leads to a nonlinear generalization of the Black–Scholes pricing equation. The model is derived for a market with transaction costs. In general, there is no nontrivial portfolio dominating a contingent claim. Following the ideas adopted from the paper by Hodges and Neuberger [63], they introduced investor's preferences into the pricing model. The investor's preferences are described by a given utility function with a constant investor's risk aversion. The resulting pricing equation is a nonlinear Black-Scholes equation with an adjusted volatility. Similarly as in the case of the Risk adjusted methodology and Frey's model the adjusted volatility is a function of the second derivative of the option price itself.

We again assume that the underlying stock price pays no dividends ($q = 0$) and it follows a geometric Brownian motion

$$dS = \rho S dt + \hat{\sigma} S dW, \tag{11.40}$$

with a drift ρ is a drift, $\hat{\sigma}$ is the volatility of the asset price and W is the standard Wiener process.

We follow the original derivation of the model due to Barles and Soner [11]. In order to model trading strategies, let us introduce two stochastic processes X_t and Y_t describing dollar holdings in the money market and the number of shares of stocks owned. By a trading strategy on the time interval $[t, T]$ we will understand a pair (L_t, M_t) of left continuous and nondecreasing functions such that $L_t = M_t = 0$. These functions can be interpreted as, respectively, the cumulative transfers (measured in the number of shares of a stock) from money market to stock (the function L_t) and vice versa (the function M_t). Let $\mu \in (0, 1)$ be a parameter describing proportional transaction costs when selling or buying assets. It means that, the ask price S_{ask} (an offer to sale) and the bid price S_{bid} are related as follows:

$$S_{ask} = (1+\mu)S, \qquad S_{bid} = (1-\mu)S,$$

where $S = (S_{ask} + S_{bid})/2$ is the average between ask and bid prices. The increments dX and dY in the dollar amount money market instruments and number of shares in the portfolio, respectively, are then expressed as follows:

$$dX = -S(1+\mu)dL + S(1-\mu)dM, \tag{11.41}$$
$$dY = dL - dM. \tag{11.42}$$

In the case when there are no option liabilities, investor's terminal wealth at $t = T$ of the portfolio will be equal to the value

$$X_T + Y_T S_T.$$

On the other hand, in the case we have sold N European call options the terminal wealth at $t = T$ of the portfolio will be

$$X_T + Y_T S_T - N(S_T - E)^+.$$

The idea how to price a European call option is based on the utility maximization approach of Hodges and Neuberger [63]. Consider a given utility function U. We will assume that the function U is an increasing and concave function and it corresponds to the so-called constant absolute risk aversion function for which we have

$$-\frac{U''(r)}{U'(r)} = \gamma, \qquad \text{for all } r > 0.$$

The constant $\gamma > 0$ is referred to as the constant absolute risk aversion parameter. In what follows, we will assume that the investor's utility function U^ε is given by

$$U^\varepsilon(\xi) = U(\xi/\varepsilon), \quad \text{for } \xi > 0,$$

where
$$U(r) = 1 - \exp(-r)$$
and the parameter
$$\varepsilon = \frac{1}{\gamma N} \tag{11.43}$$

is considered as a small parameter, $0 < \varepsilon \ll 1$. Clearly, the exponential function U^ε is a constant absolute risk aversion utility function.

Following Hodges and Neuberger [63], we shall price the European call option with maturity T and the strike price E by taking into account the goal of maximization of investor's utility function. In the first case when there are no option liabilities, investor's value (wealth) function $v^f = v^f(x,y,s,t)$ is given by a solution to the stochastic dynamic optimization problem
$$v^f(x,y,s,t) = \sup_{L,M} \mathbb{E}(U(X_T + Y_T S_T)) \tag{11.44}$$

for maximization of the terminal expected utility from the final wealth with respect to all trading strategies $(L_\tau, M_\tau), \tau \in [t,T]$, subject to the initial state conditions:
$$X_t = x, Y_t = y, S_t = s.$$

In the second case when we have sold N European call options, investor's value (wealth) function $v(x,y,s,t)$ is given by a solution to the stochastic dynamic optimization problem
$$v(x,y,s,t) = \sup_{L,M} \mathbb{E}(U(X_T + Y_T S_T - N(S_T - E)^+)). \tag{11.45}$$

The maximization of the expected utility from investor's final wealth is again taken with respect to all trading strategies $(L_\tau, M_\tau), \tau \in [t,T]$, and the initial conditions $X_t = x, Y_t = y, S_t = s$ at time t.

Hodges and Neuberger in [63] postulate that the price of the call option can be identified to the maximal solution $\Lambda = \Lambda(x,y,s,t;\gamma,N)$ of the algebraic equation
$$v(x + N\Lambda, y, s, t) = v^f(x,y,s,t). \tag{11.46}$$

It is worth noting that the state equations (11.41) are linear. As a consequence we can deduce the following scaling property of the option price $\Lambda(x,y,s,t;\gamma,N)$:
$$\Lambda(x,y,s,t;\gamma,N) = \Lambda(Nx, Ny, s, t; \gamma N, 1).$$

In what follows, we will define two auxiliary functions $z^\varepsilon(x,y,s,t)$ and $z^{f,\varepsilon}(x,y,s,t)$ using the following implicit equations:
$$v(x,y,s,t) = U^\varepsilon(x+ys-z^\varepsilon), \tag{11.47}$$
$$v^f(x,y,s,t) = U^\varepsilon(x+ys-z^{\varepsilon,f}). \tag{11.48}$$

At the terminal time $t = T$ we have
$$z^\varepsilon(x,y,s,T) = (s-E)^+, \qquad z^{\varepsilon,f}(x,y,s,T) = 0, \quad \text{for any } x,y,s.$$

Now, it follows from equation (11.46) that the option price Λ is a solution to the equation $U^\varepsilon(x+\Lambda N+ys-z^\varepsilon) = U^\varepsilon(x+ys-z^{\varepsilon,f})$. Since U^ε is an increasing function we have

$$\Lambda(x,y,s,t;\frac{1}{\varepsilon},1) = z^\varepsilon(x,y,s,t) - z^{f,\varepsilon}(x,y,s,t). \qquad (11.49)$$

Our next goal is to determine the equation for the value functions v and v^f. The idea is to adopt the Bellman optimality principle according to which we have

$$v(x,y,s,t) = \sup_{L,M} \mathbb{E}(v(X_{t+dt}, Y_{t+dt}, S_{t+dt}, t+dt)),$$

where the maximization of the expected utility is taken with respect to all trading strategies $(L_\tau, M_\tau), \tau \in [t, t+dt]$ and the initial conditions $X_t = x, Y_t = y, S_t = s$. Here $dt > 0$ is an arbitrary time increment. A similar equation holds for the value function v^f.

Using Itō's lemma for the increment $dv = v(X_{t+dt}, Y_{t+dt}, S_{t+dt}, t+dt) - v(x,y,s,t)$ over the time interval $[t, t+dt]$ we obtain

$$dv = \left(\frac{\partial v}{\partial t} + \rho s \frac{\partial v}{\partial s} + \frac{\hat{\sigma}^2 s^2}{2}\frac{\partial^2 v}{\partial s^2} + \frac{\partial v}{\partial x}\frac{dX}{dt} + \frac{\partial v}{\partial y}\frac{dY}{dt}\right)dt + \hat{\sigma}s\frac{\partial v}{\partial s}dW.$$

Taking into account equations (11.41) we have

$$\frac{dX}{dt} = -s(1+\mu)\dot{L} + s(1-\mu)\dot{M}, \qquad \frac{dY}{dt} = \dot{L} - \dot{M},$$

where $\dot{L} = \frac{dL}{dt}$ and $\dot{M} = \frac{dM}{dt}$. As the trading strategies L, M are assumed to be nondecreasing we have $\dot{L}, \dot{M} \geq 0$. With regard to the construction of Itō's integral (see Chapter 2) we have $\mathbb{E}(\hat{\sigma}s\frac{\partial v}{\partial s}dW) = 0$. Since $\mathbb{E}(v(X_{t+dt}, Y_{t+dt}, S_{t+dt}, t+dt)) = v(x,y,s,t) + \mathbb{E}(dv)$ we obtain

$$0 = \inf_{\dot{L},\dot{M}\geq 0} -\frac{\partial v}{\partial t} - \rho s\frac{\partial v}{\partial s} - \frac{\hat{\sigma}^2 s^2}{2}\frac{\partial^2 v}{\partial s^2} + \left(s(1+\mu)\frac{\partial v}{\partial x} - \frac{\partial v}{\partial y}\right)\dot{L} + \left(\frac{\partial v}{\partial y} - s(1-\mu)\frac{\partial v}{\partial x}\right)\dot{M}.$$

It is straightforward to verify the following simple implication:

$$\inf_{\dot{L},\dot{M}\geq 0} \mathcal{A} + \mathcal{B}\dot{L} + \mathcal{C}\dot{M} = 0 \quad \Rightarrow \quad \min(\mathcal{A}, \mathcal{B}, \mathcal{C}) = 0.$$

As a consequence, we obtain the governing equation for the value function v in the following form:

$$\min\left(-\frac{\partial v}{\partial t} - \rho s\frac{\partial v}{\partial s} - \frac{\hat{\sigma}^2 s^2}{2}\frac{\partial^2 v}{\partial s^2};\ s(1+\mu)\frac{\partial v}{\partial x} - \frac{\partial v}{\partial y};\ \frac{\partial v}{\partial y} - s(1-\mu)\frac{\partial v}{\partial x}\right) = 0.$$

Using the implicit equation for the auxiliary function $z(x,y,s,t)$ and taking into account the fact that the utility function U^ε is increasing we deduce the dynamic programming equation for the function z, i.e.

$$\max\left(-\frac{\partial z}{\partial t} - \frac{\hat{\sigma}^2 s^2}{2}\frac{\partial^2 z}{\partial s^2} - \frac{\hat{\sigma}^2 s^2}{2\varepsilon}\left(\frac{\partial z}{\partial s} - y\right)^2 - \rho s\left(\frac{\partial z}{\partial s} - y\right);\right.$$

$$-\frac{\partial z}{\partial y} - \mu s + s(1+\mu)\frac{\partial z}{\partial x}; \qquad (11.50)$$

$$\left.\frac{\partial z}{\partial y} - \mu s - s(1-\mu)\frac{\partial z}{\partial x}\right) = 0.$$

Since neither the terminal condition at $t = T$ for the function z nor the coefficients of (11.50) depend on the x variable one can prove that so does the solution z to (11.50), i.e., $z = z(y,s,t)$. Hence equation (11.50) can be further simplified:

$$\max\left(-\frac{\partial z}{\partial t} - \frac{\hat{\sigma}^2 s^2}{2}\frac{\partial^2 z}{\partial s^2} - \frac{\hat{\sigma}^2 s^2}{2\varepsilon}\left(\frac{\partial z}{\partial s} - y\right)^2 - \rho s\left(\frac{\partial z}{\partial s} - y\right); \left|\frac{\partial z}{\partial y}\right| - \mu s\right) = 0. \quad (11.51)$$

Similarly, for the function z^f we can prove that $z^f = z^f(y,s,t)$ and z^f is a solution to the dynamic programming equation

$$\max\left(-\frac{\partial z^f}{\partial t} - \frac{\hat{\sigma}^2 s^2}{2}\frac{\partial^2 z^f}{\partial s^2} - \frac{\hat{\sigma}^2 s^2}{2\varepsilon}\left(\frac{\partial z^f}{\partial s} - y\right)^2 - \rho s\left(\frac{\partial z^f}{\partial s} - y\right); \left|\frac{\partial z^f}{\partial y}\right| - \mu s\right) = 0. \quad (11.52)$$

Solutions $z = z^\varepsilon, z^f = z^{\varepsilon,f}$ are subject to the terminal conditions

$$z^\varepsilon(y,s,T) = (s-E)^+, \qquad z^{\varepsilon,f}(y,s,T) = 0, \quad \text{for any } y,s.$$

Concerning the transaction cost measure μ we will henceforth assume the following structural assumption

$$\mu = a\sqrt{\varepsilon}, \quad (11.53)$$

where $a > 0$ is a given constant. The next step of derivation of the Barles and Soner model is to prove the limits

$$z^{\varepsilon,f}(y,s,t) \to 0, \qquad z^\varepsilon(y,s,t) \to V(s,t) \qquad \text{as } \varepsilon \to 0^+, \quad (11.54)$$

where $V = V(s,t)$ is a solution to the nonlinear Black–Scholes equation (11.2) with the adjusted volatility σ according to (11.6). Consequently, in the limit $\varepsilon \to 0$, the call option price Λ (see (11.49)) converges to V. The rigorous proof of convergences (11.54) is rather technical and involved. We refer the reader to the paper [11] for details. A formal derivation of the limit is however rather intuitive and straightforward. Indeed, as a consequence of the terminal condition $z^{\varepsilon,f}(y,s,T) = 0$, one can prove by applying a parabolic comparison principle the following estimate for the function $z^{\varepsilon,f}$:

$$-\frac{\varepsilon\rho^2}{2\hat{\sigma}^2}(T-t) \le z^{\varepsilon,f}(y,s,t) \le \mu s|y| \equiv a\sqrt{\varepsilon}s|y|$$

from which the limit $z^{\varepsilon,f}(y,s,t) \to 0$ as $\varepsilon \to 0$ easily follows (see [11, Prop. 2.1]). The formal derivation of the limit (11.54) of the function z^ε is based on the asymptotic analysis. We assume the following asymptotic expansion of z^ε:

$$z^\varepsilon(y,s,t) = V(s,t) + \varepsilon C(r,A) + o(\varepsilon) \qquad \text{as } \varepsilon \to 0^+,$$

where

$$r = \frac{as}{\sqrt{\varepsilon}}\left(\frac{\partial V}{\partial s} - y\right), \qquad A = a^2 s^2 \frac{\partial^2 V}{\partial s^2}.$$

Then we can compute the asymptotic expansions of all the terms entering the dynamic programming equation (11.51). We obtain

$$\frac{\partial z^\varepsilon}{\partial t} = \frac{\partial V}{\partial t} + O(\sqrt{\varepsilon}), \quad \frac{\partial z^\varepsilon}{\partial s} = \frac{\partial V}{\partial s} + \sqrt{\varepsilon}\frac{A}{as}\frac{\partial C}{\partial r} + o(\sqrt{\varepsilon}),$$

$$\frac{\partial z^\varepsilon}{\partial y} = -\sqrt{\varepsilon} as \frac{\partial C}{\partial r} + o(\sqrt{\varepsilon}), \quad \frac{\partial^2 z^\varepsilon}{\partial s^2} = \frac{\partial^2 V}{\partial s^2} + \frac{A^2}{a^2 s^2}\frac{\partial^2 C}{\partial r^2} + o(\sqrt{\varepsilon}),$$

as $\varepsilon \to 0$. If we insert the above expressions into (11.51) and neglect the terms of the order $O(\sqrt{\varepsilon})$ we obtain the dynamic programming equation

$$\max\left(-\frac{\partial V}{\partial t} - \frac{\hat{\sigma}^2 s^2}{2}\frac{\partial^2 V}{\partial s^2} - \frac{\hat{\sigma}^2}{2a^2}\left(A^2 \frac{\partial^2 C}{\partial r^2} + \left[r + A\frac{\partial C}{\partial r}\right]^2\right); \left|\frac{\partial C}{\partial r}\right| - 1\right) = 0. \quad (11.55)$$

The above dynamic programming equation is fullfiled for arbitrary value of y (and consequently for all r) and therefore there must exist a function $\Psi = \Psi(A)$ such that

$$-\frac{\partial V}{\partial t} - \frac{\hat{\sigma}^2 s^2}{2}\frac{\partial^2 V}{\partial s^2} = \frac{A\hat{\sigma}^2}{2a^2}\Psi(A) \quad (11.56)$$

and the function $C(r,A)$ is a solution of the dynamic programming equation

$$\max\left(A\Psi(A) - A^2\frac{\partial^2 C}{\partial r^2} - \left[r + A\frac{\partial C}{\partial r}\right]^2; \left|\frac{\partial C}{\partial r}\right| - 1\right) = 0. \quad (11.57)$$

Finally, in [11, Appendix A] Barles and Soner have constructed an explicit solution $C = C(r,A)$ and the function $\Psi(A)$ to problem (11.57) satisfying additional boundary conditions

$$C(0,A) = \frac{\partial C}{\partial r}(0,A) = 0, \quad \lim_{|r| \to \infty}\frac{1}{|r|}\frac{\partial C}{\partial r}(r,A) = 1$$

for all A. We skip most of the details of the construction of the function C by referring to [11, Appendix A]. We only recall that the function $\Psi = \Psi(A)$ can be constructed as a solution to the nonlinear ODE:

$$\frac{d\Psi}{dA}(A) = \frac{\Psi(A) + 1}{2\sqrt{\Psi(A)A} - A}, \quad \Psi(0) = 0. \quad (11.58)$$

Having constructed the function Ψ we end up with the nonlinear Black–Scholes equation (11.56) with the term $\frac{A\hat{\sigma}^2}{2a^2}\Psi(A) \equiv \frac{\hat{\sigma}^2}{2}\Psi(a^2 s^2 \partial_s^2 V)$. It corresponds to the case when the risk-free interest rate $r = 0$. In the case when $r > 0$ one can derive the resulting nonlinear Black–Scholes equation for the call option price by taking into account a comparison of the portfolio with a risk-less bond with an interest rate $r > 0$. The final form of the pricing equation reads as

$$\frac{\partial V}{\partial t} + rs\frac{\partial V}{\partial s} + \frac{1}{2}\sigma^2(s^2\partial_s^2 V, T - t)s^2\frac{\partial^2 V}{\partial s^2} - rV = 0, \quad s > 0, \ t \in [0,T), \quad (11.59)$$

where

$$\sigma^2(s^2\partial_s^2 V, T - t) = \hat{\sigma}^2\left(1 + \Psi(a^2 e^{r(T-t)}s^2\partial_s^2 V)\right), \quad (11.60)$$

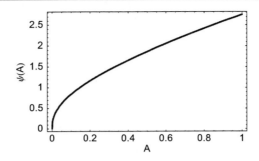

Figure 11.7. A profile of a solution $\Psi(A)$ to ODE (11.58).

is an adjusted volatility nonlinearly depending on $\partial_S^2 V$. Here where $\hat{\sigma} > 0$ is a constant volatility of the underlying stock price. The solution $V(s,t)$ is subject to the terminal payoff condition for the call option, i.e.

$$V(s,T) = (s-E)^+, \qquad s > 0. \tag{11.61}$$

A graph of the function $\Psi = \Psi(A)$ solving the ODE (11.58) is depicted in Fig. 11.7. We complete the discussion on the Barles and Soner model by pointing out the asymptotic behavior of the solution $\Psi(A)$ of the ODE (11.58). Plugging the ansatz $\Psi(A) = c|A|^{\alpha-1}A$ with $\alpha > 0$ into (11.58) we obtain

$$2\alpha c^{\frac{3}{2}} |A|^{\frac{3\alpha-1}{2}} \sim 1 + (1+c\alpha)|A|^{\alpha-1}A, \qquad \text{as } A \to 0.$$

Therefore $\alpha = \frac{1}{3}$ and $c = \left(\frac{3}{2}\right)^{\frac{2}{3}}$. It is interesting to note that the behavior of the adjusted volatility for the Barles and Soner and the Risk adjusted pricing methodology model discussed in the previous section is the same and it exhibits a cube root dependence of the volatility on the second derivative $\partial_S^2 V$ of the option price for small values of $0 \le |\partial_S^2 V| \ll 1$.

11.4.1. The Gamma Equation and a Numerical Approximation Scheme

Similarly as in the case of the RAPM model (see section 11.2.2.) we can construct a quasilinear parabolic equation for the new variable $H(x,\tau) = S\partial_S^2 V(S,t) (= S\Gamma)$.

Equation (11.59) can be rewritten in the form

$$\partial_t V + S\beta(S\Gamma, \ln(S/E), T-t) = r(V - S\partial_S V), \qquad S > 0, t \in (0,T), \tag{11.62}$$

where $\Gamma = \partial_S^2 V$ and

$$\beta(H,x,\tau) = \frac{\sigma^2}{2}\left(1 + \Psi(Ea^2 e^{r\tau+x}H)\right)H. \tag{11.63}$$

The transformed quasilinear parabolic equation for the function $H = H(x,\tau)$ where $x = \ln(S/E), x \in \mathbb{R}, \tau = T-t, \tau \in (0,T)$ reads as follows:

$$\frac{\partial H}{\partial \tau} = \frac{\partial^2 \beta}{\partial x^2} + \frac{\partial \beta}{\partial x} + r\frac{\partial H}{\partial x}, \tag{11.64}$$

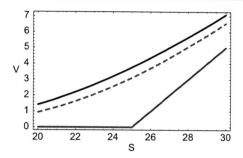
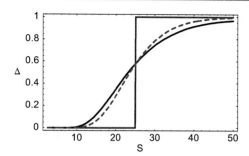

Figure 11.8. A comparison of the call option price $V(S,0)$ computed by the linear Black–Scholes equation (dashed line) and that of the nonlinear Barles and Soner model (solid line). The corresponding profile of the $\Delta = \frac{\partial V}{\partial S}$ of the call option (right).

where $\beta = \beta(H(x,\tau),x,\tau)$ and $\tau \in (0,T), x \in \mathbb{R}$. In the case of a call or put option the solution H to (11.37) is subjected to the initial condition at $\tau = 0$:

$$H(x,0) = \delta(x), \quad x \in \mathbb{R} \tag{11.65}$$

(δ is the Dirac function) and the boundary conditions

$$H(-\infty,\tau) = H(\infty,\tau) = 0, \quad \tau \in (0,T). \tag{11.66}$$

The option price $V = V(S,t)$ is given by the formula (11.24) for the call option case or (11.25) for the put option case.

In Fig. 11.8 we compare the call option price profile $V(S,0)$ and its delta $\Delta = \frac{\partial V}{\partial S}$ for the linear Black–Scholes equation and the nonlinear Barles and Soner model with the parameter $a = 0.1$.

11.5. Jumping Volatility Model and Leland's Model

In this section we present results of numerical analysis for the nonlinear generalization of the Black–Scholes equation proposed by Avellaneda, Levy and Paras [9]. Recall that the idea behind the model is to describe option pricing in incomplete markets where the volatility σ of the underlying stock process is uncertain but bounded from bellow and above by given constants $\sigma_1 < \sigma_2$. In their model the volatility can switch between σ_1 and σ_2 depending on the second derivative of the option price, i.e.

$$\sigma^2(S^2\partial_S^2 V, S, \tau) = \begin{cases} \hat{\sigma}_1^2, & \text{if } \partial_S^2 V < 0, \\ \hat{\sigma}_2^2, & \text{if } \partial_S^2 V > 0. \end{cases} \tag{11.67}$$

It is worth noting that the Leland model discussed in Chapter 5 (cf. [78]) and its generalization for more complex option strategies by Hoggard, Whaley and Wilmott in [64] can be also written as model with a jumping volatility. We remind ourselves that the volatility in Leland's model for pricing a short positioned call option is modified as follows:

$$\sigma^2(S^2\partial_S^2 V, S, \tau) = \hat{\sigma}^2(1 + \text{Le sgn}(\partial_S^2 V)),$$

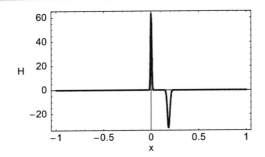 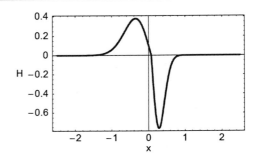

Figure 11.9. Plots of the initial approximation of the function $H(x,0)$ (left) and the solution profile $H(x,T)$ at $\tau = T$ (right).

where $\text{Le} = \sqrt{2/\pi} C/(\hat{\sigma}\sqrt{\Delta t})$ is the so-called Leland number depending on the transaction costs measure C and the time Δt between two consecutive portfolio adjustments. Clearly, the Leland volatility function can be treated as a jumping volatility model (11.67) where

$$\begin{aligned}\hat{\sigma}_1^2 &= \hat{\sigma}^2(1-\text{Le}),\\ \hat{\sigma}_2^2 &= \hat{\sigma}^2(1+\text{Le}).\end{aligned} \quad (11.68)$$

Similarly as in previously studied nonlinear Black–Scholes models, we can introduce the new variable $H(x,\tau) = S\partial_S^2 V$, where $x = \ln(S/E)$ and $\tau = T-t$. We obtain

$$\frac{\partial H}{\partial \tau} = \frac{\partial^2 \beta}{\partial x^2} + \frac{\partial \beta}{\partial x} + r\frac{\partial H}{\partial x}, \quad (11.69)$$

where $\beta = \beta(H(x,\tau))$ is given by

$$\beta(H) = \begin{cases} \frac{\hat{\sigma}_1^2}{2} H & \text{if } H < 0, \\ \frac{\hat{\sigma}_2^2}{2} H & \text{if } H > 0. \end{cases} \quad (11.70)$$

In the rest of this section we present results of numerical approximation of the jumping volatility model for the case of the bullish spread. Recall that the bullish spread strategy consists of buying one call option with a lower exercise price $E = E_1$ and selling one call option with a higher exercise price $E_2 > E_1$ (see Chapter 3). Its pay-off diagram is therefore given by the function

$$V(S,T) = (S-E_1)^+ - (S-E_2)^+.$$

The bullish spread option price $V(S,t)$ can be computed from the function $H = H(x,\tau)$ solving the Gamma equation (11.69). The initial condition $H(x,0)$ can be deduced from the bullish spread terminal pay-off diagram. As for the initial condition we have

$$H(x,0) = \delta(x-x_0) - \delta(x-x_1), \quad x \in \mathbb{R},$$

where $x_0 = 0, x_1 = \ln(E2/E1)$. We have to impose the boundary conditions

$$H(-\infty,\tau) = H(\infty,\tau) = 0, \quad \tau \in (0,T).$$

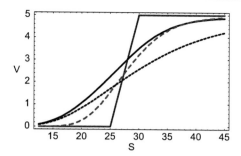
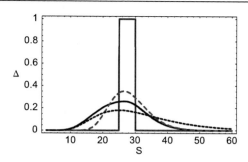

Figure 11.10. A comparison of the call option price $V(S,0)$ (left) and its delta (right) computed from the jumping volatility model (solid line) by the linear Black–Scholes. Option prices obtained from the linear Black–Scholes equation are depicted by dashed curved (for volatility σ_1) and fine-dashed curve (for volatility σ_2).

Applying the same procedure as in derivation of (11.24) we end up with the explicit formula for the bullish spread option price

$$V(S,t) = \int_{-\infty}^{\infty} (S - Ee^x)^+ H(x, T-t) dx, \qquad (11.71)$$

where $E = E_1$.

In Fig. 11.9 we plot the initial approximation of the function $H(x,0) = \delta(x-x_0) - \delta(x-x_1)$ and the solution $H(x,T)$ at $\tau = T$ corresponding to the case of bullish spread option strategy. In Fig. 11.10 we plot the bullish spread option price $V(S,0)$ computed by the jumping volatility model with volatility bounds $\sigma_1 = 0.2, \sigma_2 = 0.4$ and model parameters $r = 0.011, E_1 = 25, E_2 = 30, T = 1$. The solution $V(S,0)$ is greater than both European bullish spread prices (see Chapter 3) with constant volatilities σ_1 (dashed line) and σ_2 (fine dashed line), respectively.

11.6. Finite Difference Scheme for Solving the Γ Equation

The aim of this section is to propose an efficient numerical discretization of the Γ equation for a general function $\beta = \beta(H,x,\tau)$ including, in particular, the case of the RAPM, Frey's as well as Barles and Soner model. Our numerical scheme of the Γ equation (11.64) is based on the finite volume approximation of the partial derivatives entering (11.64). We will construct a scheme, which is semi-implicit in time.

In order to find a numerical solution to equation (11.21) we have to restrict ourselves to a finite spatial interval $x \in (-L, L)$ where $L > 0$ is sufficiently large. Since $S = Ee^x$ the restricted interval of underlying stock values to $S \in (Ee^{-L}, Ee^L)$. From a practical point of view, it is sufficient to take $L \approx 1.5$ in order to include important range of values of S. Subsequently, we also have to modify boundary conditions (11.66). Instead of (11.66) we will consider Dirichlet boundary conditions at $x = \pm L$, i.e.

$$H(-L, \tau) = H(L, \tau) = 0, \quad \tau \in (0, T).$$

We take a uniform division $\{\tau_j, j = 0, 1, \ldots, m\}$, $\tau_j = jk$, of the time interval $[0, T]$ with a time step $k = \frac{T}{m}$ and a uniform division $x_i = ih$, $i = -n, \ldots, n$, of the interval $[-L, L]$ with a step $h = \frac{L}{n}$. To construct numerical approximation of a solution H to (11.64) we derive a system of difference equations corresponding to (11.64) to be solved at every discrete time step. Difference equations involve discrete values of $H_i^j \approx H(ih, jk)$ where $j = 0, \ldots, m$.

Our numerical algorithm is semi-implicit in time. Notice that the term $\partial_x^2 \beta$, where $\beta = \beta(H(x,\tau), x, \tau)$ can be expressed in the form

$$\partial_x^2 \beta = \partial_x \left(\beta'_H(H, x, \tau) \partial_x H + \beta'_x(H, x, \tau), \right)$$

where β'_H and β'_x are partial derivatives of the function $\beta(H, x, \tau)$ with respect to H and x, respectively. In our discretization scheme, the non-linear terms $\beta'_H(H, x, \tau)$ and $\beta'_x(H, x, \tau)$ are evaluated from the previous time step τ_{j-1} whereas linear terms are solved at the current time level. Such a discretization leads to a solution of linear systems of equations at every discrete time level. Now, by replacing the time derivative by the time difference, approximating H in nodal points by the average value of neighboring segments, collecting all linear terms at the new time level j and taking all the remaining terms from the previous time level $j-1$ we obtain a *tridiagonal system* for the solution vector $H^j = (H_{-n+1}^j, \ldots, H_{n-1}^j) \in \mathbb{R}^{2n-1}$:

$$a_i^j H_{i-1}^j + b_i^j H_i^j + c_i^j H_{i+1}^j = d_i^j, \quad H_{-n}^j = 0, \quad H_n^j = 0, \tag{11.72}$$

where $i = -n+1, \ldots, n-1$ and $j = 1, \ldots, m$. The solution is subject to homogeneous Dirichlet boundary conditions imposed on new discrete values of the vector and initial condition $H_i^0 = \bar{H}(x_i)$ where $x_i = ih$. The coefficients of the tridiagonal matrix are given by

$$\begin{aligned}
a_i^j &= -\frac{k}{h^2} \beta'_H(H_{i-1}^{j-1}, x_{i-1}, \tau_{j-1}) + \frac{k}{2h} r, \\
c_i^j &= -\frac{k}{h^2} \beta'_H(H_i^{j-1}, x_i, \tau_{j-1}) - \frac{k}{2h} r, \\
b_i^j &= 1 - (a_i^j + c_i^j), \\
d_i^j &= H_i^{j-1} + \frac{k}{h} \Big(\beta(H_i^{j-1}, x_i, \tau_{j-1}) - \beta(H_{i-1}^{j-1}, x_{i-1}, \tau_{j-1}) \\
&\quad + \beta'_x(H_i^{j-1}, x_i, \tau_{j-1}) - \beta'_x(H_{i-1}^{j-1}, x_{i-1}, \tau_{j-1}) \Big).
\end{aligned}$$

It means that the vector H^j at the time level τ_j is a solution to the system of linear equations $\mathbf{A}^j H^j = d^j$, where the $(2n-1) \times (2n-1)$ matrix $\mathbf{A}^j$ is defined as

$$\mathbf{A}^j = \begin{pmatrix} b_{-n+1}^j & c_{-n+1}^j & 0 & \cdots & 0 \\ a_{-n+2}^j & b_{-n+2}^j & c_{-n+2}^j & & \vdots \\ 0 & \cdot & \cdot & \cdot & 0 \\ \vdots & \cdots & a_{n-2}^j & b_{n-2}^j & c_{n-2}^j \\ 0 & \cdots & 0 & a_{n-1}^j & b_{n-1}^j \end{pmatrix}. \tag{11.73}$$

Since tridiagonal systems admit a simple LU – matrix decomposition (see Chapter 10) we can solve the above tridiagonal system in every time step in a fast and effective way.

Table 11.1. The Mathematica source code for implicit method of the solution to the Gamma equation.

```
Needs["LinearAlgebra`Tridiagonal`"];

mu = 0.4; bfun[H_, x_, tau_] := (sigma^2/2) ( 1  + mu H^(1/3)) H;

bfunH[H_, x_, tau_] = D[bfun[H, x, tau], {H, 1}];
bfunx[H_, x_, tau_] = D[bfun[H, x, tau], {x, 1}];

n = 500;   m = 200; h = 0.005;    k = T/m;
sigma = 0.3; r = 0.011; q = 0.; T = 1; X = 25;

taustar = 0.001;
Hinit[x_] :=
Exp[- ((x + (r-q- sigma^2/2)taustar)/(sigma Sqrt[taustar]))^2/2]/
(sigma Sqrt[taustar] Sqrt[2 Pi]);

Hfn = Table[Hinit[i h], {i, -n + 1, n - 1}];   Hsol[0] = Hfn;
For[j = 1, j <= m, 1,
    {
      a = Table[
          If[i == -n + 1,
            -(k/h^2)bfunH[ 0.  , (i - 1) h, j k]  + 0.5(k/h) r,
            -(k/h^2)bfunH[ Hfn[[ i + n - 1]] , (i - 1) h, j k]
+ 0.5(k/h) r ], { i, -n + 1, n - 1} ];
      c = Table[
            -(k/h^2)bfunH[ Hfn[[ i + n]]  , i h, j k ]
- 0.5(k/h) r, { i, -n + 1, n - 1} ];
      b = Table[1 - a[[i + n]] - c[[i + n]], {i, -n + 1, n - 1}];

      a = Table[a[[i]], {i, 2, 2 n - 1}];
      c = Table[c[[i]], {i, 1, 2n - 2}];
      d = Hfn + (k/h) Table[
              If[
                i == -n + 1, (bfun[ Hfn[[1]], i h, j k ] -
                  bfun[0, i h, j k]),
                (bfun[ Hfn[[i + n ]], i h, j k] -
                  bfun[Hfn[[i - 1 + n]], i h, j k ])
              ], {i, -n + 1, n - 1}]
          + (k/h) Table[
              If[ i == -n + 1, (bfunx[ Hfn[[1]], i h, j k ] -
                  bfunx[0, i h, j k]),
                (bfunx[ Hfn[[i + n ]], i h, j k] -
                  bfunx[Hfn[[i - 1 + n]], i h, j k ])
              ], {i, -n + 1, n - 1}];
      Hfn = TridiagonalSolve[a, b, c, d]; Hsol[j] = Hfn;  j++;
    }];
ListPlot[Hfn];
V[S_]:=h Sum[Max[S-X Exp[i h], 0] Hsol[m][[i+n]], {i,-n+1,n-1}];
```

According to (11.24) and (11.25) the option price $V(S, T - \tau_j)$ can be constructed from the discrete solution H_i^j as follows:

(call option) $\quad V(S, T - \tau_j) \;=\; h \sum_{i=-n}^{n} (S - E e^{x_i})^+ H_i^j,$

(put option) $\quad V(S, T - \tau_j) \;=\; h \sum_{i=-n}^{n} (E e^{x_i} - S)^+ H_i^j,$

for $j = 1, \ldots, m$.

In Table 11.1 we present a simple source code in the Mathematica language for finite difference approximation of the Gamma equation. The function β corresponds to the RAPM model. For other nonlinear Black–Scholes models one has to modify the function $\beta = \beta(H, x, \tau)$ accordingly.

Chapter 12

Transformation Methods for Pricing American Options

In this chapter we continue our discussion on analytical and numerical methods for pricing American options initiated in Chapters 9 and 10. In these preceding chapters, we presented basic concepts of pricing American plain vanilla call and put options. In Chapter 10 we furthermore discussed the classical Projected successive over relaxation method (PSOR) for their valuation. The PSOR method is based on a solution to the variational inequality. The purpose of this chapter is to investigate the problem of pricing American options by means of the analysis of the early exercise boundary or, equivalently, the optimal stopping time. We will show how to transform the problem of valuation of American options into a solution to a parabolic partial differential equation defined on a fixed spatial domain. We will present the transformation method for call and put options as well as for nonlinear Black–Scholes models described in Chapter 11 and for American style of average strike Asian options.

12.1. Transformation Methods for Plain Vanilla Options

One of the important problems in the field of pricing derivative securities is the analysis of the early exercise boundary and the optimal stopping time for American options. In Chapter 9 we showed how it can be reduced to a problem of solving a certain free boundary problem for the Black–Scholes equation. However, the explicit analytical expression for the free boundary profile is not known yet. Many authors have investigated various approximate models leading to approximate expressions for the price and early exercise boundary position for American call and put options (see e.g. [53, 54, 68, 70, 74, 85, 97] and recent papers by Zhu [125, 126], Alobaidi et al. [5, 80, 81], Stamicar et al. [109] and the survey paper by Chadam [22]. In this section we will present results based on the fixed domain transformation method as it was presented in a series of papers by Ševčovič et al. [103, 109, 107].

Let us recall that the equation governing time evolution of the price $V(S,t)$ of the American call option is the following parabolic PDE:

$$\frac{\partial V}{\partial t} + (r-q)S\frac{\partial V}{\partial S} + \frac{\sigma^2}{2}S^2\frac{\partial^2 V}{\partial S^2} - rV = 0, \qquad 0 < S < S_f(t),\ 0 < t < T,$$

$$V(0,t) = 0,\ V(S_f(t),t) = S_f(t) - E,\ \frac{\partial V}{\partial S}(S_f(t),t) = 1, \qquad (12.1)$$

$$V(S,T) = (S-E)^+,$$

defined on a time-dependent domain $S \in (0, S_f(t))$, where $t \in (0, T)$ (see Chapter 9). As usual, $S > 0$ denotes the underlying stock price, $E > 0$ is the exercise price, $r > 0$ is the risk-free rate, $q > 0$ is the continuous stock dividend rate and $\sigma \equiv const > 0$ is a constant historical volatility of the underlying stock process.

The main purpose of this section is to derive an integral equation for the early exercise boundary position $S_f(t)$. We will show that the integral equation provides an accurate numerical method for calculating the early exercise boundary near the expiry T. The derivation of the nonlinear integral equation is based on application the Fourier and inverse Fourier transforms. Here we present a method developed by Ševčovič in [103] of reducing the free boundary problem for (12.1) to a nonlinear integral equation with a singular kernel.

Throughout this section we restrict our attention to the case when $r > q > 0$. It is well known (cf. Dewynne *et al.* [36], Kwok [75]) that, for $r > q > 0$, the free boundary $\rho(\tau) = S_f(T - \tau)$ starts at $\rho(0) = rE/q$, whereas $\rho(0) = E$ for the case $r \leq q$. Thus, the free boundary profile develops an initial jump in the case $r > q > 0$.

12.1.1. Fixed Domain Transformation for the American Call Option

In this section we will perform a fixed domain transformation of the free boundary problem (12.1) yielding a parabolic equation defined on a fixed spatial domain. As it will be shown below, imposing of the free boundary condition will result in a nonlinear time-dependent term involved in the resulting equation. To transform equation (12.1) defined on a time dependent spatial domain $(0, S_f(t))$, we introduce the following change of variables:

$$\tau = T - t, \quad x = \ln\left(\frac{\rho(\tau)}{S}\right), \qquad \text{where } \rho(\tau) = S_f(T - \tau). \qquad (12.2)$$

Clearly, $\tau \in (0, T)$ and $x \in (0, \infty)$ whenever $S \in (0, S_f(t))$. Let us furthermore define the auxiliary function $\Pi = \Pi(x, \tau)$ as follows:

$$\Pi(x, \tau) = V(S, t) - S\frac{\partial V}{\partial S}(S, t). \qquad (12.3)$$

The transformed function Π has an important financial meaning as it represents a synthetic portfolio consisting of one long positioned option and the total of $\delta = \frac{\partial V}{\partial S}$ sold underlying stocks. It is straightforward to verify the following identities:

$$\frac{\partial \Pi}{\partial x} = S^2\frac{\partial^2 V}{\partial S^2}, \quad \frac{\partial^2 \Pi}{\partial x^2} + 2\frac{\partial \Pi}{\partial x} = -S^3\frac{\partial^3 V}{\partial S^3},$$

$$\frac{\partial \Pi}{\partial \tau} + \frac{\dot{\rho}}{\rho}\frac{\partial \Pi}{\partial x} = S\frac{\partial^2 V}{\partial S \partial t} - \frac{\partial V}{\partial t}, \qquad (12.4)$$

where $\dot{\rho} = d\rho/d\tau$. Now assuming that $V = V(S,t)$ is a sufficiently smooth solution of (12.1), we may differentiate (12.1) with respect to S. Plugging expressions (12.4) into (12.1), we finally obtain that the function $\Pi = \Pi(x,\tau)$ is a solution of the parabolic equation

$$\frac{\partial \Pi}{\partial \tau} + a(\tau)\frac{\partial \Pi}{\partial x} - \frac{\sigma^2}{2}\frac{\partial^2 \Pi}{\partial x^2} + r\Pi = 0, \qquad (12.5)$$

$x \in (0,\infty)$, $\tau \in (0,T)$, with a time-dependent coefficient

$$a(\tau) = \frac{\dot{\rho}(\tau)}{\rho(\tau)} + r - q - \frac{\sigma^2}{2}.$$

It follows from the boundary condition $V(S_f(t),t) = S_f(t) - E$ and $\frac{\partial V}{\partial S}(S_f(t),t) = 1$ that

$$\Pi(0,\tau) = -E, \qquad \Pi(+\infty,\tau) = 0. \qquad (12.6)$$

The initial condition $\Pi(x,0)$ can be deduced from the pay-off diagram for $V(S,T)$. Indeed, we conclude

$$\Pi(x,0) = \begin{cases} -E & \text{for } x < \ln\left(\frac{\rho(0)}{E}\right), \\ 0, & \text{otherwise.} \end{cases} \qquad (12.7)$$

In what follows, we will show how the function $a(\tau)$ depends upon a solution Π itself. This dependence is non-local in the spatial variable x. Moreover, the initial position of the interface $\rho(0)$ enters the initial condition $\Pi(x,0)$. Therefore we have to determine the relationship between the solution $\Pi(x,\tau)$ and the free boundary function $\rho(\tau)$ first. To this end, we make use of the boundary condition imposed on V at the interface $S = S_f(t)$. Since $S_f(t) - E = V(S_f(t),t)$ we have

$$\frac{d}{dt}S_f(t) = \frac{\partial V}{\partial S}(S_f(t),t)\frac{d}{dt}S_f(t) + \frac{\partial V}{\partial t}(S_f(t),t).$$

As $\frac{\partial V}{\partial S}(S_f(t),t) = 1$ we obtain $\frac{\partial V}{\partial t}(S,t) = 0$ at $S = S_f(t)$. Assuming the function Π_x has a trace at $x = 0$ and taking into account (12.4), we may conclude that, for any $t = T - \tau \in [0,T)$,

$$S^2\frac{\partial^2 V}{\partial S^2}(S,t) \to \frac{\partial \Pi}{\partial x}(0,\tau), \quad S\frac{\partial V}{\partial S}(S,t) \to \rho(\tau) \text{ as } S \to S_f(t)^-.$$

If $\frac{\partial V}{\partial t}(S,t) \to \frac{\partial V}{\partial t}(S_f(t),t) = 0$ as $S \to S_f(t)^-$, then it follows from the Black–Scholes equation (12.1) that

$$(r-q)\rho(\tau) + \frac{\sigma^2}{2}\frac{\partial \Pi}{\partial x}(0,\tau) - r(\rho(\tau) - E)$$

$$= \lim_{S \to S_f(t)^-} \left(\frac{\partial V}{\partial t}(S,t) + (r-q)S\frac{\partial V}{\partial S}(S,t) + \frac{\sigma^2}{2}S^2\frac{\partial^2 V}{\partial S^2}(S,t) - rV(S,t)\right) = 0.$$

As a consequence, we obtain a nonlocal algebraic constraint between the free boundary function $\rho(\tau)$ and the boundary trace $\partial_x\Pi(0,\tau)$ of the derivative of the solution Π itself:

$$\rho(\tau) = \frac{rE}{q} + \frac{\sigma^2}{2q}\frac{\partial \Pi}{\partial x}(0,\tau), \quad \text{for } 0 < \tau \leq T. \qquad (12.8)$$

It remains to determine the initial position of the interface $\rho(0)$. According to Dewynne et al. [36] (see also Kwok [75]), the initial position $\rho(0)$ of the free boundary is equal to the value rE/q, for the case $0 < q < r$. Alternatively, we can derive this condition from (12.5)–(12.7) by assuming the continuity of $\partial_x \Pi$ at $(x,\tau) = (0,0)$. In this case we have

$$\lim_{\tau \to 0^+} \frac{\partial \Pi}{\partial x}(0,\tau) = \lim_{\tau \to 0^+, x \to 0^+} \frac{\partial \Pi}{\partial x}(x,\tau) = \lim_{x \to 0^+} \frac{\partial \Pi}{\partial x}(x,0) = 0$$

because $\Pi(x,0) = -E$ for x close to 0^+. From (12.8) we obtain

$$\rho(0) = \frac{rE}{q}. \tag{12.9}$$

In summary, we have shown that, under suitable regularity assumptions imposed on a solution Π to (12.5), (12.6), (12.7), the free boundary problem (12.1) can be transformed into the initial boundary value problem for parabolic PDE

$$\begin{aligned}
&\frac{\partial \Pi}{\partial \tau} + a(\tau) \frac{\partial \Pi}{\partial x} - \frac{\sigma^2}{2} \frac{\partial^2 \Pi}{\partial x^2} + r\Pi = 0, \\
&\Pi(0,\tau) = -E, \quad \Pi(+\infty, \tau) = 0, \quad x > 0, \tau \in (0,T), \\
&\Pi(x,0) = \begin{cases} -E & \text{for } x < \ln(r/q), \\ 0 & \text{otherwise,} \end{cases}
\end{aligned} \tag{12.10}$$

where $a(\tau) = \frac{\dot{\rho}(\tau)}{\rho(\tau)} + r - q - \frac{\sigma^2}{2}$ and

$$\rho(\tau) = \frac{rE}{q} + \frac{\sigma^2}{2q} \frac{\partial \Pi}{\partial x}(0,\tau), \quad \rho(0) = \frac{rE}{q}. \tag{12.11}$$

We emphasize that the problem (12.10) constitutes a nonlinear parabolic equation with a nonlocal constraint given by (12.11).

12.1.2. Reduction to a Nonlinear Integral Equation

The main purpose of this section is to show how the fully nonlinear nonlocal problem (12.10)–(12.11) can reduced to a single nonlinear integral equation for $\rho(\tau)$ giving the explicit formula for the solution $\Pi(x,\tau)$ to (12.10). The idea is to apply the Fourier sine and cosine integral transforms (cf. Stein and Weiss [117]). Let us recall that for any Lebesgue integrable function $f \in L^1(\mathbb{R}^+)$ the sine and cosine transformations are defined as follows:

$$\mathcal{F}_S(f)(\omega) = \int_0^\infty f(x) \sin \omega x \, dx, \qquad \mathcal{F}_C(f)(\omega) = \int_0^\infty f(x) \cos \omega x \, dx.$$

Their inverse transforms are given by

$$\mathcal{F}_S^{-1}(g)(x) = \frac{2}{\pi} \int_0^\infty g(\omega) \sin \omega x \, d\omega, \qquad \mathcal{F}_C^{-1}(g)(x) = \frac{2}{\pi} \int_0^\infty g(\omega) \cos \omega x \, d\omega.$$

Now we suppose that the function $\rho(\tau)$ and subsequently $a(\tau)$ are already know. Let $\Pi = \Pi(x,\tau)$ be a solution of (12.10) corresponding to a given function $a(\tau)$. Let us denote

$$p(\omega, \tau) = \mathcal{F}_S(\Pi(.,\tau))(\omega), \qquad q(\omega, \tau) = \mathcal{F}_C(\Pi(.,\tau))(\omega), \tag{12.12}$$

where $\omega \in \mathbb{R}^+, \tau \in (0, T)$. By applying the sine and cosine integral transforms, taking into account their basic properties and (12.11), we finally obtain a linear non-autonomous ω-parameterized system of ODEs

$$\frac{d}{d\tau}p(\omega,\tau) - a(\tau)\omega q(\omega,\tau) + \alpha(\omega)p(\omega,\tau) = -E\omega\frac{\sigma^2}{2},$$
$$\frac{d}{d\tau}q(\omega,\tau) + a(\tau)\omega p(\omega,\tau) + \alpha(\omega)q(\omega,\tau) = -Ea(\tau) - q\rho(\tau) + rE, \qquad (12.13)$$

where

$$\alpha(\omega) = \frac{1}{2}(\sigma^2\omega^2 + 2r).$$

The system of equations (12.13) is subject to initial conditions at $\tau = 0$, $p(\omega, 0) = \mathcal{F}_S(\Pi(.,0))(\omega)$, $q(\omega, 0) = \mathcal{F}_C(\Pi(.,0))(\omega)$. In the case of the call option, we deduce from the initial condition for Π (see (12.10)) that

$$p(\omega,0) = \frac{E}{\omega}\left(\cos\left(\omega\ln\frac{r}{q}\right) - 1\right), \quad q(\omega,0) = -\frac{E}{\omega}\sin\left(\omega\ln\frac{r}{q}\right). \qquad (12.14)$$

Taking into account (12.14) and by using the variation of constants formula for solving linear non-autonomous ODEs, we obtain an explicit formula for $p(\omega,\tau) = -E\omega^{-1} + \tilde{p}(\omega,\tau)$, where

$$\tilde{p}(\omega,\tau) = \frac{E}{\omega}e^{-\alpha(\omega)\tau}\cos(\omega(A(\tau,0) + \ln(r/q)))$$
$$+ \int_0^\tau e^{-\alpha(\omega)(\tau-s)}\left[\frac{rE}{\omega}\cos(\omega A(\tau,s)) + (rE - q\rho(s))\sin(\omega A(\tau,s))\right]ds. \qquad (12.15)$$

Here we have denoted by A the function defined as

$$A(\tau,s) = \int_s^\tau a(\xi)\,d\xi = \ln\frac{\rho(\tau)}{\rho(s)} + \left(r - q - \frac{\sigma^2}{2}\right)(\tau - s). \qquad (12.16)$$

As $\mathcal{F}_S^{-1}(\omega^{-1}) = 1$ we have $\Pi(x,\tau) = \mathcal{F}_S^{-1}(p(\omega,\tau)) = -E + \frac{2}{\pi}\int_0^\infty \tilde{p}(\omega,\tau)\sin(\omega x)\,d\omega$. From (12.11) we conclude that the free boundary function ρ satisfies the following equation:

$$\rho(\tau) = \frac{rE}{q} + \frac{\sigma^2}{q\pi}\int_0^\infty \omega\tilde{p}(\omega,\tau)\,d\omega. \qquad (12.17)$$

With regard to (12.15) and (12.17) we obtain the following nonlinear singular integral equation for the free boundary function $\rho(\tau)$:

$$\rho(\tau) = \frac{rE}{q}\left(1 + \frac{\sigma}{r\sqrt{2\pi\tau}}\exp\left(-r\tau - \frac{(A(\tau,0) + \ln(r/q))^2}{2\sigma^2\tau}\right)\right. \qquad (12.18)$$
$$\left. + \frac{1}{\sqrt{2\pi}}\int_0^\tau\left[\sigma + \frac{1}{\sigma}\left(1 - \frac{q\rho(s)}{rE}\right)\frac{A(\tau,s)}{\tau-s}\right]\frac{\exp\left(-r(\tau-s) - \frac{A(\tau,s)^2}{2\sigma^2(\tau-s)}\right)}{\sqrt{\tau-s}}\,ds\right),$$

where the function A depends on the free boundary position ρ via equation (12.16). To simplify the above integral equation, we introduce a new auxiliary function $H : [0, \sqrt{T}] \to \mathbb{R}$ as follows:

$$\rho(\tau) = \frac{rE}{q}\left(1 + \sigma\sqrt{2}H(\sqrt{\tau})\right). \qquad (12.19)$$

Using the change of variables $s = \xi^2 \cos^2 \theta$, one can rewrite the integral equation (12.17) in terms of the function H as follows:

$$H(\xi) = f_H(\xi)$$
$$+ \frac{1}{\sqrt{\pi}} \int_0^{\frac{\pi}{2}} [\xi \cos \theta - 2 \cot g \theta \, H(\xi \cos \theta) g_H(\xi, \theta)] e^{-r\xi^2 \sin^2 \theta - g_H^2(\xi, \theta)} d\theta, \quad (12.20)$$

where

$$g_H(\xi, \theta) = \frac{1}{\sigma\sqrt{2}} \frac{1}{\xi \sin \theta} \ln \left(\frac{1 + \sigma\sqrt{2} H(\xi)}{1 + \sigma\sqrt{2} H(\xi \cos \theta)} \right)$$
$$+ \frac{\Lambda}{\sqrt{2}} \xi \sin \theta, \quad \Lambda = \frac{r-q}{\sigma} - \frac{\sigma}{2}, \quad (12.21)$$

for $\xi \in [0, \sqrt{T}]$, $\theta \in (0, \pi/2)$, and

$$f_H(\xi) = \frac{1}{2r\sqrt{\pi}\xi} e^{-r\xi^2 - \left(g_H(\xi, \frac{\pi}{2}) + \frac{1}{\xi} \frac{1}{\sigma\sqrt{2}} \ln(r/q)\right)^2}. \quad (12.22)$$

Notice that equations (12.18) and (12.20) are integral equations with a singular kernel (cf. Gripenberg *et al.* [58]).

Remark 12.1. *Kwok [75] derived another integral equation for the early exercise boundary for the American call option on a stock paying continuous dividend. According to Kwok [75, Section 4.2.3], $\rho(\tau)$ satisfies the integral equation*

$$\begin{aligned}\rho(\tau) &= E + \rho(\tau) e^{-q\tau} N(d) - E e^{-r\tau} N(d - \sigma\sqrt{\tau}) \\ &+ \int_0^\tau q\rho(\tau) e^{-q\xi} N(d_\xi) - rE e^{-r\xi} N(d_\xi - \sigma\sqrt{\xi}) d\xi,\end{aligned} \quad (12.23)$$

where

$$d = \frac{1}{\sigma\sqrt{\tau}} \ln\left(\frac{\rho(\tau)}{E}\right) + \Lambda\sqrt{2\tau}, \quad d_\xi = \frac{1}{\sigma\sqrt{\xi}} \ln\left(\frac{\rho(\tau)}{\rho(\tau - \xi)}\right) + \Lambda\sqrt{2\xi}$$

and $N(u)$ is the cumulative distribution function for the normal distribution. The above integral equation covers both cases: $r \leq q$ as well as $r > q$. However, in the case $r > q$ this equation becomes singular as $\tau \to 0^+$.

In the rest of this section we derive a formula for pricing American call options based on the solution ρ to the integral equation (12.20). With regard to (12.3), we have

$$\frac{\partial}{\partial S}\left(S^{-1} V(S,t)\right) = -S^{-2} \Pi\left(\ln\left(S^{-1}\rho(T-t)\right), T-t\right).$$

Taking into account the boundary condition $V(S_f(t), t) = S_f(t) - E$ and integrating the above equation from S to $S_f(t) = \rho(T-t)$, we obtain

$$V(S, T-\tau) = \frac{S}{\rho(\tau)} \left(\rho(\tau) - E + \int_0^{\ln \frac{\rho(\tau)}{S}} e^x \Pi(x, \tau) dx \right). \quad (12.24)$$

It is straightforward to verify that V given by (12.24) is indeed a solution to the free boundary problem (12.1). Inserting the expressions (12.15) and (12.17) (recall that $\Pi(x,\tau) = -E + \frac{2}{\pi}\int_0^\infty \tilde{p}(\omega,\tau)\sin\omega x\, d\omega$) into (12.24), we end up with the formula for pricing the American call option:

$$V(S,T-\tau) = S - E + \frac{S}{\rho(\tau)} E\, I_2(A(\tau,0) + \ln(r/q), \ln(\rho(\tau)/S), \tau)$$

$$+ \frac{S}{\rho(\tau)} \int_0^\tau \Big[rE\, I_2(A(\tau,s), \ln(\rho(\tau)/S), \tau - s)$$

$$+ (rE - q\rho(s))\, I_1(A(\tau,s), \ln(\rho(\tau)/S), \tau - s) \Big] ds \qquad (12.25)$$

for any $S \in (0, S_f(t))$ and $t \in [0,T]$, where $A(\tau,s) = \ln\frac{\rho(\tau)}{\rho(s)} + (r - q - \frac{\sigma^2}{2})(\tau - s)$, and the functions I_1, I_2 are defined as follows:

$$I_1(A,L,\tau) = \frac{e^{-(r-\sigma^2/2)\tau}}{2} \left[e^A M\left(\frac{-A - \sigma^2\tau}{\sigma\sqrt{2\tau}}, \frac{L}{\sigma\sqrt{2\tau}} \right) \right. \qquad (12.26)$$

$$\left. - e^{-A} M\left(\frac{A - \sigma^2\tau}{\sigma\sqrt{2\tau}}, \frac{L}{\sigma\sqrt{2\tau}} \right) \right],$$

$$I_2(A,L,\tau) = \frac{e^{-r\tau} e^L}{2} M\left(\frac{A - L}{\sigma\sqrt{2\tau}}, \frac{2L}{\sigma\sqrt{2\tau}} \right) \qquad (12.27)$$

$$- \frac{e^{-(r-\sigma^2/2)\tau}}{2} \left[e^A M\left(\frac{-A - \sigma^2\tau}{\sigma\sqrt{2\tau}}, \frac{L}{\sigma\sqrt{2\tau}} \right) \right.$$

$$\left. + e^{-A} M\left(\frac{A - \sigma^2\tau}{\sigma\sqrt{2\tau}}, \frac{L}{\sigma\sqrt{2\tau}} \right) \right]$$

and

$$M(x,y) = \operatorname{erf}(x+y) - \operatorname{erf}(x) = \frac{2}{\sqrt{\pi}} \int_x^{x+y} e^{-\xi^2}\, d\xi.$$

We will refer to (12.25) as the semi-explicit formula for pricing the American call option. We use the term 'semi-explicit' because (12.25) contains the free boundary function $\rho(\tau) = S_f(T - \tau)$ which has to be determined first by solving the nonlinear integral equation (12.20).

12.1.3. Numerical Experiments

In this section we focus on numerical experiments. We will compute the free boundary profile

$$S_f(t) = \rho(T - t) = \frac{rE}{q}\left(1 + \sigma\sqrt{2}H(\sqrt{T-t})\right) \qquad (12.28)$$

(see (12.14)) by solving the nonlinear integral equation (12.20). We will also present a comparison of the results obtained by our methods to those obtained by other known methods for solving the American call option problem.

We first examine the leading order term of the function $H(\xi)$. Since $H(0) = 0$ we can write $H(\xi) = h\xi + O(\xi^2)$ as $\xi \to 0$. Plugging this ansatz into (12.20) and taking into account the limit

$$\lim_{\xi \to 0^+} g_H(\xi, \theta) = h\frac{1-\cos\theta}{\sin\theta}$$

we obtain the following transcendental equation for the derivative $H'(0) = h$:

$$h = \frac{1}{\sqrt{\pi}} \int_0^{\frac{\pi}{2}} \left[\cos\theta - 2h^2 \cotg^2\theta\, (1-\cos\theta)\right] e^{-h^2(1-\cos\theta)^2/\sin^2\theta}\, d\theta.$$

Solving the above equation for $h \in \mathbb{R}$ we obtain $h \approx 0.451723$. It yields the well-known first order approximation of a solution $H(\xi)$ in the form

$$H^0(\xi) = 0.451723\,\xi,$$

i.e., $\rho^0(\tau) = \frac{rE}{q}\left(1 + 0.638833\,\sigma\sqrt{\tau}\right)$ (cf. Ševčovič [103]). This asymptotics is also in agreement with that of Dewynne et al. [36].

The computation of a solution of the nonlinear integral equation is based on an iterative method. We will construct a sequence of approximate solutions to (12.20). Let $H^0(\xi) := h\xi$ be an initial approximation of a solution to (12.20). Then, for $n = 0, 1, \ldots$, we will recurrently define the $n+1$ approximation H^{n+1} as follows:

$$H^{n+1}(\xi) = f_{H^n}(\xi)$$

$$+ \frac{1}{\sqrt{\pi}} \int_0^{\frac{\pi}{2}} [\xi\cos\theta - 2\cotg\theta\, H^n(\xi\cos\theta) g_{H^n}(\xi,\theta)] e^{-r\xi^2 \sin^2\theta - g_{H^n}^2(\xi,\theta)}\, d\theta, \qquad (12.29)$$

for $\xi \in [0, \sqrt{T}]$. With regard to (12.21) and (12.22), we have

$$g_{H^n}(\xi, \theta) = \frac{1}{\sigma\sqrt{2}}\frac{1}{\xi\sin\theta} \ln\left(\frac{1+\sigma\sqrt{2}H^n(\xi)}{1+\sigma\sqrt{2}H^n(\xi\cos\theta)}\right) + \frac{\Lambda}{\sqrt{2}}\xi\sin\theta,$$

$$f_{H^n}(\xi) = \frac{1}{2r\sqrt{\pi}\xi} e^{-r\xi^2 - \left(g_{H^n}(\xi,\frac{\pi}{2}) + \frac{1}{\xi}\frac{1}{\sigma\sqrt{2}}\ln\left(\frac{r}{q}\right)\right)^2}.$$

Notice that the function g_H is bounded provided that H is nonnegative and Lipschitz continuous on $[0, \sqrt{T}]$. Recall that we have assumed $r > q > 0$. Then the function $\xi \mapsto f_H(\xi)$ is bounded for $\xi \in [0, \sqrt{T}]$ and it vanishes at $\xi = 0$. Moreover, if H is smooth then f_H is a flat function at $\xi = 0$, i.e., $f_H(\xi) = o(\xi^n)$ as $\xi \to 0^+$ for all $n \in \mathbb{N}$. From the numerical point of view such a flat function can be omitted from computations. For small values of θ, we approximate the function $\cotg\theta\, g_{H^n}(\xi, \theta)$ entering the integrand in (12.29) by its limit when $\theta \to 0^+$. It yields the approximation of the singular term in (12.29) in the form

$$\cotg\theta\, g_H(\xi,\theta) \approx \frac{1}{2}\frac{H'(\xi)}{1+\sigma\sqrt{2}H(\xi)} + \frac{\Lambda}{\sqrt{2}}\xi \qquad \text{for } 0 < \theta \ll 1,$$

where $H = H^n$.

In what follows, we present several computational examples. In Fig. 12.1 (left) we show five iterates of the free boundary function $S_f(t)$, where the auxiliary function $H(\xi)$ is constructed by means of successive iterations of the nonlinear integral operator defined by the right-hand side of (12.20). This sequence converges to a fixed point of such a map, i.e., to a solution of (12.20). Parameter values were chosen as $E = 10, r = 0.1, \sigma = 0.2, q = 0.05, T = 1$. Fig. 12.1 (right) depicts the final tenth iteration of approximation of the function $S_f(t)$. The mesh contained 100 grid points. One can observe very rapid convergence of iterates to a fixed point. In practice, no more than six iterates are sufficient to obtain the fixed point of (12.20). It is worth noting that in all our numerical simulations the convergence was monotone, i.e., the curve moves only upwards in the iteration process.

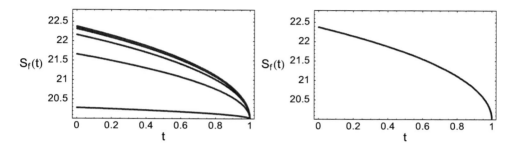

Figure 12.1. Five successive approximations of the free boundary $S_f(t)$ obtained from equation (12.20) (left). The profile of the solution $S_f(t)$. Source: Ševčovič [103] (right).

In Fig. 12.2 we show the long time behavior of the free boundary $S_f(t), t \in [0, T]$, for large values of the expiration time T. For the parameter values $T = 50, r = 0.1, q = 0.05, \sigma = 0.35$ and $E = 10$ the theoretical value of $S_f(0)$ for $T = +\infty$ is equal to 36.8179 (see Dewynne et al. [36]). This value corresponds to the price $\rho(+\infty) - E = 26.8179$ of the so-called perpetual call option.

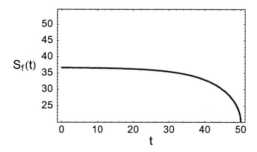

Figure 12.2. Long-time behavior of $S_f(t)$ for a large maturity. Source: Ševčovič [103].

In Table 12.1 we show a comparison of results obtained by the method based on the semi-explicit formula (12.25) and those obtained by known methods based on trinomial trees (both with the depth of the tree equal to 100), finite difference approximation (with 200 spatial and time grids) and analytic approximation of Barone-Adesi and Whalley (cf.

[12], [65, Ch. 15, p. 384]), respectively. It also turned out that the method based on solving the integral equation (12.20) is 5-10 times faster than other methods based on trees or finite differences. The reason is that the computation of $V(S,t)$ for a wide range of values of S based on the semi-explicit pricing formula (12.25) is very fast provided that the free boundary function ρ has already been computed.

Table 12.1. Comparison of the method based on formula (12.25) with other numerical methods for the parameter values $E = 10, T = 1, \sigma = 0.2, r = 0.1, q = 0.05$. The position $S_f(0) = \rho(T)$ of the free boundary at $t = 0$ (i.e., at $\tau = T$) was computed as $S_f(0) = \rho(T) = 22.3754$. Source: Ševčovič [103].

Method \ The asset value S	15	18	20	21	22.3754
Integral formula (12.25)	5.15	8.09	10.03	11.01	12.37
Trinomial tree	5.15	8.09	10.03	11.01	12.37
Finite differences	5.49	8.48	10.48	11.48	12.48
Analytic approximation	5.23	8.10	10.04	11.02	12.38

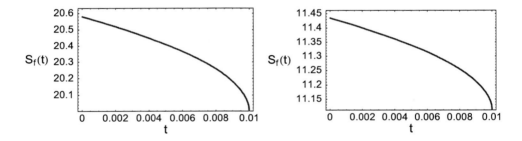

Figure 12.3. The early exercise boundary $S_f(t) = \rho(T-t)$ for parameters $r = 0.1, q = 0.05$ (left) and for $r = 0.1, q = 0.09$ (right). Source: Ševčovič [103].

In Fig. 12.3 the early exercise boundary $S_f(t)$ is computed for various values of the dividend yield parameter q. In these computations we chose $E = 10, T = 0.01, \sigma = 0.45$. Of interest is the case where q is close to r ($q = 0.09$ and $r = 0.1$).

12.1.4. Early Exercise Boundary for the American Put Option

In this section, we will review the fixed domain transformation method applied to construction of the early exercise boundary for American style of a put option. The method was derived and analyzed by Stamicar, Ševčovič and Chadam in [109] (see also the survey paper [107] by Ševčovič).

According to Chapter 9, the early exercise boundary problem for American put option can be formulated in terms of a solution to the parabolic partial differential equation

$$\frac{\partial V}{\partial t} + rS\frac{\partial V}{\partial S} + \frac{\sigma^2}{2}S^2\frac{\partial^2 V}{\partial S^2} - rV = 0, \quad 0 < t < T, \ S_f(t) < S < \infty,$$

$$V(+\infty, t) = 0, \ V(S_f(t), t) = E - S_f(t), \ \frac{\partial V}{\partial S}(S_f(t), t) = -1, \quad (12.30)$$

$$V(S, T) = (E - S)^+,$$

defined on a time-dependent domain $S \in (S_f(t), \infty)$, where $t \in (0, T)$ (cf. Chapter 9 or Kwok [75]). Again $S > 0$ stands for the underlying stock price, $E > 0$ is the exercise price, $r > 0$ is the risk-free rate and $\sigma > 0$ is the volatility of the underlying stock process. Throughout this section we shall assume that the underlying stock pays no dividends, i.e., $q = 0$. In order to perform a fixed domain transformation of the free boundary problem (12.31) we introduce the following change of variables

$$x = \ln\left(\frac{S}{\rho(\tau)}\right), \quad \text{where } \tau = T - t, \rho(\tau) = S_f(T - \tau).$$

Similarly as in the case of a call option we define a synthetised portfolio Π for the put option $\Pi(x, \tau) = V(S, t) - S\frac{\partial V}{\partial S}(S, t)$. Then it is easy to verify that Π is a solution to the following the parabolic equation

$$\frac{\partial \Pi}{\partial \tau} - a(\tau)\frac{\partial \Pi}{\partial x} - \frac{\sigma^2}{2}\frac{\partial^2 \Pi}{\partial x^2} + r\Pi = 0, \quad x > 0, \tau \in (0, T),$$

$$\Pi(0, \tau) = E, \quad \Pi(\infty, \tau) = 0, \quad (12.31)$$

$$\frac{\sigma^2}{2}\frac{\partial \Pi}{\partial x}(0, \tau) = -rE \quad \text{for } \tau \in (0, T), \quad (12.32)$$

$$\Pi(x, 0) = 0 \quad \text{for } x > 0,$$

where $a(\tau) = \frac{\dot{\rho}(\tau)}{\rho(\tau)} + r - \frac{\sigma^2}{2}$. Following calculations described in the previous sections 12.1.1. and 12.1.2. one can construct the Fourier image of the function Π in terms of the free boundary position ρ. The resulting equation for the free boundary position reads as $\frac{\sigma^2}{2}\frac{\partial \Pi}{\partial x}(0, \tau) = -rE$, from which the expression for the function ρ can be found by using the inverse Fourier transform. We refer the reader to the paper by Stamicar et al. [109] for details. After straightforward calculations, it turns out that the function $\rho(\tau)$ can be computed from the equation:

$$\rho(\tau) = Ee^{-(r-\frac{\sigma^2}{2})\tau + \sigma\sqrt{2\tau}\eta(\tau)}, \quad (12.33)$$

where the auxiliary function $\eta(\tau)$ is a solution to the following nonlinear integral equation

$$\eta(\tau) = -\sqrt{-\ln\left[\frac{r\sqrt{2\pi\tau}}{\sigma}e^{r\tau}\left(1 - \frac{F_\eta(\tau)}{\sqrt{\pi}}\right)\right]}, \quad \text{for } \tau \in [0, T], \quad (12.34)$$

with the function F_η depending on η through the expression

$$F_\eta(\tau) = 2\int_0^{\pi/2} e^{-r\tau\cos^2\theta - g_\eta^2(\tau,\theta)}\left(\frac{\sigma\sqrt{\tau}}{\sqrt{2}}\sin\theta + g_\eta(\tau,\theta)\tan\theta\right)d\theta, \quad (12.35)$$

$$g_\eta(\tau, \theta) = \frac{1}{\cos\theta}\left[\eta(\tau) - \eta(\tau\sin^2\theta)\sin\theta\right], \quad (12.36)$$

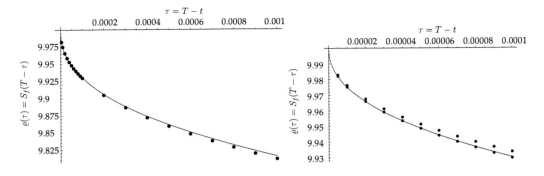

Figure 12.4. Left: asymptotic approximation vs. binomial method for $\sigma = 0.25, r = 0.1, E = 10$ and $T - t = 8.76$ hrs. (0.001 of a year), MBW approximation vs. the asymptotic solution (12.37) for $T - t = 0.876$ hrs (right). Source: Stamicar, Ševčovič and Chadam [109].

for $\tau \in [0, T], \theta \in [0, \frac{\pi}{2}]$. According to [109], the asymptotic analysis of the above integral equation for the unknown function $\eta(\tau)$ enables us to conclude the asymptotic approximation formula for $\eta(\tau)$ as $\tau \to 0$. The early exercise behavior of $\rho(\tau)$ for $\tau \to 0$ can be then deduced from the second order iteration to the system (12.34) and (12.35) when starting from the initial guess $\eta_0(\tau) = (r - \frac{\sigma^2}{2}) \frac{\tau^{\frac{1}{2}}}{\sigma\sqrt{2}}$ corresponding to the constant early exercise boundary $S_{f0}(t) \equiv E$. One can iteratively compute F_{η_0}, η_1 and F_{η_1}, η_2. It turned out from calculation performed in [109] that the second consecutive iterate η_2 is the lowest order (in τ) approximation of η. Namely,

$$\eta(\tau) = -\sqrt{-\ln\left[\frac{2r}{\sigma}\sqrt{2\pi\tau}e^{r\tau}\right]} + o(\sqrt{\tau}\sqrt{-\ln\tau}) \quad \text{as } \tau \to 0^+. \quad (12.37)$$

Interestingly enough, it has been shown just recently by Chen *et al.* [24] that the early exercise boundary function ρ is convex (see also Chadam [22], Ekström and Tysk [40], [41]). Moreover, the approximation formula (12.37) derived by Stamicar, Ševčovič and Chadam [109] provides the right asymptotic behavior for $\tau \to 0^+$. Furthermore, Chen *et al.* in [24] derived sixth-th order Taylor expansion of the function α,

$$\alpha(\tau) = -\xi - \frac{1}{2\xi} + \frac{1}{8\xi^2} + \frac{17}{24\xi^3} - \frac{51}{64\xi^4} - \frac{287}{120\xi^5} + \frac{199}{32\xi^6} + O(\xi^{-7}), \quad (12.38)$$

expressed in terms of the variable $\xi = \ln\sqrt{\frac{8\pi r^2 \tau}{\sigma^2}} \to -\infty$ as $\tau \to 0^+$ where the function α is given by the formula:

$$\rho(\tau) = E e^{-\sigma\sqrt{2\tau\alpha(\tau)}}. \quad (12.39)$$

In Fig. 12.4 (left) we examine how accurately our asymptotic approximation matches the data from the binomial method (cf. Kwok [75]). Near expiry at about one hour, the asymptotic approximation matches the data from the binomial method. With $\sigma = 0.25, r = 0.1, E = 10$ at 8.76 hours the approximation gives an overestimate but of only 0.4 cents (see [109]). In Fig. 12.4 (right) we also compared our asymptotic solution with MacMillan, Barone-Adesi, and Whalley's [12], [79] numerical approximation of the American put free boundary. We remind ourselves that the method due to MacMillan et al. is based on a transformation that results in a Cauchy-Euler equation that can be solved analytically. For times very close to expiry, one can see that approximation of the free boundary by Stamicar, Ševčovič and Chadam developed in [109] matches the data from the binomial and trinomial methods more accurately.

12.1.5. Analytical Approximation Valuation Formula by Zhu

In this section we show that the numerical and analytical results obtained by the transformation method due to Stamicar, Ševčovič and Chadam (see [109] and the previous section) are in agreement with those obtained recently by Zhu [125, 126]. In [125] Zhu derived a new analytical approximation formula of the early exercise boundary by application of the Laplace and inverse Laplace integral transforms to a dimensionless form of the governing parabolic PDE. He obtained a closed analytic approximation formula for the early exercise boundary position as a sum of a perpetual option and integral that valuates early exercise boundary position (see also [126]). The resulting formula for the early exercise boundary $S_f(t) = \rho(T-t)$ reads as follows:

$$\rho^{Zhu}(\tau) = \frac{\gamma E}{1+\gamma} + \frac{2E}{\pi} \int_0^\infty \frac{\zeta e^{-\tau \frac{\sigma^2}{2}(a^2+\zeta^2)}}{a^2+\zeta^2} e^{-f_1^*(\zeta)} \sin(f_2^*(\zeta)) d\zeta, \quad (12.40)$$

where $\gamma = \frac{2r}{\sigma^2}$, $a = \frac{1+\gamma}{2}$, $b = \frac{1-\gamma}{2}$ and

$$\begin{aligned}
f_1^*(\zeta) &= \frac{1}{b^2+\zeta^2}\left[b\ln\left(\frac{1}{\gamma}\sqrt{a^2+\zeta^2}\right) + \zeta\arctan(\zeta/a)\right], \\
f_2^*(\zeta) &= \frac{1}{b^2+\zeta^2}\left[\zeta\ln\left(\frac{1}{\gamma}\sqrt{a^2+\zeta^2}\right) - b\arctan(\zeta/a)\right].
\end{aligned} \quad (12.41)$$

Notice that the first summand in (12.40) represents the constant value of a perpetual put option i.e., the limit $\lim_{\tau \to \infty} \rho(\tau) = \gamma E/(1+\gamma)$.

In a recent paper by Lauko and Ševčovič [77] we presented qualitative and quantitative comparison of analytical and numerical approximation methods for computation of the early exercise boundary of the American put option paying zero dividends. We compared asymptotic formulae by Evans, Kuske and Keller [44], Stamicar, Ševčovič and Chadam [109] and the analytic approximation formula by Zhu [125, 125]. In Fig. 12.5 (left) we present a comparison of the analytic solution $\rho^{Zhu}(\tau)$ (12.40) for $\tau \in [0,T]$ and $E = 100, \sigma = 0.3, r = 0.1, T = 10^{-4}$. In [77] we furthermore proposed a new local iterative numerical scheme for construction of the entire early exercise boundary, which is based on a solution to the nonlinear integral equation (12.34). A comparison of numerical results obtained by the new method to those of the projected successive over relaxation method and the analytical approximation formula by Zhu is depicted in Fig. 12.5 (right).

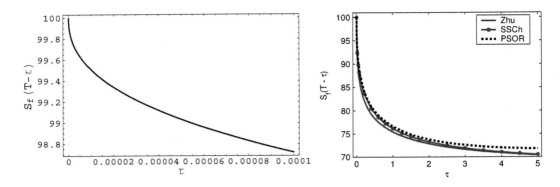

Figure 12.5. The profile the function ρ^{Zhu} (left) close to expiry. A comparison of the early exercise boundary position computed by various methods in the long time horizon (right). Source: Lauko and Ševčovič [77].

12.2. Transformation Method for a Class of Nonlinear Equations of the Black–Scholes Type

In Chapter 11 we have recalled and discused various nonlinear generalizations of the Black–Scholes equation. The main purpose of this section is to present the fixed domain transformation methodology adopted for pricing American style of options whose price is described by a solution to a generalized Black–Scholes equation with a volatility nonlinearly depending on a solution itself. We shall assume that the governing nonlinear Black–Scholes equation has the form

$$\frac{\partial V}{\partial t} + (r-q)S\frac{\partial V}{\partial S} + \frac{\sigma^2}{2}S^2\frac{\partial^2 V}{\partial S^2} - rV = 0, \qquad (12.42)$$

where

$$\sigma = \sigma(S^2 \partial_S^2 V, S, T-t)$$

is the volatility function which may depend on the second derivative of the option price $V = V(S,t)$, $S > 0$ is the price of an underlying asset and $T-t$ stands for the time to expiry. As usual, $r > 0$ is the interest rate of a zero-coupon bond, $q \geq 0$ is the dividend yield rate.

In the case of an American call option a solution to equation (12.42) is defined on a time dependent domain $0 < S < S_f(t)$, $0 < t < T$. It is subject to the boundary conditions

$$V(0,t) = 0, \quad V(S_f(t),t) = S_f(t) - E, \quad \frac{\partial V}{\partial S}(S_f(t),t) = 1, \qquad (12.43)$$

and terminal pay-off condition at expiry $t = T$,

$$V(S,T) = (S-E)^+, \qquad (12.44)$$

where $E > 0$ is a strike price. Analogously, one can construct boundary and terminal pay-off condition corresponding to the American put option (see Chapter 9).

The goal of this section is to present an iterative numerical algorithm for solving the free boundary problem. The algorithm was derived in the paper [104] by Ševčovič. Similarly as in the case of American style of plain vanilla options, the key idea of this method consists in transformation of the free boundary problem into a semilinear parabolic equation defined on a fixed spatial domain coupled with a nonlocal algebraic constraint equation for the free boundary position. It has been proposed and analyzed by Ševčovič in [109, 103, 104, 107]. The method has been also investigated and analyzed by Ehrhardt and Ankudinova in [8, 7]. In contrast to the transformation method presented in previous sections 12.1.1. and 12.1.4. we are unable to solve the resulting parabolic equation explicitly. This is due to the nonconstant character of the volatility function $\sigma = \sigma(S^2 \partial_S^2 V, S, T - t)$. Instead we have to compute the solution numerically. Since the resulting parabolic equation contains a strong convective term we make use of the operator splitting method in order to overcome numerical difficulties. A full space-time discretization of the problem leads to a system of semi-linear algebraic equations that can be solved by an iterative procedure at each time level. We apply the transformation method for calculations of the early exercise boundary for various nonlinear generalization of the Black–Scholes equation, including, in particular, the Risk adjusted pricing methodology model, Frey's model and the model by Barles and Soner studied in Chapter 11.

For the sake of simplicity we will present a detailed derivation of an equation only for the case of an American call option. Derivation of the corresponding equation for the American put option is similar. We shall again consider the following change of variables:

$$\tau = T - t, \quad x = \ln(\rho(\tau)/S), \quad \text{where } \rho(\tau) = S_f(T - \tau).$$

Clearly, $\tau \in (0, T)$ and $x \in (0, \infty)$ iff $S \in (0, S_f(t))$. The boundary value $x = 0$ corresponds to the free boundary position $S = S_f(t)$ whereas $x = +\infty$ corresponds to the default value $S = 0$ of the underlying asset. Following ideas from section 12.1.1. we construct the transformed function $\Pi = \Pi(x, \tau)$ defined as follows:

$$\Pi(x, \tau) = V(S, t) - S \frac{\partial V}{\partial S}(S, t). \tag{12.45}$$

Again, it represents a synthetic portfolio consisting of one long positioned option and $\Delta = \frac{\partial V}{\partial S}$ underlying short stocks. Similarly as in section 12.1.1. we have

$$\frac{\partial \Pi}{\partial x} = S^2 \frac{\partial^2 V}{\partial S^2}, \quad \frac{\partial \Pi}{\partial \tau} + \frac{\dot\rho}{\rho} \frac{\partial \Pi}{\partial x} = -\frac{\partial}{\partial t}\left(V - S \frac{\partial V}{\partial S}\right),$$

where we have denoted $\dot\rho = d\rho/d\tau$. Assuming sufficient smoothness of a solution $V = V(S,t)$ to (12.42) we can deduce from (12.42) the following parabolic equation for the synthetic portfolio function $\Pi = \Pi(x, \tau)$:

$$\frac{\partial \Pi}{\partial \tau} + \left(b(\tau) - \frac{\sigma^2}{2}\right)\frac{\partial \Pi}{\partial x} - \frac{1}{2}\frac{\partial}{\partial x}\left(\sigma^2 \frac{\partial \Pi}{\partial x}\right) + r\Pi = 0,$$

defined on a fixed domain $x \in \mathbb{R}, t \in (0, T)$, with a time-dependent coefficient

$$b(\tau) = \frac{\dot\rho(\tau)}{\rho(\tau)} + r - q, \tag{12.46}$$

and a diffusion coefficient σ^2 given by: $\sigma^2 = \sigma^2(\partial_x \Pi(x,\tau), \rho(\tau)e^{-x}, \tau)$. It may depend on τ, x and the gradient $\partial_x \Pi$ of a solution Π. Now the boundary conditions $V(0,t) = 0$, $V(S_f(t),t) = S_f(t) - E$ and $\frac{\partial V}{\partial S}(S_f(t),t) = 1$ imply

$$\Pi(0,\tau) = -E, \quad \Pi(+\infty,\tau) = 0, \quad 0 < \tau < T, \tag{12.47}$$

and, from the terminal pay-off diagram for $V(S,T)$, we deduce

$$\Pi(x,0) = \begin{cases} -E & \text{for } x < \ln\left(\frac{\rho(0)}{E}\right), \\ 0, & \text{otherwise.} \end{cases} \tag{12.48}$$

In order to close the system of equations determining the value of a synthetic portfolio Π we have to construct an equation for the free boundary position $\rho(\tau)$. Indeed, both the coefficient b as well as the initial condition $\Pi(x,0)$ depend on the function $\rho(\tau)$. Similarly as in the case of a constant volatility σ (see [103, 109]) we proceed as follows: since $S_f(t) - E = V(S_f(t),t)$ and $\partial_S V(S_f(t),t) = 1$ we have $\frac{d}{dt}S_f(t) = \partial_S V(S_f(t),t)\frac{d}{dt}S_f(t) + \partial_t V(S_f(t),t)$ and so $\partial_t V(S,t) = 0$ along the free boundary $S = S_f(t)$. Moreover, assuming $\partial_x \Pi$ is continuous up to the boundary $x = 0$ we obtain $S^2 \partial_S^2 V(S,t) \to \partial_x \Pi(0,\tau)$ and $S \partial_S V(S,t) \to \rho(\tau)$ as $S \to S_f(t)^-$. Now, by taking the limit $S \to S_f(t)^-$ in the Black–Scholes equation (11.2) we obtain $(r-q)\rho(\tau) + \frac{1}{2}\sigma^2 \partial_x \Pi(0,\tau) - r(\rho(\tau) - E) = 0$. Therefore

$$q\rho(\tau) = rE + \frac{1}{2}\sigma^2(\partial_x \Pi(0,\tau), \rho(\tau), \tau)\frac{\partial \Pi}{\partial x}(0,\tau),$$

for $0 < \tau \leq T$. The value of $\rho(0)$ can be derived from the smoothness assumption made on $\partial_x \Pi$ at the origin $(x,\tau) = (0,0)$ under the structural assumption

$$0 < q \leq r$$

made on the interest and dividend yield rates r, q (cf. [103, 104]). The assumption on continuity of $\partial_x \Pi$ at the origin $(0,0)$ implies $\lim_{\tau \to 0^+} \partial_x \Pi(0,\tau) = \partial_x \Pi(0,0) = \lim_{x \to 0^+} \partial_x \Pi(x,0) = 0$ because $\Pi(x,0) = -E$, for x close to 0^+ provided $\ln(r/q) > 0$. From the above equation for $\rho(\tau)$ we deduce $\rho(0) = \frac{rE}{q}$ by taking the limit $\tau \to 0^+$. Putting all the above equations together we end up with a closed system of equations for $\Pi = \Pi(x,\tau)$ and $\rho = \rho(\tau)$

$$\frac{\partial \Pi}{\partial \tau} + \left(b(\tau) - \frac{\sigma^2}{2}\right)\frac{\partial \Pi}{\partial x} - \frac{1}{2}\frac{\partial}{\partial x}\left(\sigma^2 \frac{\partial \Pi}{\partial x}\right) + r\Pi = 0,$$
$$\Pi(0,\tau) = -E, \quad \Pi(+\infty,\tau) = 0, \, x > 0, \tau \in (0,T),$$
$$\Pi(x,0) = \begin{cases} -E & \text{for } x < \ln(r/q), \\ 0, & \text{otherwise,} \end{cases} \tag{12.49}$$

where $\sigma^2 = \sigma^2(\partial_x \Pi(x,\tau), \rho(\tau)e^{-x}, \tau)$, $b(\tau) = \frac{\dot{\rho}(\tau)}{\rho(\tau)} + r - q$ and the free boundary position $\rho(\tau) = S_f(T-\tau)$ satisfies the implicit algebraic equation:

(Algebraic part)

$$\rho(\tau) = \frac{rE}{q} + \frac{\sigma^2(\partial_x \Pi(0,\tau), \rho(\tau), \tau)}{2q}\frac{\partial \Pi}{\partial x}(0,\tau), \quad \text{with } \rho(0) = \frac{rE}{q}, \tag{12.50}$$

where $\tau \in (0,T)$. In order to guarantee parabolicity of equation (12.49) we have to assume that the function $p \mapsto \sigma^2(p,\rho(\tau)e^{-x},\tau)p$ is strictly increasing. More precisely, we shall assume that there exists a positive constant $\gamma > 0$ such that

$$\sigma^2(p,\xi,\tau) + p\partial_p \sigma^2(p,\xi,\tau) \geq \gamma > 0 \qquad (12.51)$$

for any $\xi > 0, \tau \in (0,T)$ and $p \in \mathbb{R}$. Notice that condition (12.51) is satisfied for the RAPM model in which $\sigma^2 = \hat{\sigma}^2(1 + \mu p^{\frac{1}{3}} \xi^{-\frac{1}{3}})$ for any $\mu \geq 0$ and $p \geq 0$. Clearly $p = S^2 \partial_S^2 V > 0$ for the case of plain vanilla call or put options. As far as the Barles and Soner model is concerned, we have $\sigma^2 = \hat{\sigma}^2(1 + \Psi(a^2 e^{rt} p))$ and condition (12.51) is again satisfied because the function Ψ is a positive and increasing function in the Barles and Soner model.

Finally, by following exactly the same argument as in (12.24) one can derive an explicit expression for the option price $V(S,t)$:

$$V(S, T-\tau) = \frac{S}{\rho(\tau)} \left(\rho(\tau) - E + \int_0^{\ln \frac{\rho(\tau)}{S}} e^x \Pi(x,\tau) \, dx \right).$$

12.2.1. Alternative Representation of the Early Exercise Boundary

Although equation (12.50) provides an algebraic formula for the free boundary position $\rho(\tau)$ in terms of the derivative $\partial_x \Pi(0,\tau)$ such an expression is not quite suitable for construction of a robust numerical approximation scheme. The reason is that any small inaccuracy in approximation of the value $\partial_x \Pi(0,\tau)$ is transferred in to the entire computational domain $x \in (0,\infty)$ making thus a numerical scheme very sensitive to the value of the derivative of a solution evaluated in one point $x = 0$. In what follows, we present an equivalent equation for the free boundary position $\rho(\tau)$, which is more robust from the numerical approximation point of view.

By Integration the governing equation (12.49) with respect to $x \in (0,\infty)$ we obtain

$$\frac{d}{d\tau} \int_0^\infty \Pi dx + \int_0^\infty \left(b(\tau) - \frac{\sigma^2}{2} \right) \frac{\partial \Pi}{\partial x} dx - \frac{1}{2} \int_0^\infty \frac{\partial}{\partial x} \left(\frac{\sigma^2}{2} \frac{\partial \Pi}{\partial x} \right) dx + r \int_0^\infty \Pi dx = 0.$$

Now, taking into account the boundary conditions $\Pi(0,\tau) = -E, \Pi(\infty,\tau) = 0$ and consequently $\partial_x \Pi(\infty,\tau) = 0$ we obtain, by applying condition (12.50), the following differential equation:

(Integral form of the algebraic part)

$$\frac{d}{d\tau} \left(E \ln \rho(\tau) + \int_0^\infty \Pi(x,\tau) dx \right) + q\rho(\tau) - qE$$

$$+ \int_0^\infty \left(-\frac{1}{2} \sigma^2(\partial_x \Pi(x,\tau), \rho(\tau)e^{-x}, \tau) \frac{\partial \Pi}{\partial x}(x,\tau) + r\Pi(x,\tau) \right) dx = 0. \qquad (12.52)$$

12.2.2. Numerical Iterative Algorithm for Approximation of the Early Exercise Boundary

The idea of the iterative numerical algorithm for solving the problem (12.49), (12.50) is rather simple: we use the backward Euler method of finite differences in order to discretize the parabolic equation (12.49) in time. At each time level we find a new approximation of a solution pair (Π, ρ). First we determine a new position of ρ from the algebraic equation (12.50). We remind ourselves that (even in the case σ is constant) the free boundary function $\rho(\tau)$ behaves like $rE/q + O(\tau^{1/2})$ for $\tau \to 0^+$ (see e.g. [36] or [103]) and so $b(\tau) = O(\tau^{-1/2})$. Hence the convective term $b(\tau)\partial_x \Pi$ becomes a dominant part of equation (12.49) for small values of τ. In order to overcome this difficulty we employ the operator splitting technique for successive solving of the convective and diffusion parts of equation (12.49).

Now we present our algorithm in more details. We restrict the spatial domain $x \in (0, \infty)$ to a finite interval of values $x \in (0, L)$ where $L > 0$ is sufficiently large. For practical purposes one can take $L \approx 3$ as it corresponds to the interval $S \in (S_f(t)e^{-L}, S_f(t))$ in the original asset price variable S. The value $S_f(t)e^{-L}$ is then could be a good approximation for the default value $S = 0$ if $L \approx 3$. Let us denote by $k > 0$ the time step, $k = T/m$ and by $h > 0$ the spatial step, $h = L/n$ where $m, n \in \mathbb{N}$ stand for the number of time and space discretization steps, respectively. We denote by Π_i^j an approximation of $\Pi(x_i, \tau_j)$, $\rho^j \approx \rho(\tau_j)$, $b^j \approx b(\tau_j)$ where $x_i = ih, \tau_j = jk$. We approximate the value of the volatility σ at the node (x_i, τ_j) by the finite difference approximation as follows:

$$\sigma_i^j = \sigma_i^j(\rho^j, \Pi^j) = \sigma((\Pi_{i+1}^j - \Pi_i^j)/h, \rho^j e^{-x_i}, \tau_j).$$

Then for the Euler backward in time finite difference approximation of equation (12.49) we have

$$\frac{\Pi^j - \Pi^{j-1}}{k} + \left(b^j - \frac{1}{2}(\sigma^j)^2\right)\frac{\partial}{\partial x}\Pi^j - \frac{1}{2}\frac{\partial}{\partial x}\left((\sigma^j)^2 \frac{\partial}{\partial x}\Pi^j\right) + r\Pi^j = 0 \quad (12.53)$$

and the solution $\Pi^j = \Pi^j(x)$ is subject to the Dirichlet boundary conditions at $x = 0$ and $x = L$. We set $\Pi^0(x) = \Pi(x, 0)$. Next we decompose the above problem into two parts - a convection part and a diffusive part by introducing an auxiliary intermediate step $\Pi^{j-\frac{1}{2}}$:

(*Convective part*)
$$\frac{\Pi^{j-\frac{1}{2}} - \Pi^{j-1}}{k} + b^j \frac{\partial}{\partial x}\Pi^{j-\frac{1}{2}} = 0, \quad (12.54)$$

(*Diffusive part*)
$$\frac{\Pi^j - \Pi^{j-\frac{1}{2}}}{k} - \frac{(\sigma^j)^2}{2}\frac{\partial}{\partial x}\Pi^j - \frac{1}{2}\frac{\partial}{\partial x}\left((\sigma^j)^2 \frac{\partial}{\partial x}\Pi^j\right) + r\Pi^j = 0. \quad (12.55)$$

The idea of the operator splitting technique consists in comparison the sum of solutions to convective and diffusion part to a solution of (12.53). Indeed, if $\partial_x \Pi^j \approx \partial_x \Pi^{j-\frac{1}{2}}$ then it is reasonable to assume that Π^j computed from the system (12.54)–(12.55) is a good approximation of the system (12.53).

The convective part can be approximated by an explicit solution to the transport equation:

$$\partial_\tau \tilde{\Pi} + b(\tau)\partial_x \tilde{\Pi} = 0, \quad \text{for } x > 0, \ \tau \in (\tau_{j-1}, \tau_j], \quad (12.56)$$

subject to the boundary condition $\tilde{\Pi}(0, \tau) = -E$ and initial condition $\tilde{\Pi}(x, \tau_{j-1}) = \Pi^{j-1}(x)$. For American style of call option the free boundary $\rho(\tau) = S_f(T - \tau)$ must be an increasing function in τ. Since we have assumed $0 < q < r$ we conclude $b(\tau) = \dot{\rho}(\tau)/\rho(\tau) + r - q > 0$ and so prescribing the in-flowing boundary condition $\tilde{\Pi}(0,\tau) = -E$ is consistent with the transport equation. Let us denote by $B(\tau)$ the primitive function to $b(\tau)$, i.e., $B(\tau) = \ln\rho(\tau) + (r-q)\tau$. Equation (12.56) can be integrated to obtain its explicit solution:

$$\tilde{\Pi}(x,\tau) = \begin{cases} \Pi^{j-1}(x - B(\tau) + B(\tau_{j-1})), & \text{if } x - B(\tau) + B(\tau_{j-1}) > 0, \\ -E, & \text{otherwise.} \end{cases} \quad (12.57)$$

Thus the spatial approximation $\Pi_i^{j-\frac{1}{2}}$ can be constructed from the formula

$$\Pi_i^{j-\frac{1}{2}} = \begin{cases} \Pi^{j-1}(\xi_i) & \text{if } \xi_i = x_i - \ln\rho_j + \ln\rho_{j-1} - (r-q)k > 0, \\ -E, & \text{otherwise,} \end{cases} \quad (12.58)$$

where a piecewise linear interpolation between discrete values $\Pi_i^{j-1}, i = 0, 1, \ldots, n$, is being used to compute the value $\Pi^{j-1}(x_i - \ln\rho_j + \ln\rho_{j-1} - (r-q)k)$.

The diffusive part can be solved numerically by means of finite differences. Using central finite difference for approximation of the derivative $\partial_x \Pi^j$ we obtain

$$\frac{\Pi_i^j - \Pi_i^{j-\frac{1}{2}}}{k} + r\Pi_i^j - \frac{(\sigma_i^j)^2}{2}\frac{\Pi_{i+1}^j - \Pi_{i-1}^j}{2h}$$
$$- \frac{1}{2h}\left((\sigma_i^j)^2\frac{\Pi_{i+1}^j - \Pi_i^j}{h} - (\sigma_{i-1}^j)^2\frac{\Pi_i^j - \Pi_{i-1}^j}{h}\right) = 0.$$

Hence, the vector of discrete values $\Pi^j = \{\Pi_i^j, i = 1, 2, \ldots, n\}$ at the time level $j \in \{1, 2, \ldots, m\}$ satisfies the tridiagonal system of equations:

$$\alpha_i^j \Pi_{i-1}^j + \beta_i^j \Pi_i^j + \gamma_i^j \Pi_{i+1}^j = \Pi_i^{j-\frac{1}{2}}, \quad (12.59)$$

for $i = 1, 2, \ldots, n$, where

$$\begin{aligned} \alpha_i^j &\equiv \alpha_i^j(\rho^j, \Pi^j) = -\frac{k}{2h^2}(\sigma_{i-1}^j)^2 + \frac{k}{2h}\frac{(\sigma_i^j)^2}{2}, \\ \gamma_i^j &\equiv \gamma_i^j(\rho^j, \Pi^j) = -\frac{k}{2h^2}(\sigma_i^j)^2 - \frac{k}{2h}\frac{(\sigma_i^j)^2}{2}, \\ \beta_i^j &\equiv \beta_i^j(\rho^j, \Pi^j) = 1 + rk - (\alpha_i^j + \gamma_i^j). \end{aligned} \quad (12.60)$$

The initial and boundary conditions at $\tau = 0$ and $x = 0, L$, respectively., can be approximated as follows:

$$\Pi_i^0 = \begin{cases} -E & \text{for } x_i < \ln(r/q), \\ 0 & \text{for } x_i \geq \ln(r/q), \end{cases}$$

for $i = 0, 1, \ldots, n$ and $\Pi_0^j = -E$, $\Pi_n^j = 0$.

Next we proceed by approximation of equation (12.50) which introduces a nonlinear constraint condition between the early exercise boundary function $\rho(\tau)$ and the trace of the solution Π at the boundary $x = 0$ ($S = S_f(t)$ in the original variable). Taking a finite difference approximation of $\partial_x \Pi$ at the origin $x = 0$ we obtain

(*Algebraic part*)

$$\rho^j = \frac{rE}{q} + \frac{1}{2q}\sigma^2\left((\Pi_1^j - \Pi_0^j)/h, \rho^j, \tau_j\right)\frac{\Pi_1^j - \Pi_0^j}{h}. \qquad (12.61)$$

Recall that the above discretization of the algebraic constraint (12.50) can be replaced by its integral form derived in section 12.2.1.. The formula (12.52) can be discretized as follows:

(*Integral form of the algebraic part*)

$$E\ln\rho^j = E\ln\rho^{j-1} + I_0(\Pi^{j-1}) - I_0(\Pi^j) + k\left(qE - q\rho^j - I_1(\rho^j, \Pi^j)\right), \qquad (12.62)$$

where $I_0(\Pi)$ stands for numerical trapezoid quadrature of the integral $\int_0^\infty \Pi(\xi)d\xi$ whereas $I_1(\rho^j,\Pi)$ is a trapezoid quadrature of the second integral in (12.52), i.e.

$$I_1(\rho^j,\Pi) \approx \int_0^\infty \left(-\frac{1}{2}\sigma^2(\partial_x\Pi(x),\rho^j e^{-x},\tau_j)\frac{\partial\Pi}{\partial x}(x) + r\Pi(x)\right)dx.$$

Now, equations (12.58), (12.59) and (12.61) can be rewritten in the abstract operator form:

$$\begin{aligned}
\rho^j &= \mathcal{F}(\Pi^j, \rho^j), \\
\Pi^{j-\frac{1}{2}} &= \mathcal{T}(\Pi^j, \rho^j), \\
\mathcal{A}(\Pi^j, \rho^j)\Pi^j &= \Pi^{j-\frac{1}{2}},
\end{aligned} \qquad (12.63)$$

where $\mathcal{F}(\Pi^j,\rho^j)$ is the right-hand side of the algebraic equation (12.61). Alternatively, $\mathcal{F}(\Pi^j,\rho^j)$ can be computed by using (12.62). The operator $\mathcal{T}(\Pi^j,\rho^j)$ is the transport equation solver given by the right-hand side of (12.58) and $\mathcal{A} = \mathcal{A}(\Pi^j,\rho^j)$ is a tridiagonal matrix with coefficients given by (12.60). The system (12.63) can be approximately solved by means of successive iterations procedure. We define, for $j \geq 1$, $\Pi^{j,0} = \Pi^{j-1}, \rho^{j,0} = \rho^{j-1}$. Then the $(p+1)$-th approximation of Π^j and ρ^j is obtained as a solution to the system:

$$\begin{aligned}
\rho^{j,p+1} &= \mathcal{F}(\Pi^{j,p}, \rho^{j,p}), \\
\Pi^{j-\frac{1}{2},p+1} &= \mathcal{T}(\Pi^{j,p}, \rho^{j,p+1}), \\
\mathcal{A}(\Pi^{j,p}, \rho^{j,p+1})\Pi^{j,p+1} &= \Pi^{j-\frac{1}{2},p+1}.
\end{aligned} \qquad (12.64)$$

Notice that the last equation represents a tridiagonal system of linear equation for the vector $\Pi^{j,p+1}$ whereas $\rho^{j,p+1}$ and $\Pi^{j-\frac{1}{2},p+1}$ can be directly computed from (12.61) and (12.58), respectively. If the sequence of approximate solutions $\{(\Pi^{j,p},\rho^{j,p})\}_{p=1}^\infty$ converges to some limiting value $(\Pi^{j,\infty},\rho^{j,\infty})$ as $p \to \infty$ then this limit is a solution to a nonlinear system of equations (12.63) at the time level j and we can proceed by computing the approximate solution the next time level $j+1$.

12.2.3. Numerical Approximations of the Early Exercise Boundary

In this section we focus on numerical experiments based on the iterative scheme described in the previous section. All numerical examples are borrowed from the paper [107] by Ševčovič. The main purpose is to compute the free boundary profile $S_f(t) = \rho(T-t)$ for different nonlinear generalizations of the Black–Scholes models discussed in Chapter 11. A solution (Π, ρ) has been computed by our iterative algorithm for the following basic model parameters: $E = 10, T = 1$ (one year), $r = 0.1$ (10% p.a) , $q = 0.05$ (5% p.a.) and $\hat{\sigma} = 0.2$. We used $n = 750$ spatial points and $m = 225000$ time discretization steps. Such a time step $k = T/m$ corresponds to 140 seconds between consecutive time levels when expressed in real time scale. In average, we needed $p \leq 6$ micro-iterates (12.64) in order to solve the nonlinear system (12.63) with the precision 10^{-7}.

The Case of a Constant Volatility. A Comparison Study

In our first numerical experiment we make attempt to compare our iterative approximation scheme for solving the free boundary problem for an American call option to known schemes in the case when the volatility $\sigma > 0$ is constant. We compare our solution to the one computed by means of a solution to a nonlinear integral equation for $\rho(\tau)$ (see also [103, 109]). This comparison can be also considered as a benchmark or test example for which we know a solution that can be computed by another justified algorithm. In Fig. 12.6, part a), we show the function ρ computed by our iterative algorithm for $E = 10, T = 1, r = 0.1, q = 0.05, \sigma = 0.2$. At the expiry $T = 1$, the value of $\rho(T)$ was computed as: $\rho(T) = 22.321$. The corresponding value $\rho(T)$ computed from the integral equation (12.18) (cf. [103]) was $\rho(T) = 22.375$. The relative error is less than 0.25%. In the part b) we present 7 approximations of the free boundary function $\rho(\tau)$ computed for different mesh sizes h (see Tab. 12.2 for details). The sequence of approximate free boundaries $\rho_h, h = h_1, h_2, \ldots$, converges monotonically from below to the free boundary function ρ as $h \downarrow 0$. The next part c) of Fig. 12.6 depicts various solution profiles of a function $\Pi(x, \tau)$. In order to achieve a reasonable approximation to equation (12.61) we need very accurate approximation of $\Pi(x, \tau)$ for x close to the origin 0. The parts d) and e) of Fig. 12.6 depict the contour and 3D plots of the function $\Pi(x, \tau)$.

In Tab. 12.2 we present the numerical error analysis for the distance $\|\rho_h - \rho\|_p$ measured in two different norms (L^∞ and L^2) of a computed free boundary position ρ_h corresponding to the mesh size h and the solution ρ computed from the integral equation described in (12.18) (cf. [103]). The time step k has been adjusted to the spatial mesh size h in order to satisfy CFL condition $\hat{\sigma}^2 k/h^2 \approx 1/2$. We also computed the experimental order of convergence $\text{EOC}(L^p)$ for $p = 2, \infty$. It is defined as the ratio:

$$\text{EOC}(L^p) = \frac{\ln(\|\rho_{h_i} - \rho\|_p) - \ln(\|\rho_{h_{i-1}} - \rho\|_p)}{\ln h_i - \ln h_{i-1}}.$$

It can be interpreted as such an exponent $\alpha = \text{EOC}(L^p)$ for which we have $\|\rho_h - \rho\|_p = O(h^\alpha)$ for $h \to 0$. It turns out from Tab. 12.2 that the conjecture on the order of convergence $\|\rho_h - \rho\|_\infty = O(h)$ whereas $\|\rho_h - \rho\|_2 = O(h^{3/2})$ as $h \to 0^+$ can be reasonable.

Risk Adjusted Pricing Methodology Model

In the next example we computed the position of the free boundary $\rho(\tau)$ in the case of the Risk Adjusted Pricing Methodology model. It is a nonlinear Black–Scholes type model derived by Jandačka and Ševčovič in [66] and recalled in Chapter 11. In this model the

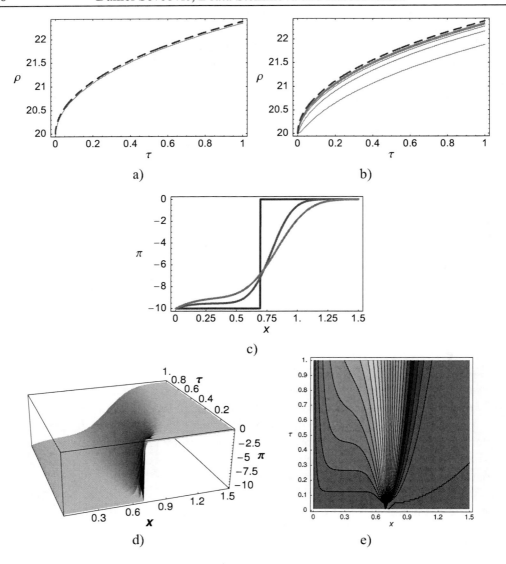

Figure 12.6. a) A comparison of the free boundary function $\rho(\tau)$ computed by the iterative algorithm (solid curve) to the integral equation based approximation (dashed curve); b) free boundary positions computed for various mesh sizes; c) solution profiles $\Pi(x,\tau)$ for $\tau=0$, $\tau=T/2$, $\tau=T$; d) 3D plot and e) contour plot of the function $\Pi(x,\tau)$. Source: Ševčovič [107].

volatility σ is a nonlinear function of the asset price S and the second derivative $\partial_S^2 V$ of the option price. The volatility function is given by formula (11.8). In Fig. 12.7 we present results of numerical approximation of the free boundary position $\rho^R(\tau) = S_f^R(T-\tau)$ in the case when the coefficient of transaction costs $C = 0.01$ is fixed and the risk premium measure R varies from $R = 5, 15, 40, 70$, up to $R = 100$. We compare the position of the free boundary $\rho^R(\tau)$ to the case when there are no transaction costs and no risk from volatile

Table 12.2. Experimental order of convergence of the iterative algorithm for approximating the free boundary position. Source: Ševčovič [107].

h	err(L^∞)	EOC(L^∞)	err(L^2)	EOC(L^2)
0.03	0.5	-	0.808	-
0.012	0.215	0.92	0.227	1.39
0.006	0.111	0.96	0.0836	1.44
0.004	0.0747	0.97	0.0462	1.46
0.003	0.0563	0.98	0.0303	1.47
0.0024	0.0452	0.98	0.0218	1.48
0.002	0.0378	0.98	0.0166	1.48

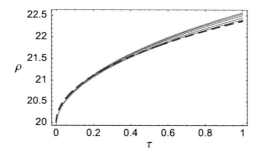

Figure 12.7. A comparison of the free boundary function $\rho^R(\tau)$ computed for the Risk Adjusted Pricing Methodology model. Dashed curve represents a solution corresponding to $R = 0$, whereas the solid curves represent a solution $\rho^R(\tau)$ for different values of the risk premium coefficients $R = 5, 15, 40, 70, 100$. Source: Ševčovič [107].

portfolio, i.e., we compare it with the free boundary position $\rho^0(\tau)$ for the linear Black–Scholes equation. An increase in the risk premium coefficient R resulted in an increase of the free boundary position as it can be expected.

In Tab. 12.3 and Fig. 12.8 we summarize results of comparison of the free boundary position ρ^R for various values of the risk premium coefficient to the reference position $\rho = \rho^0$ computed from the Black–Scholes model with a constant volatility $\sigma = \hat\sigma$, i.e., $R = 0$. The experimental order α_p of the distance function $\|\rho^R - \rho^0\|_p = O(R^{\alpha_p})$ has been computed for $p = 2, \infty$, as follows:

$$\alpha_p = \frac{\ln(\|\rho^{R_i} - \rho^0\|_p) - \ln(\|\rho^{R_{i-1}} - \rho^0\|_p)}{\ln R_i - \ln R_{i-1}}.$$

According to the values presented in Tab. 12.3 it turns out that a reasonable conjecture on the order of convergence is that $\|\rho^R - \rho^0\|_p = O(R^{1/3})$ for both norms $p = 2$ and $p = \infty$. Since the transaction cost coefficient C and risk premium measure R enter the expression for the RAPM volatility (11.8) only in the product $C^2 R$ we can conjecture that $\|\rho^{R,C} - \rho^{0,0}\|_p = O(C^{2/3} R^{1/3})$ as either $C \to 0^+$ or $R \to 0^+$.

Table 12.3. The distance $\|\rho^R - \rho^0\|_p$ ($p=2,\infty$) of the free boundary position ρ^R from the reference free boundary position ρ^0 and experimental orders α_∞ and α_2 of convergence. Source: Ševčovič [107].

R	$\|\rho^R - \rho^0\|_\infty$	α_∞	$\|\rho^R - \rho^0\|_2$	α_2
1	0.0601	-	0.0241	-
2	0.0754	0.33	0.0303	0.328
5	0.102	0.33	0.0408	0.326
10	0.128	0.33	0.0511	0.324
15	0.145	0.32	0.0582	0.323
20	0.16	0.32	0.0639	0.322
30	0.182	0.32	0.0727	0.321
40	0.2	0.32	0.0798	0.32
50	0.214	0.32	0.0856	0.319
60	0.227	0.32	0.0907	0.318
70	0.239	0.32	0.0953	0.317
80	0.249	0.32	0.0994	0.317
90	0.259	0.32	0.103	0.316
100	0.268	0.32	0.107	0.316

Barles and Soner Model

Our next example is devoted to the nonlinear Black–Scholes model due to Barles and Soner (see [11]). In this model the volatility is given by equation (11.6). Numerical results are depicted in Fig. 12.9. Choosing a larger value of the risk aversion coefficient $a > 0$ resulted in increase of the free boundary position $\rho^a(\tau)$. The position of the early exercise boundary $\rho^a(\tau)$ has considerably increased in comparison to the linear Black–Scholes equation with constant volatility $\sigma = \hat{\sigma}$. In contrast to the case of constant volatility as well as the RAPM model, there is, at least a numerical evidence (see Fig.12.9 and ρ^a for the largest value $a = 0.35$) that the free boundary profile $\rho^a(\tau)$ need not be necessarily convex. Recall that that convexity of the free boundary profile has been proved analytically by Ekström *et al.* and Chen *et al.* in a recent papers [24, 40, 41] in the case of a American put option and constant volatility $\sigma = \hat{\sigma}$.

Similarly as in the RAPM model we have also investigated the dependence of the free boundary position $\rho = \rho^a(\tau)$ on the risk aversion parameter $a > 0$. In Fig. 12.10 we present results of comparison of the free boundary position ρ^a for various values of the risk aversion coefficient a to the reference position $\rho = \rho^0$.

12.3. Early Exercise Boundary for American Style of Asian Options

The fixed domain transformation method discussed in this chapter can be successfully adopted for construction of the early exercise boundary for American style of Asian average strike options. It is the purpose of this section to review recent results in this field due

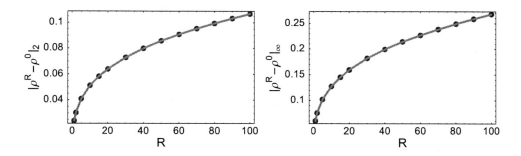

Figure 12.8. The dependence of the norms $\|\rho^R - \rho^0\|_p$ ($p = \infty, 2$) of the deviation of the free boundary $\rho = \rho^R(\tau)$ for the RAPM model on the risk premium coefficient R. Source: Ševčovič [107].

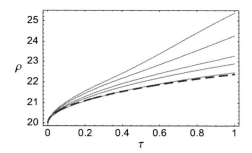

Figure 12.9. A comparison of the free boundary function $\rho(\tau)$ computed for the Barles and Soner model. Dashed curve represents a solution corresponding to $a = 0$, whereas the solid curves represents a solution $\rho(\tau)$ for different values of the risk aversion coefficient $a = 0.01, 0.07, 0.13, 0.25, 0.35$. Source: Ševčovič [107].

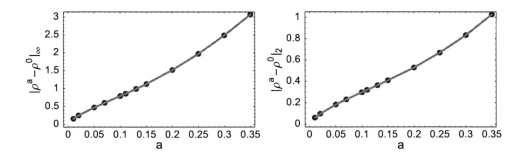

Figure 12.10. Dependence of the norms $\|\rho^a - \rho^0\|_p$ ($p = \infty, 2$) of the deviation of the free boundary $\rho = \rho^a(\tau)$ for the Barles-Soner model on the risk aversion parameter a. Source: Ševčovič [107].

to Ševčovič and Bokes in [107] and [15]. Recall that Asian options belong to a group of the so-called path-dependent options (cf. Chapter 6). Their pay-off diagrams depend on the value of the underlying asset price during the whole or some part(s) of the life span of the option. Usually Asian options may depend on the (arithmetic or geometric) average of the spot price of the underlying. In recent years many authors have investigated the problem of construction of the early exercise boundary for Asian options by means of integral transformation methods or adopted PSOR algorithms (cf. Hansen and Jørgensen [59], Pascucci [91], Wu, Kwok and Yu [123], Wu and Fu [124] and others).

In this section we focus on the so-called the floating strike Asian call option which exercise price depends on the averaged path history of the underlying asset. More precisely, we are interested in pricing American-style Asian call and put options having the pay-off functions $V(S,A,T) = (S-A)^+$ and $V(S,A,T) = (A-S)^+$, respectively. The strike price A is given as an average of the underlying over the time history $[0,T]$ (cf. [103, 15, 59, 33, 76]). Our goal is to propose an efficient numerical algorithm for determining the free boundary position for American-style of Asian options. Similarly as in the previous section, construction of the algorithm will be based on a solution to a nonlocal parabolic partial differential equation. We will also present numerical results obtained by the transformation method and compare them to those of the Projected successive over relaxation method recently developed and adopted for Asian options by Dai and Kwok in [33].

Recall that in Chapter 6 we derived a partial differential equation for pricing the Asian options in the form of a parabolic equation:

$$\frac{\partial V}{\partial t} + \frac{\sigma^2}{2}S^2\frac{\partial^2 V}{\partial S^2} + (r-q)S\frac{\partial V}{\partial S} + Af\left(\frac{A}{S},t\right)\frac{\partial V}{\partial A} - rV = 0, \qquad (12.65)$$

for the price of the Asian option price $V(S,A,t)$, where $0 < t < T$, $S,A > 0$ (see Chapter 6 and [33, 75]). For the Asian call option the above equation is subject to the terminal pay-off condition $V(T,S,A) = (S-A)^+, S,A > 0$. Here A denotes the continuously averaged underlying asset price S (see Chapter 6 for details).

We also remind ourselves that for arithmetic or geometric averaging we have $dA/A = f(A/S,t)dt$ where the function $f = f(x,t)$ is defined as follows:

$$f(x,t) = \begin{cases} \frac{x^{-1}-1}{t} & \text{arithmetic averaging of the exercise price,} \\ \frac{-\ln x}{t} & \text{geometric averaging of the exercise price,} \end{cases} \qquad (12.66)$$

(cf. Chapter 6). It is well known (see Chapter 6 or [75, 33]) that for Asian options with a floating strike we can perform dimension reduction by introducing the following similarity variable:

$$x = A/S, \qquad W(x,\tau) = V(S,A,t)/A,$$

where $\tau = T - t$. It is straightforward to verify that $V(S,A,t) = AW(A/S, T-t)$ is a solution of (12.65) iff $W = W(x,\tau)$ is a solution to the following parabolic PDE:

$$\frac{\partial W}{\partial \tau} - \frac{\sigma^2}{2}\frac{\partial}{\partial x}\left(x^2\frac{\partial W}{\partial x}\right) + (r-q)x\frac{\partial W}{\partial x} - f(x, T-\tau)\left(W + x\frac{\partial W}{\partial x}\right) + rW = 0, \qquad (12.67)$$

where $x > 0$ and $0 < \tau < T$. The initial condition for W immediately follows from the terminal pay-off diagram for the call option, i.e., $W(x,0) = (x^{-1} - 1)^+$.

12.3.1. American-Style of Asian Call Options

According to Dai and Kwok [33] the set

$$\mathcal{E} = \{(S,A,t) \in [0,\infty) \times [0,\infty) \times [0,T], \ V(S,A,t) = V(S,A,T)\}$$

is the exercise region for American-style of Asian call options. In the case of a call option this region can be described by the early exercise boundary function $S_f = S_f(A,t)$ such that $\mathcal{E} = \{(S,A,t) \in [0,\infty) \times [0,\infty) \times [0,T], \ S \geq S_f(A,t)\}$. It means that the Black–Scholes equation (12.65) holds true in the time dependent domain $0 < t < T, A > 0, 0 < S < S_f(A,t)$.

For American-style of an Asian call option we have to impose a homogeneous Dirichlet boundary condition $V(0,A,t) = 0$. According to [33] the C^1 continuity condition at the contact point $(S_f(A,t), A, t)$ of a solution V with its pay-off diagram implies the following boundary condition at the free boundary position $S_f(A,t)$:

$$\frac{\partial V}{\partial S}(S_f(A,t),A,t) = 1, \quad V(S_f(A,t),A,t) = S_f(A,t) - A, \tag{12.68}$$

for any $A > 0$ and $0 < t < T$. It is important to emphasize that the free boundary function S_f can be also reduced to a function of one variable by introducing a new state function x_t^* as follows:

$$S_f(A,t) = A/x_t^*.$$

The function $t \mapsto x_t^*$ is a free boundary function for the transformed state variable $x = A/S$. For American-style of Asian call options the spatial domain for the reduced equation (12.67) is given by $1/\rho(\tau) < x < \infty$, $\tau \in (0,T)$, where $\rho(\tau) = 1/x_{T-\tau}^*$. Taking into account boundary conditions (12.68) for the option price V we end up with corresponding boundary conditions for the function W:

$$W(\infty,\tau) = 0, \quad W(x,\tau) = \frac{1}{x} - 1, \quad \frac{\partial W}{\partial x}(x,\tau) = -\frac{1}{x^2} \text{ at } x = \frac{1}{\rho(\tau)}, \tag{12.69}$$

for any $0 < \tau < T$ and the initial condition

$$W(x,0) = \max(x^{-1} - 1, 0), \quad \text{for any } x > 0. \tag{12.70}$$

12.3.2. Fixed Domain Transformation

In order to apply the fixed domain transformation for the free boundary problem (12.67), (12.69), (12.70) we introduce a new state variable ξ and an auxiliary function $\Pi = \Pi(\xi,\tau)$ again representing the synthetic portfolio. Let us define:

$$\xi = \ln(\rho(\tau)x), \quad \Pi(\xi,\tau) = W(x,\tau) + x\frac{\partial W}{\partial x}(x,\tau).$$

Clearly, $x \in (\rho(\tau)^{-1}, \infty)$ iff $\xi \in (0,\infty)$ for $\tau \in (0,T)$. The value $\xi = \infty$ of the transformed variable corresponds to the value $x = \infty$, i.e., $S = 0$ when expressed in the original variable. On the other hand, the value $\xi = 0$ corresponds to the free boundary position $x = x_t^*$, i.e.,

$S = S_f(A,t)$. After straightforward calculations we conclude that the function $\Pi = \Pi(\xi,\tau)$ is a solution to the following parabolic PDE:

$$\frac{\partial \Pi}{\partial \tau} + a(\xi,\tau)\frac{\partial \Pi}{\partial \xi} - \frac{\sigma^2}{2}\frac{\partial^2 \Pi}{\partial \xi^2} + b(\xi,\tau)\Pi = 0,$$

where the term $a(\xi,\tau)$ depends on the free boundary position ρ. The functions a,b are given by

$$a(\xi,\tau) = \frac{\dot\rho(\tau)}{\rho(\tau)} + r - q - \frac{\sigma^2}{2} - f(e^\xi/\rho(\tau), T-\tau),$$

$$b(\xi,\tau) = r - \frac{\partial}{\partial x}(xf(x,T-\tau))\Big|_{x=\frac{e^\xi}{\rho(\tau)}}. \tag{12.71}$$

Notice that $b(\xi,\tau) = r + 1/(T-\tau)$ in the case of arithmetic averaging, i.e., $f(x,t) = (x^{-1} - 1)/t$. The initial condition for the solution Π can be determined from (12.70)

$$\Pi(\xi,0) = \begin{cases} -1 & \xi < \ln\rho(0), \\ 0 & \xi > \ln\rho(0). \end{cases}$$

Since $\partial_x W(x,\tau) = -\frac{1}{x^2}$ and $W(x,\tau) = \frac{1}{x} - 1$ for $x = \frac{1}{\rho(\tau)}$ and $W(\infty,\tau) = 0$ we conclude the Dirichlet boundary conditions for the transformed function $\Pi(\xi,\tau)$

$$\Pi(0,\tau) = -1, \qquad \Pi(\infty,\tau) = 0.$$

It remains to determine an algebraic constraint between the free boundary function $\rho(\tau)$ and the solution Π. Similarly as in the case of a linear or nonlinear Black–Scholes equation (cf. [104]) we obtain, by differentiating the condition $W(\frac{1}{\rho(\tau)},\tau) = \rho(\tau) - 1$ with respect to τ, the following identity:

$$\frac{d\rho}{d\tau}(\tau) = \frac{\partial W}{\partial x}(\rho(\tau)^{-1},\tau)(-\rho(\tau)^{-2})\frac{d\rho}{d\tau}(\tau) + \frac{\partial W}{\partial \tau}(\rho(\tau)^{-1},\tau).$$

Since $\partial_x W(\rho(\tau)^{-1},\tau) = -\rho(\tau)^2$ we have $\frac{\partial W}{\partial \tau}(x,\tau) = 0$ at $x = \rho(\tau)$. Assuming continuity of the function $\Pi(\xi,\tau)$ and its derivative $\Pi_\xi(\xi,\tau)$ up to the boundary $\xi = 0$ we obtain

$$x^2\frac{\partial^2 W}{\partial x^2}(x,\tau) \to \frac{\partial \Pi}{\partial \xi}(0,\tau) + 2\rho(\tau), \quad x\frac{\partial W}{\partial x}(x,\tau) \to -\rho(\tau) \quad \text{as } x \to \rho(\tau)^{-1}.$$

Passing to the limit $x \to \rho(\tau)^{-1}$ in (12.67) we end up with the algebraic equation

$$q\rho(\tau) - r + f(\rho(\tau)^{-1}, T-\tau) = \frac{\sigma^2}{2}\frac{\partial \Pi}{\partial \xi}(0,\tau), \qquad \tau \in (0,T], \tag{12.72}$$

for the free boundary position $\rho(\tau)$. Notice that, in the case of arithmetic averaging where $f(\rho(\tau)^{-1}, T-\tau) = (\rho(\tau) - 1)/(T-\tau)$, we can derive the following explicit expression for the free boundary position $\rho(\tau)$:

$$\rho(\tau) = \frac{1 + r(T-\tau) + \frac{\sigma^2}{2}(T-\tau)\frac{\partial \Pi}{\partial \xi}(0,\tau)}{1 + q(T-\tau)}, \qquad 0 < \tau < T,$$

as a function of the derivative $\partial_\xi \Pi(0,\tau)$ evaluated at $\xi = 0$.

The initial value $\rho(0)$ can be deduced from the smoothness of the solution Π at $(\xi,\tau) = (0,0)$. We can proceed in the same way as in section 12.1.1.. We have $\partial_\xi \Pi(0,0) = 0$. In the case of arithmetic averaging we obtain from the above equation for $\rho(\tau)$ by passing to the limit $\tau \to 0$ that $\rho(0) = (1+rT)/(1+qT)$ provided $0 \leq q < r$. If $r \leq q$ we have $\rho(0) = 1$. Summarizing, for the arithmetically averaged Asian call option we have the following expression:

$$\text{(arithmetic averaging)} \quad \rho(0) = \max\left(\frac{1+rT}{1+qT}, 1\right), \tag{12.73}$$

(cf. Dai and Kwok [33] and Bokes and Ševčovič [15]). In the case of geometrically averaged Asian call option the initial value $\rho(0)$ is a solution to the transcendental equation:

$$\text{(geometric averaging)} \quad \rho(0) = \max\left(1/\tilde{x}_T, 1\right), \quad \text{where } \ln\tilde{x}_T = \frac{qT}{\tilde{x}_T} - rT. \tag{12.74}$$

(cf. Wu, Kwok and Yu [123] and Bokes and Ševčovič [15]).

In summary, we have derived the following nonlocal parabolic equation for the synthesized portfolio $\Pi(\xi,\tau)$:

$$\frac{\partial \Pi}{\partial \tau} + a(\xi,\tau)\frac{\partial \Pi}{\partial \xi} - \frac{\sigma^2}{2}\frac{\partial^2 \Pi}{\partial \xi^2} + b(\xi,\tau)\Pi = 0, \quad 0 < \tau < T, \ \xi > 0,$$

with the algebraic constraint

$$q\rho(\tau) - r + f(\rho(\tau)^{-1}, T-\tau) = \frac{\sigma^2}{2}\frac{\partial \Pi}{\partial \xi}(0,\tau), \quad 0 < \tau < T.$$

A solution Π is subject to the boundary and initial conditions: (12.75)

$$\Pi(0,\tau) = -1, \quad \Pi(\infty,\tau) = 0,$$

$$\Pi(\xi,0) = \begin{cases} -1, & \text{for } \xi < \ln(\rho(0)), \\ 0, & \text{for } \xi > \ln(\rho(0)), \end{cases}$$

where $a(\xi,\tau)$ and $b(\xi,\tau)$ are given by (12.71),
and the starting point $\rho(0)$ is given by (12.73) or (12.74), respectively.

An Equivalent Form of the equation for the Free Boundary

In section 12.2.1. we presented an idea how to overcome the problem of implementing the algebraic constraint (12.72). Notice that (12.72) provides a formula for the free boundary position $\rho(\tau)$ in terms of the derivative $\partial_\xi \Pi(0,\tau)$. Again, such an expression is not suitable for construction of a robust numerical approximation scheme because of the sensitivity of the entire parabolic equation with respect to the approximation of $\partial_\xi \Pi(0,\tau)$ evaluated at the single point $\xi = 0$.

Integrating the governing equation (12.75) with respect to $\xi \in (0,\infty)$ yields

$$\frac{d}{d\tau}\int_0^\infty \Pi d\xi + \int_0^\infty a(\xi,\tau)\frac{\partial \Pi}{\partial \xi}d\xi - \frac{\sigma^2}{2}\int_0^\infty \frac{\partial^2 \Pi}{\partial \xi^2}d\xi + \int_0^\infty b(\xi,\tau)\Pi d\xi = 0.$$

Now, taking into account the boundary conditions $\Pi(0,\tau) = -1, \Pi(\infty,\tau) = 0$, and consequently $\partial_\xi \Pi(\infty,\tau) = 0$ we obtain, by applying condition (12.72), the following differential equation:

$$\frac{d}{d\tau}\left(\ln\rho(\tau) + \int_0^\infty \Pi(\xi,\tau)d\xi\right) + q\rho(\tau) - q - \frac{\sigma^2}{2}$$
$$+ \int_0^\infty \left[r - f\left(\frac{e^\xi}{\rho(\tau)}, T - \tau\right)\right]\Pi(\xi,\tau)d\xi = 0.$$

In the case of arithmetic averaging where $f(x,t) = (x^{-1} - 1)/t$ we obtain

$$\frac{d}{d\tau}\left(\ln\rho(\tau) + \int_0^\infty \Pi(\xi,\tau)d\xi\right) + q\rho(\tau) - q - \frac{\sigma^2}{2}$$
$$+ \int_0^\infty \left[r - \frac{\rho(\tau)e^{-\xi} - 1}{T - \tau}\right]\Pi(\xi,\tau)d\xi = 0. \quad (12.76)$$

12.3.3. A Numerical Approximation Operator Splitting Scheme

Our numerical approximation scheme is based on a solution to the transformed system (12.75). For the sake of simplicity, the scheme will be derived for the case of arithmetically averaged American style Asian call option. Derivation of the scheme for geometric or weighted arithmetic averaging is similar and therefore omitted.

We restrict the spatial domain $\xi \in (0,\infty)$ to a finite interval of values $\xi \in (0,L)$ where $L > 0$ is sufficiently large. For practical purposes it sufficient to take $L \approx 2$. Let $k > 0$ denote by the time step, $k = T/m$ and by $h = L/n > 0$ the spatial step. Here $m, n \in \mathbb{N}$ denote the number of time and space discretization steps, respectively. We let denote by $\Pi^j = \Pi^j(\xi)$ the time discretization of $\Pi(\xi,\tau_j)$ and $\rho^j \approx \rho(\tau_j)$ where $\tau_j = jk$. By Π_i^j we shall denote the full space–time approximation for the value $\Pi(\xi_i,\tau_j)$. Then for the Euler backward in time finite difference approximation of equation (12.75) we have

$$\frac{\Pi^j - \Pi^{j-1}}{k} + c^j \frac{\partial \Pi^j}{\partial \xi} - \left(\frac{\sigma^2}{2} + \frac{\rho^j e^{-\xi} - 1}{T - \tau_j}\right)\frac{\partial \Pi^j}{\partial \xi} - \frac{\sigma^2}{2}\frac{\partial^2 \Pi^j}{\partial^2 \xi} + \left(r + \frac{1}{T - \tau_j}\right)\Pi^j = 0$$

where c^j is an approximation of the value $c(\tau_j)$ where the $c(\tau) = \frac{\dot\rho(\tau)}{\rho(\tau)} + r - q$. The solution $\Pi^j = \Pi^j(x)$ is subject to Dirichlet boundary conditions at $\xi = 0$ and $\xi = L$. We set $\Pi^0(\xi) = \Pi(\xi,0)$ (see (12.75)). In what follows, we shall again make use of the time step operator splitting method. We split the above problem into a convection part and a diffusive part by introducing an auxiliary intermediate step solution $\Pi^{j-\frac{1}{2}}$:

(*Convective part*)
$$\frac{\Pi^{j-\frac{1}{2}} - \Pi^{j-1}}{k} + c^j \partial_\xi \Pi^{j-\frac{1}{2}} = 0, \quad (12.77)$$

(*Diffusive part*)
$$\frac{\Pi^j - \Pi^{j-\frac{1}{2}}}{k} - \left(\frac{\sigma^2}{2} + \frac{\rho^j e^{-\xi} - 1}{T - \tau_j}\right)\frac{\partial \Pi^j}{\partial \xi} - \frac{\sigma^2}{2}\frac{\partial^2 \Pi^j}{\partial^2 \xi} + \left(r + \frac{1}{T - \tau_j}\right)\Pi^j = 0. \quad (12.78)$$

Similarly as in [104] we shall approximate the convective part by the explicit solution to the transport equation $\partial_\tau \tilde{\Pi} + c(\tau)\partial_\xi \tilde{\Pi} = 0$ for $\xi > 0$ and $\tau \in (\tau_{j-1}, \tau_j]$ subject to the boundary condition $\tilde{\Pi}(0,\tau) = -1$ and the initial condition $\tilde{\Pi}(\xi, \tau_{j-1}) = \Pi^{j-1}(\xi)$. It is known that the free boundary function $\rho(\tau)$ need not be monotonically increasing (see e.g. [33, 107] or [59]). Therefore depending whether the value of $c(\tau)$ is positive or negative the boundary condition $\tilde{\Pi}(0,\tau) = -1$ at $\xi = 0$ is either in-flowing ($c(\tau) > 0$) or out-flowing ($c(\tau) < 0$). Hence the boundary condition $\Pi(0,\tau) = -1$ can be prescribed only if $c(\tau_j) \geq 0$. Let us denote by $C(\tau)$ the primitive function to $c(\tau)$, i.e., $C(\tau) = \ln\rho(\tau) + (r-q)\tau$. Solving the transport equation $\partial_\tau \tilde{\Pi} + c(\tau)\partial_\xi \tilde{\Pi} = 0$ for $\tau \in [\tau_{j-1}, \tau_j]$ subject to the initial condition $\Pi(\xi, \tau_{j-1}) = \Pi^{j-1}(\xi)$ we obtain: $\tilde{\Pi}(\xi, \tau) = \Pi^{j-1}(\xi - C(\tau) + C(\tau_{j-1}))$ if $\xi - C(\tau) + C(\tau_{j-1}) > 0$ and $\tilde{\Pi}(\xi, \tau) = -1$ otherwise. Hence the full time-space approximation of the half-step solution $\Pi_i^{j-\frac{1}{2}}$ can be obtained from the formula

$$\Pi_i^{j-\frac{1}{2}} = \begin{cases} \Pi^{j-1}(\eta_i) & \text{if } \eta_i = \xi_i - \ln\rho^j + \ln\rho^{j-1} - (r-q)k > 0, \\ -1 & \text{otherwise.} \end{cases} \quad (12.79)$$

In order to compute the value $\Pi^{j-1}(\eta_i)$ we make use of a linear interpolation between discrete values $\Pi_i^{j-1}, i = 0, 1, \ldots, n$.

Using central finite differences for approximation of the derivative $\partial_\xi \Pi^j$ we can approximate the diffusive part of a solution of (12.78) as follows:

$$\frac{\Pi_i^j - \Pi_i^{j-\frac{1}{2}}}{k} + \left(r + \frac{1}{T - \tau_j}\right)\Pi_i^j - \left(\frac{\sigma^2}{2} + \frac{\rho^j e^{-\xi_i} - 1}{T - \tau_j}\right)\frac{\Pi_{i+1}^j - \Pi_{i-1}^j}{2h} - \frac{\sigma^2}{2}\frac{\Pi_{i+1}^j - 2\Pi_i^j + \Pi_{i-1}^j}{h^2} = 0.$$

Hence the vector of discrete values $\Pi^j = \{\Pi_i^j, i = 1, 2, \ldots, n\}$ at the time level $j \in \{1, 2, \ldots, m\}$ is a solution of a tridiagonal system of linear equations

$$\alpha_i^j \Pi_{i-1}^j + \beta_i^j \Pi_i^j + \gamma_i^j \Pi_{i+1}^j = \Pi_i^{j-\frac{1}{2}}, \quad \text{for } i = 1, 2, \ldots, n, \quad \text{where} \quad (12.80)$$

$$\alpha_i^j(\rho^j) = -\frac{k}{2h^2}\sigma^2 + \frac{k}{2h}\left(\frac{\sigma^2}{2} + \frac{\rho^j e^{-\xi_i} - 1}{T - \tau_j}\right),$$

$$\gamma_i^j(\rho^j) = -\frac{k}{2h^2}\sigma^2 - \frac{k}{2h}\left(\frac{\sigma^2}{2} + \frac{\rho^j e^{-\xi_i} - 1}{T - \tau_j}\right), \quad (12.81)$$

$$\beta_i^j(\rho^j) = 1 + \left(r + \frac{1}{T - \tau_j}\right)k - (\alpha_i^j + \gamma_i^j).$$

The initial and boundary conditions at $\tau = 0$ and $x = 0, L$, can be approximated as follows:

$$\Pi_i^0 = \begin{cases} -1, & \text{for } \xi_i < \ln((1+rT)/(1+qT)), \\ 0, & \text{for } \xi_i \geq \ln((1+rT)/(1+qT)), \end{cases}$$

for $i = 0, 1, \ldots, n$ and $\Pi_0^j = -1$, $\Pi_n^j = 0$, for $j = 1, \ldots, m$.

Finally, we employ the differential equation (12.76) to determine the free boundary position ρ. Taking the Euler finite difference approximation of $\frac{d}{d\tau}(\ln \rho + \int_0^\infty \Pi d\xi)$ we obtain

(*Algebraic part*)

$$\ln \rho^j = \ln \rho^{j-1} + I_0(\Pi^{j-1}) - I_0(\Pi^j) + k\left(q + \frac{\sigma^2}{2} - q\rho^{j-1} - I_1(\rho^{j-1}, \Pi^j)\right), \quad (12.82)$$

where $I_0(\Pi)$ stands for a numerical trapezoid quadrature of the integral $\int_0^\infty \Pi(\xi) d\xi$ whereas $I_1(\rho^{j-1}, \Pi)$ is a trapezoid quadrature of the second integral $\int_0^\infty \left(r - \frac{\rho^{j-1}e^{-\xi}-1}{T-\tau_j}\right) \Pi(\xi) d\xi$.

We formally rewrite discrete equations (12.79), (12.80) and (12.82) in the operator form:

$$\rho^j = \mathcal{F}(\Pi^j), \qquad \Pi^{j-\frac{1}{2}} = \mathcal{T}(\rho^j), \qquad \mathcal{A}(\rho^j)\Pi^j = \Pi^{j-\frac{1}{2}}, \quad (12.83)$$

where $\ln \mathcal{F}(\Pi^j)$ is the right-hand side of equation (12.82), $\mathcal{T}(\rho^j)$ is the transport equation solver given by the right-hand side of (12.79) and $\mathcal{A} = \mathcal{A}(\rho^j)$ is a tridiagonal matrix with coefficients given by (12.81). The system (12.83) can be approximately solved by means of successive iterations procedure. We define, for $j \geq 1$, $\Pi^{j,0} = \Pi^{j-1}, \rho^{j,0} = \rho^{j-1}$. Then the $(p+1)$-th approximation of Π^j and ρ^j is obtained as a solution to the system:

$$\begin{aligned}
\rho^{j,p+1} &= \mathcal{F}(\Pi^{j,p}), \\
\Pi^{j-\frac{1}{2},p+1} &= \mathcal{T}(\rho^{j,p+1}), \\
\mathcal{A}(\rho^{j,p+1})\Pi^{j,p+1} &= \Pi^{j-\frac{1}{2},p+1}.
\end{aligned} \quad (12.84)$$

Supposing the sequence of approximate discretized solutions $\{(\Pi^{j,p}, \rho^{j,p})\}_{p=1}^\infty$ converges to the limiting value $(\Pi^{j,\infty}, \rho^{j,\infty})$ as $p \to \infty$ then this limit is a solution to a nonlinear system of equations (12.83) at the time level j and we can proceed by computing the approximate solution in the next time level $j+1$.

12.3.4. Computational Examples of the Free Boundary Approximation

Finally we present several computational examples of application of the numerical approximation scheme (12.84) for the solution $\Pi(\xi, \tau)$ and the free boundary position $\rho(\tau)$ of (12.75). We consider American-style of Asian arithmetically averaged floating strike call options.

In Fig. 12.11 we show the behavior of the early exercise boundary function $\rho(\tau)$ and the function $x_t^* = 1/\rho(T-t)$. In this numerical experiment we chose $r = 0.06, q = 0.04, \sigma = 0.2$ and very long expiration time $T = 50$ years. These parameters correspond to the example presented by Dai and Kwok in [33]. As far as other numerical parameters are concerned, we chose the mesh of $n = 200$ spatial grid points and we have chosen the number of time steps $m = 10^5$ in order to achieve very fine time stepping corresponding to 260 minutes between consecutive time steps when expressed in the original time scale of the problem.

In Fig. 12.12 we can see the behavior of the transformed function Π in both 3D as well as contour plot perspectives. We also plot the initial condition $\Pi(\xi, 0)$ and five time steps of the function $\xi \mapsto \Pi(\xi, \tau_j)$ for $\tau_j = 0.1, 1, 5, 25, 50$.

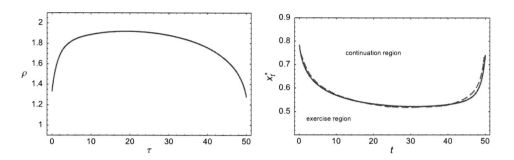

Figure 12.11. The function $\rho(\tau)$ (left). A comparison of the free boundary position $x_t^* = 1/\rho(T-t)$ (right) obtained by our method (solid curve) and that of the PSOR algorithm by Dai and Kwok (dashed curve). Source: Bokes and Ševčovič [15].

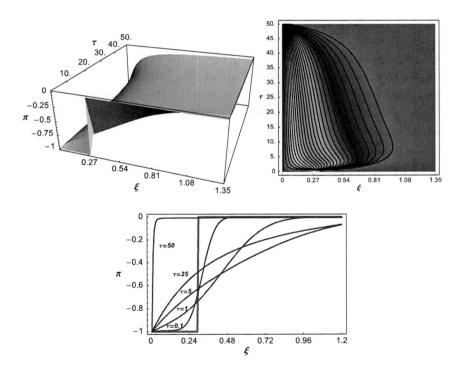

Figure 12.12. A 3D plot (left) and contour plot (right) of the function $\Pi(\xi,\tau)$. Profiles of the function $\Pi(\xi,\tau)$ for various times $\tau \in [0,T]$. Source: Bokes and Ševčovič [15].

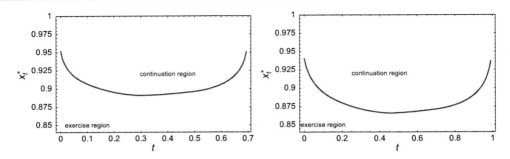

Figure 12.13. The free boundary position for expiration times $T = 0.7$ (left) and $T = 1$ (right). Source: Bokes and Ševčovič [15].

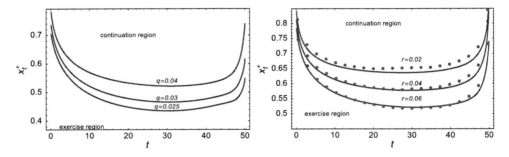

Figure 12.14. A comparison of the free boundary position x_t^* for various dividend yield rates $q = 0.04, 0.03, 0.025$ and fixed interest rate $r = 0.06$ (left). Comparison of x_t^* for various interest rates $r = 0.06, 0.04, 0.02$ and dividend yield $q = 0.04$. Dots represents the solution obtained by Dai and Kwok (right). Source: Bokes and Ševčovič [15].

Table 12.4. A comparison of PSOR method due to Dai and Kwok and our transformation method for $T = 50, \sigma = 0.2, q = 0.04$. Source: Bokes and Ševčovič [15].

	$r = 0.06$	$r = 0.04$	$r = 0.02$
$\|x_t^{*,trans} - x_t^{*,psor}\|_\infty$	0.09769	0.03535	0.05359
$\|x_t^{*,trans} - x_t^{*,psor}\|_1$	0.00503	0.00745	0.01437
$\min x_t^{*,trans}$	0.52150	0.57780	0.63619

A comparison of early exercise boundary profiles with respect to varying interest rates r and dividend yields q is shown in Fig. 12.14. A comparison of the free boundary position $x_t^* = 1/\rho(T - t)$ obtained by our method (solid curve) and that of the projected successive over relaxation algorithm by Dai and Kwok [33] (dashed curve) for different values of the interest rate r is shown in Fig. 12.14 (right). The algorithm due to Dai and Kwok is based on a numerical solution to the variational inequality for the function $W = W(x, \tau)$ which a solution to (12.67) in the continuation region and it is smoothly pasted to its pay-off diagram (12.69). It is clear that our method and that of [33] give almost the same results. A quantitative comparison of both methods is given in Table 12.4 for model parameters $T = 50, \sigma = 0.2, q = 0.04$ and various interest rates $r = 0.02, 0.04, 0.06$. We evaluated discrete $L^\infty(0,T)$ and $L^1(0,T)$ norms of the difference $x_t^{*,trans} - x_t^{*,psor}$ between the numerical solution $x_t^{*,trans}, t \in [0,T]$, obtained by our method and that of Dai and Kwok denoted by $x_t^{*,psor}$. We also show the minimal value $\min_{t \in [0,T]} x_t^{*,trans}$ of the early exercise boundary. Finally, in Fig. 12.13 we present numerical experiments for shorter expiration times $T = 0.7$ and $T = 1$ (one year) with zero dividend rate $q = 0$ and $r = 0.06, \sigma = 0.2$.

Chapter 13

Calibration of Interest Rate and Term Structure Models

The main goal of this chapter is to review calibration techniques and results for one factor interest rate and term structure models including, in particular, the Vasicek and Cox–Ingersoll–Ross models. In comparison to previous Chapter 7, we present, in a more detail, the generalized method of moments due to Chan, Karolyi, Longstaff and Sanders [23] and the so-called Nowman's type of Gaussian parameter estimates. These methods are based on the statistical analysis of the short rate time series. In the second part of this chapter we focus our attention to calibration methods based on the entire term structure information. We present the two phase min-max method of parameters calibration introduced and analyzed by Urbánová-Csajková and Ševčovič in [106, 102].

13.1. Generalized Method of Moments

One of important results in the area of calibrating interest rate models is the paper [23] by Chan, Karolyi, Longstaff and Sanders. Their main result is that for one-factor models, the form of the volatility term is a very important feature distinguishing the models. It turns out that the most successfull models are those that allow the volatility to be highly sensitive function with respect to the level of the short rate. In [23], Chan *et al.* considered a general short rate model expressed in terms of a single stochastic differential equation

$$dr = (\alpha + \beta r)dt + \sigma r^\gamma dw. \tag{13.1}$$

A comparison of qualitative and quantitative properties of short rate interest rate models of the form (13.1) became a topic of a wide range of papers dealing with interest rate modeling. Special cases of (13.1), considered in [23] are summarized in Table 13.1. The short rate Model 1, i.e. the Brownian motion with drift, was first used in [84] in order to derive a model of discount bond prices. Model 2 is a classical Vasicek model derived in [120]. Notice that the Vasicek model has been already studied in Chapter 7. Model 3 the Cox-Ingersoll-Ross model from [31], which we also discussed in Chapter 7. Model 4 was first used by Dothan in [37] in the problem of valuing discount bonds and, subsequently, by Brennan and Schwartz in [18]. In the latter paper the authors developed numerical models of savings, retractable, and callable bonds. Model 5 is similar to the geometric Brownian motion process used in the Black–Scholes option pricing model [14]. As an interest rate model it was proposed by Marsh and Rosenfeld in [82]. Model 6 was treated by Brennan and Schwartz [19] in the context of deriving a numerical model for convertible bond prices and, by Courtadon [28], for developing a model of discount bond option prices. Model 7 was introduced by Cox, Ingersoll and Ross in their seminal paper [30] of variable-rate securities. Model 8 represents the so-called constant elasticity of variance process introduced by Cox in [29] and by Cox and Ross in [32]. Marsh and Rosenfeld discussed its application to interest rate modeling in [82].

Table 13.1. Overview of one-factor short rate models. Source: Chan, Karolyi, Longstaff, Sanders [23].

Model	Equation for the short rate
1. Merton	$dr = \alpha dt + \sigma dw$
2. Vasicek	$dr = (\alpha + \beta r)dt + \sigma dw$
3. CIR SR	$dr = (\alpha + \beta r)dt + \sigma r^{1/2} dw$
4. Dothan	$dr = \sigma r dw$
5. GBM	$dr = \beta r dt + \sigma r dw$
6. Brennan-Schwartz	$dr = (\alpha + \beta r)dt + \sigma r dw$
7. CIR VR	$dr = \sigma r^{3/2} dw$
8. CEV	$dr = \beta r dt + \sigma r^{\gamma} dw$

The parameter estimation method proposed in [23] (see also Hansen [60]) is referred to as the generalized method of moments (GMM). This methodology takes into account a discrete-time econometric version of the short term interest rate process. In the case of the stochastic differential equation (13.1) the authors consider the discrete-time form of the stochastic process:

$$r_{t+1} - r_t = \alpha + \beta r_t + \varepsilon_{t+1},$$

where

$$E(\varepsilon_{t+1}) = 0, \qquad E((\varepsilon_{t+1})^2) = \sigma^2 (r_t)^{2\gamma}.$$

Here $E(.)$ is the sample mean of the process. The unknown parameters are $\theta = (\alpha, \beta, \sigma^2, \gamma)$. The idea of the GMM method is to estimate unknown model parameters from the equation $E(f_t(\theta)) = 0$, where

$$f_t(\theta) = (\varepsilon_{t+1}, \, \varepsilon_{t+1} r_t, \, \varepsilon_{t+1}^2 - \sigma^2 r_t^{2\gamma}, \, (\varepsilon_{t+1}^2 - \sigma^2 r_t^{2\gamma}) r_t)^T \in \mathbb{R}^4.$$

The GMM is based on replacing $E(f_t(\theta))$ by its sample average counterpart, i.e.

$$g_T(\theta) = \frac{1}{T} \sum_{t=1}^{T} f_t(\theta),$$

where T is the number of observations. The vector $\theta \in \mathbb{R}^4$ of model parameters is then estimated by means of minimization of the quadratic form:

$$J_T(\theta) = g_T^T(\theta) W_T(\theta) g_T(\theta)$$

for some positive definite weight matrix $W_T(\theta)$. Here g_T^T stands for transposition of the vector $g_T \in \mathbb{R}^4$. It means that the estimate $\hat{\theta}$ of the vector of model parameters is given by

$$\hat{\theta} = \arg\min_{\theta} J_T(\theta).$$

If there are no restrictions on model parameters then the quadratic form $J_T(\theta)$ has the zero minimum any choice of the weight matrix $W_T(\theta)$. For the nested interest rate models, the parameters are overidentified and the GMM estimates depend on the choice of W_T. In [60], Hansen showed that the choice $W_T(\theta) = S^{-1}(\theta)$, where $S(\theta)$ is the 4×4 matrix,

$$S(\theta) = E(g_T(\theta)g_T^T(\theta)),$$

yields the GMM estimator of θ with the smallest asymptotic covariance matrix.

To test the hypothesis on parameters, determined by restriction imposed by models, authors often use the method of Newey and West [86]. In this procedure, a general null hypothesis of the form, $a(\theta) = 0$, where $a(\theta)$ is a vector of order k, is tested with the statistic

$$R = T\left[J_T(\tilde{\theta}) - J_T(\hat{\theta})\right],$$

where $\tilde{\theta}$ is the unrestricted and $\hat{\theta}$ is the restricted estimate. If the null hypothesis is confirmed, this statistic is an asymptotically distributed χ^2 distribution with k degrees of freedom.

Next we will present an empirical example of parameter estimation. The results, obtained by using the annualized one-month U.S, treasury bill yield from June 1964 to December 1989 (306 observations), are summarized in Table 13.2. Considering the usual 5 percent significance level, the model is rejected if the P value is less than 0.05, which means that in this case we reject Merton, Vasicek and CIR SR model. Rejecting models of Vasicek and Cox-Ingersoll-Ross, which have analytical formulas for bond prices (see Chapter 8), motivates the study of bond prices also in other models. In Chapter 14 we study the analytical approximation for a more general model when compared to (13.1). For the further comments on these results and more tests of the model we refer the reader to the paper [23] by Chen *et al.*

Let us note that these results are not universal. For different term structures we can obtain different estimations of parameters including, in particular, the parameter γ. For example, in [2] the authors estimated the parameter γ to be less than one by using the same estimation methodology but for the LIBOR term structure. It means that the volatility is less than the one estimated by Chan, Karolyi, Longstaff and Sanders.

We recall that a modification of the generalized method of moments, which is robust with resecpect to presence of outliers, was developed in [35]. It is refereed to as the robust generalized method of moments.

13.2. Nowman's Parameter Estimates

Nowman's estimates are based on approximating the likelihood function of the model. In Chapter 7 we considered the maximum likelihood estimation methodology for the Vasicek

Table 13.2. Estimated parameters and corresponding P values obtained by the GMM estimation the respective models as restrictions of (13.1). Source: Chan, Karolyi, Longstaff, Sanders [23].

Model	α	β	σ^2	γ	P value
unrestricted parameters	0.0408	-0.5921	1.6704	1.4999	-
Merton	0.0055	0	0.0004	0	0.0341
Vasicek	0.0154	-0.1779	0.0004	0	0.0029
CIR SR	0.0189	-0.2339	0.0073	0.5	0.0131
Dothan	0	0	0.1172	1	0.1327
GBM	0	0.1101	0.1185	1	0.2066
Brennan-Schwartz	0.0242	-0.3142	0.1185	1	0.1364
CIR VR	0	0	1.5778	1.5	0.1019
CEV	0	0.1026	0.5207	1.2795	0.0793

model. Because of the constant volatility, we were able to derive the exact likelihood function. The main idea of Nowman's estimation methodology is to approximate the volatility by a piece-wise constant function, which remains constant between two observations (see [87] and [43]).

An equation for the short rate is the same as in the previous section, i.e.

$$dr_s = (\alpha + \beta r_s)ds + \sigma r_s^\gamma dw_s.$$

Next we multiply it by the term $e^{-\beta s}$. It yields

$$e^{-\beta s}dr_s - \beta e^{-\beta s}r_s ds = \alpha e^{-\beta s}ds + \sigma e^{-\beta s}r_s^\gamma dw_s,$$

$$\frac{d}{ds}\left(e^{\beta s}r_s\right) = \alpha e^{-\beta s}ds + \sigma e^{-\beta s}r_s^\gamma dw_s,$$

from which we obtain, by integration over the time interval $[t-1,t]$ to time t, the equation:

$$e^{-\beta t}r_t - e^{-\beta(t-1)}r_{t-1} = \frac{\alpha}{\beta}\left(e^{-\beta(t-1)} - e^{-\beta t}\right) + \int_{t-1}^t \sigma r_s^\gamma e^{-\beta s}dw_s.$$

Using approximation according to which the volatility is constant on the interval $[t-1,t)$ and it is equal to the value at the beginning of the interval, we deduce

$$\int_{t-1}^t \sigma r_s^\gamma e^{-\beta s}dw_s = \sigma r_{t-1}^\gamma \int_{t-1}^t e^{-\beta s}dw_s,$$

and hence

$$e^{-\beta t}r_t - e^{-\beta(t-1)}r_{t-1} = \frac{\alpha}{\beta}\left(e^{-\beta(t-1)} - e^{-\beta t}\right) + \sigma r_{t-1}^\gamma \int_{t-1}^t e^{-\beta s}dw_s.$$

Multiplying the above equation by the term $e^{\beta t}$ and denoting

$$\varepsilon_t = \sigma r_{t-1}^\gamma e^{\beta t}\int_{t-1}^t e^{-\beta s}dw_s$$

we obtain a discrete short rate model

$$r_t = e^\beta r_{t-1} + \frac{\alpha}{\beta}\left(e^\beta - 1\right) + \varepsilon_t, \quad \text{for } t = 2,\ldots,N. \tag{13.2}$$

The conditional distribution of ε_t for a given value of r_{t-1} follows from properties of Itō's integral: ε_t are normally distributed and independent for $t = 1, 2, \ldots$, with a zero expected value and the variance v_t^2 satisfying

$$\begin{aligned}
v_t^2 &:= Var(\varepsilon_t) = \sigma^2 r_{t-1}^{2\gamma} e^{2\beta t} Var\left(\int_{t-1}^t e^{-\beta s} dw_s\right) \\
&= \sigma^2 r_{t-1}^{2\gamma} \int_{t-1}^t e^{-2\beta s} ds = \sigma^2 r_{t-1}^{2\gamma} \frac{e^{2\beta} - 1}{2\beta},
\end{aligned}$$

where we have used Itō's isometry (see Chapter 2).

The likelihood function L for this model is equal, up to an additive constant, to the following expression:

$$\ln L = -\frac{1}{2} \sum_{t=2}^N \left(\log v_t^2 + \frac{\varepsilon_t^2}{v_t^2}\right), \tag{13.3}$$

where

$$v_t^2 = \frac{\sigma^2}{2\beta}\left(e^\beta - 1\right) r_{t-1}^{2\gamma}, \quad \varepsilon_t = r_t - \frac{\alpha}{\beta}\left(e^\beta - 1\right) - e^\beta r_{t-1} \tag{13.4}$$

(see [87], [43]). Nowman's estimates of the parameters are then the arguments of maximum of the function $\log L$.

These results are obtained in the case when the length of the interval $[t-1,t]$ between two consecutive values of r_t is taken to be a unit of time. In a case of a different time scale in which the length of the interval equals Δt we can derive the model

$$r_k = e^{\beta \Delta t} r_{k-1} + \frac{\alpha}{\beta}\left(e^{\beta \Delta t} - 1\right) + \varepsilon_k, \quad k = 2,\ldots,N, \tag{13.5}$$

where k is a number of observations[1] with ε_k normally distributed, uncorrelated and having the zero expected value and variance $\sigma^2 r_{k-1}^{2\gamma} \frac{e^{2\beta \Delta t}-1}{2\beta}$. It can be written as follows:

$$r_k = e^{\tilde\beta} r_{k-1} + \frac{\tilde\alpha}{\tilde\beta}\left(e^{\tilde\beta} - 1\right) + \tilde\varepsilon_k, \quad k = 2,\ldots,N, \tag{13.6}$$

where $\tilde\varepsilon_k$ are normally distributed, uncorrelated, with a zero expected value and the variance $\tilde\sigma^2 r_{k-1}^{2\gamma} \frac{e^{2\tilde\beta}-1}{2\tilde\beta}$, where

$$\tilde\alpha = \alpha \Delta t, \quad \tilde\beta = \beta \Delta t, \quad \tilde\sigma^2 = \sigma^2 \Delta t. \tag{13.7}$$

When we are investigating the existence of maximum of the likelihood function, we can study a model in the form (13.6), which is equivalent to (13.2). On the other hand, when we are estimating the parameters α, β, σ^2, we have to divide estimates of $\tilde\alpha$, $\tilde\beta$ and $\tilde\sigma^2$ by the factor Δt. If we are estimating the volatility σ, we have to divide $\tilde\sigma$ by $\sqrt{\Delta t}$).

---

[1] In order to simplify the notation, instead of time, we are indexing observations by their number

As an example of estimation results obtained by this method we review the estimates from [43], where the models was estimated using monthly data of 1-month interest rates from 10 countries, see Table 13.3. Results of unconstrained estimation are presented in Table 13.4. Various models were tested by the likelihood ratio test. If θ is a vector of parameters, $\tilde{\theta}$ then its unconstrained and $\hat{\theta}$ its constrained estimate, the null hypothesis is tested by the likelihood ratio statistic

$$LR = (-2)\left[\ln L(\hat{\theta}) - \ln L(\tilde{\theta})\right].$$

This statistics asymptotically has a χ^2 distribution with the degrees of freedom equal to the number of restrictions. Models, tested in this way, formed a subset of models considered in Chen *et al.* paper [23]. In particular, these are the models labeled by 2,3,6,7,8 in Table 13.1. Estimation results are presented in Table 13.5. The reader is referred to the paper by Episcopos [43] for further details and estimation results.

Table 13.3. Data used in parameter estimation. Source: Episcopos [43].

Country	Range of the data	Number of observations
Australia	March 1986 - April 1998	146
Belgium	October 1989 - April 1998	103
Germany	November 1990 - April 1998	90
Japan	December 1985 - April 1998	149
Netherlands	January 1979 - April 1998	232
New Zealand	April 1986 - April 1998	145
Singapore	April 1986 - April 1998	145
Switzerland	January 1986 - April 1998	148
United Kingdom	January 1975 - April 1998	280
USA	January 1986 - April 1998	148

Finally, let us recall that, in [119], Treeponggama and Grey studied robustness of the parameter estimates with respect to a sample period and use of interest rates with different maturities as short rate proxies. They considered a similar sample of countries as Episcopos in the paper [43]. Different types of interest rates as short rate proxies were used in [88] for the Japanese financial market. Parameter estimations for UK and USA data and a subsequent study of the forecasting power of the models were performed in the paper [21] by Byers and Nowman.

13.3. Method Based on Comparison with Entire Market Term Structures

In this section we presents a min-max calibration method for Vasicek and CIR models. It is based on results presented in papers [106] and [102] by Ševčovič and Urbánová-Csajková.

The idea of the phase min-max method is rather simple. In the first step we minimize the sum of squares of differences of theoretical yield curve computed from the models and real market yield curve. The minimum is attained on a one dimensional curve in the four

Table 13.4. Estimated parameters. Source: Episcopos, [43].

Model	α	β	σ²	γ
Australia	0.0008	-0.0170	0.0354	1.5174
Belgium	0.0007	-0.0192	0.1147	1.5617
Germany	0.0002	-0.0133	0.0001	0.5501
Japan	0.0001	-0.0148	0.0002	0.4143
Netherlands	0.0007	-0.0126	0.0072	1.0245
New Zealand	0.0045	-0.048	0.0034	0.7815
Singapore	0.0043	-0.109	0.0002	0.1976
Switzerland	0.0007	-0.019	0.0001	0.2064
United Kingdom	0.0023	-0.0238	0.0008	0.5663
USA	0.0013	-0.0234	0.0001	0.4239

Table 13.5. P values from testing the models as restrictions of (13.1). Source: Episcopos [43].

Country / Model	Vasicek	CIR SR	BR-SC	CIR VR	CEV
Australia	0.0000	0.0000	0.0003	0.5069	0.3401
Belgium	0.0000	0.0000	0.0089	0.7511	0.4029
Germany	0.0037	0.7885	0.0179	0.0000	0.7118
Japan	0.0000	0.1960	0.0000	0.0000	0.7084
Netherlands	0.0000	0.0000	0.8201	0.0002	0.3494
New Zealand	0.0000	0.0092	0.0477	0.0000	0.0345
Singapore	0.0971	0.0102	0.0000	0.0000	0.0059
Switzerland	0.0596	0.0059	0.0000	0.0000	0.3858
United Kingdom	0.0001	0.6286	0.0011	0.0000	0.0825
USA	0.0154	0.6539	0.0006	0.0000	0.1740

dimensional parameter space of Vasicek or CIR model parameters. Then, by maximization of the likelihood function over this curve, we obtain estimation of the four parameters of the model.

13.3.1. Parameters Reduction Principle

Case of the Cox-Ingersoll-Ross Model

In the Cox–Ingersoll–Ross model the price of a zero coupon bond is a solution to the following partial differential equation

$$-\frac{\partial P}{\partial \tau} + (\kappa(\theta - r) - \lambda r)\frac{\partial P}{\partial r} + \frac{\sigma^2}{2}r\frac{\partial^2 P}{\partial r^2} - rP = 0, \quad t \in (0, T), \, r > 0.$$

In contrast to Chapter 7, we consider a slightly modified market price of risk function having the form $\frac{\lambda}{\sigma}\sqrt{r}$. As it has been already pointed out by Pearson and Sun in [92], the adjustment speed κ and the risk premium λ appear in the CIR bond price only in the summation $\kappa +$

λ. This is why four CIR parameters can be reduced to three essential parameters fully describing the behavior of the bond prices.

The parameter reduction for the CIR model consists of introduction of the following set of new variables:

$$\beta = e^{-\eta}, \quad \xi = \frac{\kappa + \lambda + \eta}{2\eta}, \quad \rho = \frac{2\kappa\theta}{\sigma^2}, \tag{13.8}$$

where $\eta = \sqrt{(\kappa+\lambda)^2 + 2\sigma^2}$. Returning back to the original CIR parameters $(\kappa, \sigma, \theta, \lambda)$ we have

$$\kappa = \eta(2\xi - 1) - \lambda, \quad \sigma = \eta\sqrt{2\xi(1-\xi)}, \quad \theta = \frac{\rho\sigma^2}{2\kappa}, \tag{13.9}$$

where $\eta = -\ln\beta$.

Proposition 13.1. *In terms of transformed parameters, the value of a bond $P = P(T - \tau, T, r)$ can be expressed as $P = Ae^{-Br}$, where $\tau = T - t \in [0, T]$ and functions $A = A(\beta, \xi, \rho, \tau)$, $B = B(\beta, \xi, \rho, \tau)$ satisfy*

$$B = -\frac{1}{\ln\beta} \frac{1-\beta^\tau}{\xi(1-\beta^\tau) + \beta^\tau}, \quad A = \left(\frac{\beta^{(1-\xi)\tau}}{\xi(1-\beta^\tau) + \beta^\tau}\right)^\rho. \tag{13.10}$$

Moreover, $(\beta, \xi, \rho) \in \Omega = (0,1) \times (0,1) \times \mathbb{R}^+ \subset \mathbb{R}^3$.

It is convenient to introduce the transformation $T : \mathcal{D} \to \Omega$ defined as in (13.8) where $\mathcal{D} = (0,\infty)^3 \times \mathbb{R} \subset \mathbb{R}^4$. Then $T(\kappa, \sigma, \theta, \lambda) = (\beta, \xi, \rho)$, is a smooth mapping and, for any $(\check{\beta}, \check{\xi}, \check{\rho}) \in \Omega$, the preimage

$$T^{-1}(\check{\beta}, \check{\xi}, \check{\rho}) = \{(\kappa_\lambda, \sigma_\lambda, \theta_\lambda, \lambda) \in \mathbb{R}^4, \lambda \in \check{J}\}, \quad \check{J} = (-\infty, -(2\check{\xi}-1)\ln\check{\beta}),$$

is a smooth one-dimensional λ-parameterized curve in $\mathcal{D} \subset \mathbb{R}^4$ where

$$\kappa_\lambda = -\lambda - (2\check{\xi}-1)\ln\check{\beta},$$
$$\sigma_\lambda = -\sqrt{2\check{\xi}(1-\check{\xi})}\ln\check{\beta}, \tag{13.11}$$
$$\theta_\lambda = \frac{\check{\rho}\sigma_\lambda^2}{2\kappa_\lambda}, \tag{13.12}$$

where $\lambda \in \check{J}$.

Case of the Vasicek Model

As far as the Vasicek model is considered we put

$$\beta = e^{-\kappa}, \quad \xi = \theta - \frac{\sigma^2}{2\kappa^2} - \frac{\sigma\lambda}{\kappa}, \quad \rho = \frac{\sigma^2}{4\kappa}. \tag{13.13}$$

Then for the original Vasicek parameters we have:

$$\kappa = -\ln\beta, \quad \sigma = 2\sqrt{\rho\kappa}, \quad \theta = \xi + \frac{\sigma^2}{2\kappa^2} + \frac{\sigma\lambda}{\kappa}. \tag{13.14}$$

Proposition 13.2. *In terms of transformed parameters the value of a bond $P = P(\tau, r)$ can be expressed as $P = Ae^{-Br}$, where $\tau = T - t \in [0, T]$ and functions $A = A(\beta, \xi, \rho, \tau)$, $B = B(\beta, \xi, \rho, \tau)$ satisfy*

$$B = -\frac{1-\beta^\tau}{\ln\beta}, \quad A = \exp\left(\xi(B(\tau) - \tau) - \rho B^2(\tau)\right), \quad (13.15)$$

where $(\beta, \xi, \rho) \in \Omega = (0,1) \times \mathbb{R} \times \mathbb{R}^+ \subset \mathbb{R}^3$.

The transformation $T : \mathcal{D} \to \Omega$ defined as in (13.13), i.e. $T(\kappa, \sigma, \theta, \lambda) = (\beta, \xi, \rho)$, where $\mathcal{D} = (0, \infty)^3 \times \mathbb{R} \subset \mathbb{R}^4$, is a smooth mapping too and, for any $(\check\beta, \check\xi, \check\rho) \in \Omega$, the preimage

$$T^{-1}(\check\beta, \check\xi, \check\rho) = \{(\kappa_\lambda, \sigma_\lambda, \theta_\lambda, \lambda) \in \mathbb{R}^4, \ \lambda \in \check{J}\}, \quad \check{J} = \mathbb{R},$$

is a smooth one-dimensional λ-parameterized curve in $\mathcal{D} \subset \mathbb{R}^4$. In this case

$$\kappa_\lambda = -\ln\check\beta, \quad \sigma_\lambda = 2\sqrt{\check\rho\kappa_\lambda}, \quad \theta_\lambda = \check\xi + \frac{\sigma_\lambda^2}{2\kappa_\lambda^2} + \frac{\sigma_\lambda \lambda}{\kappa_\lambda}. \quad (13.16)$$

Summarizing, in both studied one factor models the yield curve depends only on three transformed parameters β, ξ and ρ defined in (13.8) and (13.13), respectively.

13.3.2. The Loss Functional

In this section we introduce the loss functional measuring the quality of approximation of the set of real market yield curves by computed yield curves from each model.

Definition 13.1. *The loss functional is the time-weighted distance of the real market yield curves $\{R_j^i, j = 1, \ldots, m\}$ and the set of computed yield curves $\{\bar{R}_j^i, j = 1, \ldots, m\}$ at time $i = 1, \ldots, n$, determined from the bond price - yield curve relationship*

$$A_j e^{-B_j R_0^i} = e^{-\bar{R}_j^i \tau_j}, \quad (13.17)$$

where $r^i = R_0^i$ is the overnight interest rate at time $i = 1, \ldots, n$, $A_j = A(\tau_j)$ and $B_j = B(\tau_j)$ where $0 = \tau_0 < \tau_1 < \tau_2 < \cdots < \tau_m$ stand for maturities of bonds forming the yield curve, is defined as follows:

$$U(\beta, \xi, \rho) = \frac{1}{m}\sum_{j=1}^{m}\frac{1}{n}\sum_{i=1}^{n}(R_j^i - \bar{R}_j^i)^2 \tau_j^2. \quad (13.18)$$

Recall that $A(\tau)$ and $B(\tau)$ are defined by (13.10) and (13.15).

Proposition 13.3. *In terms of the averaged term structure values and their covariance values the loss functional can be expressed in form:*

$$\begin{aligned} U(\beta, \xi, \rho) &= \frac{1}{m}\sum_{j=1}^{m}((\tau_j E(R_j) - B_j E(R_0) + \ln A_j)^2 \\ &+ Var(\tau_j R_j - B_j R_0)), \end{aligned} \quad (13.19)$$

where $E(X_j)$ and $Var(X_j)$ denote the mean value and variance of the vector $X_j = \{X_j^i, i = 1, \ldots, n\}$.

Expression (13.19) for the loss functional is much more suitable for computational purposes because it contains aggregated time series information from the yield curve only, the cumulative statistics like the mean and covariance of term structure R_j series. These statistical informations can be pre-processed prior to optimization.

13.3.3. Non-linear Regression Problem for the Loss Functional

Introducing the short form of the loss functional (13.19) is prerequisition to the next steps. The core of the estimation method is to minimize the function $U(\beta, \xi, \rho)$, i.e.:

$$\min_{(\beta,\xi,\rho)\in\Omega} U(\beta,\xi,\rho),$$

where $\Omega = (0,1) \times (0,1) \times (0,\rho_{max})$ is a bounded domain[2] in $\mathbb{R}^3$. During this step of our approach we obtain the vector of $(\check{\beta},\check{\xi},\check{\rho})$ for any given λ. This problem is highly non-linear. For that reason we discuss different numerical procedures in the next section. Having identified the curve of global minimizers of the loss functional we proceed by the second step which will be discussed later.

For the CIR as well as for the Vasicek model we have first order necessary conditions for the minimizer of the loss functional. These conditions can be used either for further parameter reduction of the problem (2D problem for the CIR model and even 1D problem for the Vasicek model) or for testing whether a numerical approximation is close to a minimizer. Latter property has been used in practical implementation of the minimization method.

Case of the Cox-Ingersoll-Ross Model

Proposition 13.4. *Given β and ξ, an optimal value for the parameter ρ in the CIR model $\rho_c^{opt} = \rho_c^{opt}(\beta, \xi)$ can be found as a function of β and ξ. Solving the first order optimality condition $\frac{\partial U}{\partial \rho} = 0$ we have:*

$$\sum_{j=1}^{m} (\ln A_j)^2 = -\sum_{j=1}^{m} (\tau_j E(R_j) - B_j E(R_0)) \ln A_j \quad (13.20)$$

and the optimal ρ_c is determined as follows:

$$\rho_c^{opt} = -\frac{\sum_{j=1}^{m} (\tau_j E(R_j) - B_j E(R_0)) \ln A_j(\beta,\xi,1)}{\sum_{j=1}^{m} (\ln A_j(\beta,\xi,1))^2}. \quad (13.21)$$

Case of the Vasicek Model

Proposition 13.5. *Given β, a pair of optimal values for the parameter (ρ, ξ) in the Vasicek model $\rho_v^{opt} = \rho_v^{opt}(\beta)$, $\xi_v^{opt} = \xi_v^{opt}(\beta)$ can be found. Solving the system of first order*

---
[2] ρ_{max} is sufficiently large. In our computation we chose $\rho_{max} = 5$.

optimality conditions $\frac{\partial U}{\partial \rho} = 0$ *and* $\frac{\partial U}{\partial \xi} = 0$ *we have:*

$$0 = \sum_{j=1}^{m} (\tau_j E(R_j) - B_j E(R_0) + \xi(B_j - \tau_j) - \rho B_j^2) B_j^2 \qquad (13.22)$$

$$0 = \sum_{j=1}^{m} (\tau_j E(R_j) - B_j E(R_0) + \xi(B_j - \tau_j) - \rho B_j^2)(B_j - \tau_j)$$

and the pair of optimal values $(\rho_v^{opt}, \xi_v^{opt})$ can be determined from the system of linear equations:

$$\rho_v^{opt} = \frac{\sum_{j=1}^{m} (\tau_j E(R_j) - B_j E(R_0) + \xi_v^{opt}(B_j - \tau_j)) B_j^2}{\sum_{j=1}^{m} B_j^4}, \qquad (13.23)$$

$$\xi_v^{opt} = -\frac{\sum_{j=1}^{m} (\tau_j E(R_j) - B_j E(R_0) - \rho_v^{opt} B_j^2)(B_j - \tau_j)}{\sum_{j=1}^{m} (B_j - \tau_j)^2}.$$

13.3.4. Evolution Strategies

It is well known fact that steepest-descent gradient methods of Newton-Kantorovich type (cf. [4]) may converge to a local minimum only. This is why we have to consider a different and more robust numerical method generically converging to a global minimum of the functional U. There is a wide range of optimization methods based on stochastic optimization algorithms.

These methods are often referred to as evolution strategies (ES) (see e.g. [94, 100, 101]). The main concept of this strategy is based on the survival of the fitness. There exist many different types of this stochastic algorithm like the two membered $(1+1)$ ES, multi-membered (p,c) ES, $(p+c)$ ES (see [94, 100, 101]).

In our case we used a slight modification of the well known $(p+c)$ ES [94]. Recall that the $(p+c)$ ES has p parents and c children (offsprings) per population among which the p best individuals are selected to be next generation parents by their fitness value. The procedure is repeated until some termination criterion is satisfied.

The mathematical description of the modification of $(p+c)$ ES called $(p+c+d)$ ES is as follows:

The problem is defined as finding the real valued vector $x \in \Omega$, which is a global minimum of objective function U in $\Omega \subset R^n$.

1. The initial population of parent vectors $x_k \in \Omega$, $k = 1, \ldots, p$ is generated randomly from bounded three dimensional space $\Omega_b = \{(\beta, \xi, \rho) \in \Omega, 0 \le \rho \le \rho_{max}\}$ where ρ_{max} is large enough. Ω_b is a subset of the domain Ω.

2. In each step of the ES algorithm we generate a set of c offsprings from the parent population $(c \le p)$. Each vector of children (offspring) $\bar{x}_l$, $l = 1, \ldots, c$ is created from parents x_k, $k = 1, \ldots, p$ by mutation and recombination. Mutation means perturbation of parent generation x_k, $k = 1, \ldots, p$ by Gaussian noise with zero mean and preselected standard deviation σ_{gauss}. Recombination means crossing over parts of randomly chosen vectors of children.

3. The modification $(p+c+d)$ ES comprise selection on a wider set. It means that we include a randomly generated set of d wild type individuals forming the so-called wild population. The procedure of generation of the wild type population x_o, $o = 1,\ldots,d$, from bounded space Ω_b is the same as for the initial population.

4. Every member of the population (parents, children, wild population) is characterized with its fitness value, which is the value of the loss functional U.

5. Selection chooses p best vectors from the population by their fitness value to be next generation parents. A set of p intermediate parents is obtained.

6. Next we include a corrector step consisting of improving the set of p intermediate parents by NK iterates of the Newton-Kantorovich gradient minimization method. As a result we obtain a set of p improved parents.

7. The best p individuals from the set of p parents, p improved parents, c offsprings and d wild type individuals are selected to be the next generation of parents.

8. We repeat this procedure until the overall number of steps is less than N. We also perform the first order necessity test as described in Chapter 5.

In our computations we have chosen $N = 300$, $p = c = d = 10^5$, $NK = 30$ and $\sigma_{gauss} = 0.01$. We have not update the standard deviation according to Rechenberger's rule (see [94]) as it turned to be ineffective.

13.3.5. Calibration Based on Maximization of the Restricted Likelihood Function

Recall that in the first step, as it was described in the section 13.3.2. we identify one dimensional curve of the model parameters by minimizing the loss functional. Having identified the curve of global minimizers of the loss functional we proceed by the second step. This step consists of maximization of the likelihood function restricted to that curve so the global maximum is attained in a unique point, which is the estimation of the model parameters.

Notice that the aim of the first "minimization" step of the method was to find a point $(\check{\beta},\check{\xi},\check{\rho})$ - a unique global minimum of the loss functional $U = U(\beta,\xi,\rho)$. Bearing in mind parameter reduction described in the previous section, there exists a C^∞ smooth one dimensional curve of original model parameters $(\kappa_\lambda,\theta_\lambda,\sigma_\lambda,\lambda) \in R^4$ parameterized by $\lambda \in \check{J}$ corresponding to the same transformed triple $(\check{\beta},\check{\xi},\check{\rho})$ for which the minimum of U (in terms of transformed variables β,ξ,ρ) is attained. In order to construct estimation of the model parameters $\kappa,\theta,\sigma,\lambda$ we proceed with the second optimization step in which we find a global maximum of the standard Gaussian likelihood function (LF) over the above mentioned λ-parameterized curve representing of global minimizers of the loss functional U. The two step optimization method combines the maximum likelihood estimation with minimization of the loss functional U. In the case of parameter estimation of a stand-alone short rate process having the form (13.1) the LF is:

$$\ln L(\kappa,\sigma,\theta) = -\frac{1}{2} \sum_{t=2}^{n} \left(\ln v_t^2 + \frac{\varepsilon_t^2}{v_t^2} \right), \qquad (13.24)$$

where $v_t^2 = \frac{\sigma^2}{2\kappa}\left(1-e^{-2\kappa\Delta t}\right)r_{t-1}^{2\gamma}$, $\varepsilon_t = r_t - e^{-\kappa\Delta t}r_{t-1} - \theta\left(1-e^{-\kappa\Delta t}\right)$ (see previous section). Here $\Delta t > 0$ denotes the time step between observations $r_t, r_{t-1}, t = 1,..,n$, evaluated on the yearly basis, e.g. $\Delta t = 1/365$. If estimation of model parameters (κ, σ, θ) is realized by maximization of the likelihood function over the whole set $\mathbb{R}_+^3$ then the maximum is unrestricted. The value of the unrestricted maximum likelihood function is:

$$\ln L^u = \ln L(\kappa^u, \sigma^u, \theta^u) = \max_{\kappa,\sigma,\theta>0} \ln L(\kappa, \sigma, \theta). \qquad (13.25)$$

In our approach we make use of restricted maximization of $\ln L$ over the λ-parameterized curve $\{(\kappa_\lambda, \theta_\lambda, \sigma_\lambda), \lambda \in \check{J}\}$. This can be expressed in original model parameters as follows:

$$\ln L^r = \ln L(\kappa_{\bar{\lambda}}, \sigma_{\bar{\lambda}}, \theta_{\bar{\lambda}}) = \max_{\lambda \in \check{J}} \ln L(\kappa_\lambda, \sigma_\lambda, \theta_\lambda), \qquad (13.26)$$

where $\check{J} = (-\infty, -(2\check{\xi}-1)\ln\check{\beta})$ in the case of the CIR model and $\check{J} = \mathbb{R}$ for the Vasicek model. The argument $\bar{\kappa} = \kappa_{\bar{\lambda}}$, $\bar{\sigma} = \sigma_{\bar{\lambda}}$, $\bar{\theta} = \theta_{\bar{\lambda}}$ of the maximum of the restricted likelihood function $\ln L^r$ is adopted as a result of two step optimization method for calibrating the model parameters. A global maximizer of the unrestricted likelihood function $\ln L^u$ has been computed by the same variant of the ES algorithm described in section 13.3.4. Since maximization of the restricted likelihood function $\ln L^r$ is performed over one dimensional parameter λ and the function $\lambda \mapsto \ln L(\kappa_\lambda, \sigma_\lambda, \theta_\lambda)$ is smooth we could apply a standard optimization software package Mathematica in order to find a global maximizer of the restricted likelihood function. For measuring of accuracy of calibration we introduce the maximum likelihood ratio (MLR) as a ratio of the maximum values of the restricted $\ln L^r$ and unrestricted $\ln L^u$ likelihood functions. We have MLR ≤ 1 and if MLR is close to 1 then the restricted maximum likelihood value is close to the unrestricted one. In this case one can therefore expect that the estimated values $(\bar{\kappa}, \bar{\sigma}, \bar{\theta})$ of the model parameters are close to the argument $(\kappa^u, \sigma^u, \theta^u)$ of the unique global maximum of the unrestricted likelihood function. It may indicate that a simple estimation of parameters based on the mean reversion equation (13.1) for the short rate process r_t is also suitable for estimation of the whole term structure.

13.3.6. Qualitative Measure of Goodness of Fit and Non-linear $\mathcal{R}^2$ Ratio

In linear regression statistical methods, the appropriateness of linear regression function is measured by the $\mathcal{R}^2$ ratio. If the value of $\mathcal{R}^2$ ratio is close to one, it indicates that the given data set can be regressed by a linear function. In the case of non-linear regression, there is no unique way how to define the equivalent concept of the linear $\mathcal{R}^2$ ratio. The non-linear $\mathcal{R}^2$ ratio essentially depends on the choice of the reference value. We take this value of the loss functional (13.18) by taking the argument $(\beta, \xi, \rho) = (1, 1, 1)$. Since $\lim_{\beta \to 1} B_j = \tau_j$ and $\ln A_j = 0$ for $\beta = 1$ it is easy to calculate that

$$U(1,1,1) = \frac{1}{m}\sum_{j=1}^{m} \tau_j^2 E((R_j - R_0)^2),$$

and, moreover, $U(1,1,1) = U(1,\xi,\rho)$ for any $\xi \in [0,1]$ and $\rho \in \mathbb{R}$.

Now we are able to define the non-linear $\mathcal{R}^2$ ratio measuring the quality of non-linear regression as follows:

$$R^2 = 1 - \frac{U(\check{\beta},\check{\xi},\check{\rho})}{U(1,1,1)}, \quad (13.27)$$

where $(\check{\beta},\check{\xi},\check{\rho})$ is the argument of the unique global minimum of the loss functional U. Then $0 \leq \mathcal{R}^2 \leq 1$. The value of $\mathcal{R}^2$ close to one indicates perfect matching of the yield curve computed for parameters $(\check{\beta},\check{\xi},\check{\rho})$ and that of the given real market data set.

13.3.7. Results of Calibration

The results of calibration for the CIR model parameters as well as the $\mathcal{R}^2$ ratios are summarized in Table 13.6 for term structures with maturities up to one year. It reports quarterly results for PRIBOR and EURIBOR in the year 2003. Estimated parameters $\kappa, \sigma, \theta, \lambda$, the value of the loss functional (U) and the non-linear $\mathcal{R}^2$ ratio are presented. Behavior of the expected long-term interest rate θ is in accordance with the expectancy of the market in the long-term run. It predicts interest rates close to 1.7% for EURIBOR as well as for PRIBOR.

Table 13.6. Numerical results of calibration for short term structures (up to one year) for PRIBOR and EURIBOR. Results cover 4 quarters of 2003.

	κ	σ	θ	λ	U ($\times 10^{-6}$)	$\mathcal{R}^2$
PRIBOR						
1/4 2003	0.674	0.007	0.004	-0.483	0.134	0.633
2/4 2003	41.3	0.728	0.018	-9.07	0.238	0.428
3/4 2003	3.78	0.066	0.015	-1.15	0.028	0.897
4/4 2003	3.385	0.097	0.019	-0.626	0.088	0.924
EURIBOR						
1/4 2003	47.7	1.030	0.017	-15.5	0.506	0.783
2/4 2003	0.925	0.028	0.021	0.145	0.319	0.746
3/4 2003	43.2	0.644	0.016	-10.9	0.143	0.807
4/4 2003	16.2	0.39	0.017	-4.06	0.145	0.941

Risk Premium Analysis

In this section we discuss and analyze results of parameter estimation for the parameter λ representing the market price of risk in the Cox–Ingersoll–Ross model. We remind ourselves that the price of a zero coupon bond $P = P(t,T,r)$ computed by means of CIR model satisfies the parabolic equation

$$-\frac{\partial P}{\partial \tau} + (\kappa(\theta - r) - \lambda r)\frac{\partial P}{\partial r} + \frac{\sigma^2}{2}r\frac{\partial^2 P}{\partial r^2} - rP = 0, \quad t \in (0,T), r > 0. \quad (13.28)$$

A solution P for the CIR model can be expressed by the explicit formula $P(t,T,r) = A(T-t)e^{-B(T-t)r}$. Thus $\partial_r P = -BP$. As a consequence, equation (13.28) for the bond price P can

Calibration of Interest Rate and Term Structure Models   245

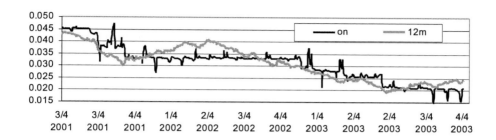

(a)

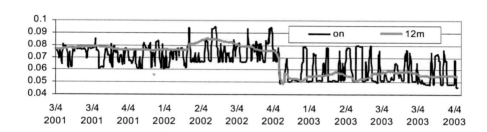

(b)

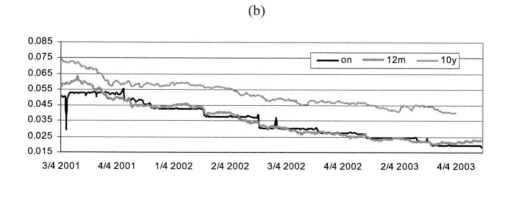

(c)

Figure 13.1. Graphical description of overnight (short rate) interest rates and those of bond with longer maturity. Daily data are plotted for EURO-LIBOR (a), BRIBOR (b) and PRIBOR (c). The 10y PRIBOR stands for the 10 year yield on government bonds. Source: Ševčovič and Csajková, [102].

be rewritten as

$$\frac{\partial P}{\partial t} + \kappa(\theta - r)\frac{\partial P}{\partial r} + \frac{1}{2}\sigma^2 r \frac{\partial^2 P}{\partial r^2} = r^* P, \quad t \in (0,T), r > 0, \tag{13.29}$$

where $r^* = (1 - \lambda B)r$. Indeed, both equations (13.28) as well as (13.29) posses the same solution $P(t,T,r) = A(T-t)e^{-B(T-t)r}$. Hence the multiplier $1 - \lambda B$ can be interpreted as the risk premium factor and r^* as the expected rate of return on the bond (cf. Pearson and Sun [92]). It is easy calculus to show $B(\tau) \geq \tau > 0$ and therefore we have $r^* > r$ if and only if $\lambda < 0$. On the other hand, if $\lambda > 0$ market bond return r^* is less than riskless return rate r.

13.4. Overview of Other Methods

Let us briefly mention other approaches to estimating short rate models. There are other moment-based estimators, besides the generalized method of moments used by [23], such as simulation-based efficient method of moments by Dai and Singleton [34]. Also a large number of methods based on likelihood function exist: quasi maximum likelihood (for example Duffee [38]), estimating the likelihood using simulations in [93] by Pedersen, Ait-Sahalia's series expansions of likelihood function [3]. Another approach to estimating the models uses Bayesian methodology. A comprehensive overview of Bayesian methods and Markov Chain Monte Carlo algorithms in finance can be found in [67] by Johannes and Poulson, which includes also a chapter on estimating the interest rate models.

Chapter 14

Advanced Topics in the Term Structure Modeling

The last chapter of this book is devoted to advanced topics in modeling of term structures with a focus on two factor interest rate models and generalized one factor models. In the first part, we deal with a general one factor model in which the stochastic part is proportional to a power of the short rate. We derive and analyze approximate formulae for a solution to the partial differential equation for pricing zero coupon bonds. In the second part of this chapter, we turn our attention to the problem of averaging of the bond price with respect to stochastic factors (e.g., stochastic volatility). Most of the results contained in this chapter were obtained by Stehlíková and Ševčovič in a series of papers [113, 114, 115, 112].

14.1. Approximate Analytical Solution for a Class of One-factor Models

In this section we discuss the approximate analytical solution for bond prices derived by Choi and Wirjanto in [25]. They considered a class of one-factor models of the form (13.1) and proposed an approximate formula for prices of zero coupon bonds given by a solution to (14.2). We will recall a method how to derive the order of accuracy of their approximation as it was presented recently in the paper [114] by Stehlíková and Ševčovič. Moreover, we will also derive a new approximation of a higher order of accuracy from the paper [114].

Interest rate models for short rate, considered by Choi and Wirjanto in [25], have the following form:

$$dr = (\alpha + \beta r)dt + \sigma r^\gamma dw. \tag{14.1}$$

The short rate equation is written under the risk-neutral measure. It corresponds to the real measure process:

$$dr = (\alpha + \beta r + \lambda(t,r)\sigma r^\gamma)dt + \sigma r^\gamma dw,$$

where $\lambda(t,r)$ is the so-called market price of risk. Let us recall that for a general market price of risk function $\lambda(t,r)$, the price P of a zero-coupon bond can be obtained from a solution to the following partial differential equation:

$$-\frac{\partial P}{\partial \tau} + \frac{1}{2}\sigma^2 r^{2\gamma}\frac{\partial^2 P}{\partial r^2} + (\alpha+\beta r)\frac{\partial P}{\partial r} - rP = 0, \; r>0, \; \tau \in (0,T), \qquad (14.2)$$

satisfying the initial condition $P(0,r) = 1$, for all $r > 0$. In what follows, we use the notation $\partial_\tau P$ for $\partial P/\partial \tau$, similarly $\partial_r P$ for $\partial P/\partial r$ and $\partial_r^2 P$ for $\partial^2 P/\partial r^2$.

In the paper [25] Choi and Wirjanto derived the following approximation P^{ap} for the exact solution P^{ex}:

Theorem 14.1. *[25, Theorem 2] The approximate analytical solution P^{ap} is given by*

$$\begin{aligned}\ln P^{ap}(\tau,r) &= -rB + \frac{\alpha}{\beta}(\tau-B) + (r^{2\gamma}+q\tau)\frac{\sigma^2}{4\beta}\left[B^2 + \frac{2}{\beta}(\tau-B)\right] \\ &\quad - q\frac{\sigma^2}{8\beta^2}\left[B^2(2\beta\tau-1) - 2B\left(2\tau - \frac{3}{\beta}\right) + 2\tau^2 - \frac{6\tau}{\beta}\right],\end{aligned} \qquad (14.3)$$

where

$$q(r) = \gamma(2\gamma-1)\sigma^2 r^{2(2\gamma-1)} + 2\gamma r^{2\gamma-1}(\alpha+\beta r) \qquad (14.4)$$

and

$$B(\tau) = (e^{\beta\tau}-1)/\beta. \qquad (14.5)$$

The derivation of formula (14.3) is based on calculation of the price as an expected value under a risk neutral measure. The tree property of conditional expectation was used and the integral appearing in the exact price was approximated to obtain a closed form approximation. The reader is referred to [25] for more details of derivation of (14.3).

Choi and Wirjanto furthermore showed that such an approximation coincides with the exact solution in the case of the Vasicek model [120]. Moreover, they compared the above approximation with the exact solution of the CIR model, which is also known in a closed form (cf. [31]). Graphical and tabular descriptions of the relative error in the bond prices have been also provided in [25].

Our next goal is to derive the order of accuracy of the approximation formula (14.3) by estimating the difference $\ln P^{ap} - \ln P^{ex}$ of logarithms of approximative and exact solutions of the bond valuation equation (14.2). Then, we give recall a new higher order accurate approximation formula derived in [114] and we analyze its order of convergence analytically and numerically.

14.1.1. Uniqueness of a Solution to the PDE for Bond Prices

It is worth noting that comparison of approximate and exact solutions is meaningful only if the uniqueness of the exact solution can be guaranteed. In the next theorem we provide a proof of uniqueness of a solution to (14.2) satisfying the following definition 14.1.

Definition 14.1 ([114, Def. 1]). *By a complete solution to (14.2) we mean a function $P = P(\tau,r)$ having continuous partial derivatives $\partial_\tau P$, $\partial_r P$, $\partial_r^2 P$ on $Q_T = [0,\infty) \times (0,T)$, satisfying equation (14.2) on Q_T, the initial condition $P(0,r) = 1$ for $r \in [0,\infty)$ and fulfilling the following growth conditions: $|P(\tau,r)| \le Me^{-mr^\delta}$ and $|P_r(\tau,r)| \le M$ for any $r > 0, t \in (0,T)$, where $M,m,\delta > 0$ are constants.*

Now we can state the theorem on uniqueness of a solution to the bond pricing equation.

Theorem 14.2 ([114, Thm. 1]). *Assume $\frac{1}{2} < \gamma < \frac{3}{2}$ or $\gamma = \frac{1}{2}$ and $2\alpha \geq \sigma^2$. Then there exists a unique complete solution to (14.2).*

Notice that if the condition $2\alpha \geq \sigma^2$ for $\gamma = \frac{1}{2}$ is not satisfied then the solution need not be unique. Indeed, Heston in [61] gave an example of a solution to the CIR model having multiple solutions of the bond pricing equation. Based on the existence of a cheaper solution to the CIR model, Heston presented an interpretation of the so-called pricing bubbles in bond markets.

Proof. Our aim is to prove that the inequality

$$\frac{d}{d\tau} \int_0^\infty r^\omega P^2 dr \leq K \int_0^\infty r^\omega P^2 dr \tag{14.6}$$

is satisfied by any solution of (14.2) with some constants K and $\omega \geq 0$. It implies the uniqueness of a solution to the PDE (14.2). Indeed, if P_1 and P_2 are two complete solutions of (14.2) with the same initial condition $P(0,r) = 1$, then $P = P_1 - P_2$ is also a solution to (14.2) with $P(0,r) = 0$. Let us define a function

$$y(\tau) = \int_0^\infty r^\omega P^2(\tau, r) dr.$$

Then the inequality (14.6) means $\frac{dy(\tau)}{d\tau} \leq Ky(\tau)$ for $\tau > 0$. It implies:

$$\frac{d}{d\tau}\left(e^{-K\tau}y(\tau)\right) = -Ke^{-K\tau}y(\tau) + e^{-K\tau}\frac{dy(\tau)}{d\tau} \leq 0.$$

Since $y(0) = 0$ and $y(\tau) \geq 0$, it follows that $y(\tau) = 0$ for all τ. Therefore $P(\tau, r) = 0$ for all $\tau \geq 0, r \geq 0$ and hence $P_1 \equiv P_2$, as claimed.

Now let us derive inequality (14.6). Multiplying the equation by $r^\omega P$, where $\omega > 0$ and $2\gamma + \omega - 1 > 0$ using the identity $\frac{1}{2}\frac{d}{d\tau}\int_0^\infty r^\omega P^2 dr = \int_0^\infty r^\omega P \partial_\tau P dr$, and integrating with respect to r from 0 to infinity we obtain:[1]

$$\frac{1}{2}\frac{d}{d\tau}\int_0^\infty r^\omega P^2 = \frac{\sigma^2}{2}\int_0^\infty r^{2\gamma+\omega}\partial_r^2 PP + \int_0^\infty (\alpha + \beta r)r^\omega \partial_r PP - \int_0^\infty r^{\omega+1}P^2. \tag{14.7}$$

First, we make use integration by parts for the following integrals from the above equation:

$$\int_0^\infty r^{2\gamma+\omega}P\partial_r^2 P = -(2\gamma+\omega)\int_0^\infty r^{2\gamma+\omega-1}P\partial_r P - \int_0^\infty r^{2\gamma+\omega}(\partial_r P)^2$$

$$= \frac{1}{2}(2\gamma+\omega)(2\gamma+\omega-1)\int_0^\infty r^{2\gamma+\omega-2}P^2 - \int_0^\infty r^{2\gamma+\omega}(\partial_r P)^2,$$

where we have used the identity $\int_0^\infty r^{\omega+\xi}P\partial_r P = -\frac{\omega+\xi}{2}\int_0^\infty r^{\omega+\xi-1}P^2$ valid for any $\omega, \xi \geq 0$, $\omega + \xi > 0$, and a function P satisfying the decay estimates from Definition 14.1. Substituting

---
[1] Henceforth, we shall omit the differential dr from the notation.

this to (14.7), we end up with the identity

$$\frac{1}{2}\frac{d}{d\tau}\int_0^\infty r^\omega P^2 = \frac{\sigma^2}{4}(2\gamma+\omega)(2\gamma+\omega-1)\int_0^\infty r^{2\gamma+\omega-2}P^2 - \frac{\sigma^2}{2}\int_0^\infty r^{2\gamma+\omega}(\partial_r P)^2$$
$$- \frac{\alpha\omega}{2}\int_0^\infty r^{\omega-1}P^2 - \frac{(\omega+1)\beta}{2}\int_0^\infty r^\omega P^2 - \int_0^\infty r^{\omega+1}P^2. \quad (14.8)$$

Case 1: $\gamma = \frac{1}{2}$ and $2\alpha \geq \sigma^2$. In the case of CIR model ($\gamma = \frac{1}{2}$) we recall that the condition $2\alpha \geq \sigma^2$ is very well understood as it almost surely guarantees the strict positivity of the stochastic processes r_t, $t \geq 0$, satisfying the stochastic differential equation: $dr = (\alpha + \beta r)dt + \sigma\sqrt{r}dw$ (see e.g., [75]).

Subcase 1a: $2\alpha > \sigma^2$. We use the identity (14.8) with $\gamma = 1/2$ and $\omega = \frac{2\alpha}{\sigma^2} - 1 > 0$ to obtain the desired inequality (14.6) with $K = (\omega+1)\beta$.

Subcase 1b: $2\alpha = \sigma^2$. Using the identity (14.8) with $\omega = 0$ (or simply by multiplying the PDE with P and integrating over $(0,\infty)$) we obtain the inequality (14.6) with $K = \beta$.

Case 2: $\gamma \in (\frac{1}{2}, 1)$. We use equation (14.7) with $\omega = 2$ and estimate the integral $\int_0^\infty r^{2\gamma}P^2$ by using Hölder's inequality:

$$\int_0^\infty r^{2\gamma}P^2 = \int_0^\infty \left(r^{4\gamma-2}P^{4\gamma-2}\right)\left(r^{2-2\gamma}P^{4-4\gamma}\right) \leq \left(\int_0^\infty r^2 P^2\right)^{2\gamma-1}\left(\int_0^\infty rP^2\right)^{2-2\gamma}.$$

Now it follows from the Young's inequality $ab \leq \frac{1}{p\varepsilon^p}a^p + \frac{1}{q}\varepsilon^q b^q$ valid for $p,q \geq 1$, $\frac{1}{p}+\frac{1}{q}=1$, that for any $\varepsilon > 0$ we obtain

$$\int_0^\infty r^{2\gamma}P^2 \leq (2\gamma-1)\left(\frac{1}{\varepsilon}\right)^{\frac{1}{2\gamma-1}}\int_0^\infty r^2 P^2 + (2-2\gamma)\varepsilon^{\frac{1}{2-2\gamma}}\int_0^\infty rP^2.$$

Again using (14.8) with $\omega = 2$ and the above estimate we obtain

$$\frac{1}{2}\frac{d}{d\tau}\int_0^\infty r^2 P^2 \leq \frac{\sigma^2}{2}(\gamma+1)(2\gamma+1)\int_0^\infty r^{2\gamma}P^2 - \alpha\int_0^\infty rP^2 - \frac{3\beta}{2}\int_0^\infty r^2 P^2$$
$$\leq K\int_0^\infty r^2 P^2 + \left(\sigma^2(\gamma+1)(2\gamma+1)(1-\gamma)\varepsilon^{\frac{1}{2-2\gamma}} - \alpha\right)\int_0^\infty rP^2.$$

where $K = \frac{\sigma^2}{2}(\gamma+1)(2\gamma+1)(2\gamma-1)\left(\frac{1}{\varepsilon}\right)^{\frac{1}{2\gamma-1}} - \frac{3\beta}{2}$. By choosing $\varepsilon > 0$ sufficiently small such that $\sigma^2(\gamma+1)(2\gamma+1)(1-\gamma)\varepsilon^{\frac{1}{2-2\gamma}} - \alpha < 0$, we finally obtain the desired inequality $\frac{1}{2}\frac{d}{d\tau}\int_0^\infty r^2 P^2 \leq K\int_0^\infty r^2 P^2$.

Case 3: $\gamma = 1$. We again use equation (14.8) with $\omega = 2$. We obtain (14.6) with $K = 3(2\sigma^2 - \beta)$.

Case 4: $\gamma \in (1, \frac{3}{2})$. Similarly as in the case $\frac{1}{2} < \gamma < 1$ we make use of the Hölder inequality. We obtain:

$$\int_0^\infty r^{2\gamma}P^2 = \int_0^\infty \left(r^{6-4\gamma}P^{6-4\gamma}\right)\left(r^{6\gamma-6}P^{4\gamma-4}\right) \leq \left(\int_0^\infty r^2 P^2\right)^{3-2\gamma}\left(\int_0^\infty r^3 P^2\right)^{2\gamma-2}$$

and, by Young's inequality, we have, for any $\varepsilon > 0$,

$$\int_0^\infty r^{2\gamma} P^2 \leq (3-2\gamma)\left(\frac{1}{\varepsilon}\right)^{\frac{1}{3-2\gamma}} \int_0^\infty r^2 P^2 + (2\gamma-2)\varepsilon^{\frac{1}{2\gamma-2}} \int_0^\infty r^3 P^2.$$

By (14.8) with $\omega = 2$ we have

$$\frac{1}{2}\frac{d}{d\tau}\int_0^\infty r^2 P^2 \leq \frac{\sigma^2}{2}(\gamma+1)(2\gamma+1)\int_0^\infty r^{2\gamma} P^2 - \frac{3\beta}{2}\int_0^\infty r^2 P^2 - \int_0^\infty r^3 P^2$$

$$\leq K\int_0^\infty r^2 P^2 + \left(\sigma^2(\gamma+1)(2\gamma+1)(\gamma-1)\varepsilon^{\frac{1}{2\gamma-2}} - 1\right)\int_0^\infty r^3 P^2,$$

where $K = \frac{\sigma^2}{2}(\gamma+1)(2\gamma+1)(3-2\gamma)\left(\frac{1}{\varepsilon}\right)^{\frac{1}{3-2\gamma}} - \frac{3\beta}{2}$. By choosing $\varepsilon > 0$ sufficiently small such that $\sigma^2(\gamma+1)(2\gamma+1)(\gamma-1)\varepsilon^{\frac{1}{2\gamma-2}} - 1 < 0$ we end up with the desired inequality $\frac{1}{2}\frac{d}{d\tau}\int_0^\infty r^2 P^2 \leq K\int_0^\infty r^2 P^2$. $\square$

14.1.2. Error Estimates for the Approximate Analytical Solution

In this part we present derivation the order of accuracy for the approximative solution proposed by Choi and Wirjanto in [25].

Theorem 14.3 ([114, Thm. 3]). *Let P^{ap} be the approximative solution given by (14.3) and P^{ex} be the exact bond price given as a unique complete solution to (14.2). Then*

$$\ln P^{ap}(\tau,r) - \ln P^{ex}(\tau,r) = c_5(r)\tau^5 + o(\tau^5)$$

as $\tau \to 0^+$, where

$$c_5(r) = -\frac{1}{120}\gamma r^{2(\gamma-2)}\sigma^2\left[2\alpha^2(-1+2\gamma)r^2 + 4\beta^2\gamma r^4 - 8r^{3+2\gamma}\sigma^2\right.$$
$$+2\beta(1-5\gamma+6\gamma^2)r^{2(1+\gamma)}\sigma^2 + \sigma^4 r^{4\gamma}(2\gamma-1)^2(4\gamma-3)$$
$$\left.+2\alpha r\left(\beta(-1+4\gamma)r^2 + (2\gamma-1)(3\gamma-2)r^{2\gamma}\sigma^2\right)\right]. \quad (14.9)$$

Convergence is uniform with respect to r on compact subintervals $[r_1, r_2] \subset (0, \infty)$.

Proof. Recall that the exact bond price $P^{ex}(\tau, r)$ for the model (14.1) is given by a solution of PDE (14.2). Let us define the following auxiliary function: $f^{ex}(\tau, r) = \ln P^{ex}(\tau, r)$. Clearly, $\partial_\tau P^{ex} = P^{ex}\partial_\tau f^{ex}$, $\partial_r P^{ex} = P^{ex}\partial_r f^{ex}$ and $\partial_r^2 P^{ex} = P^{ex}\left[(\partial_r f^{ex})^2 + \partial_r^2 f^{ex}\right]$. Hence the PDE for the function f^{ex} reads as follows:

$$-\partial_\tau f^{ex} + \frac{1}{2}\sigma^2 r^{2\gamma}\left[(\partial_r f^{ex})^2 + \partial_r^2 f^{ex}\right] + (\alpha + \beta r)\partial_r f^{ex} - r = 0. \quad (14.10)$$

Substitution of $f^{ap} = \ln P^{ap}$ into equation (14.10) yields a nontrivial right-hand side $h(\tau, r)$ for the equation for the approximative solution f^{ap}:

$$-\partial_\tau f^{ap} + \frac{1}{2}\sigma^2 r^{2\gamma}\left[(\partial_r f^{ap})^2 + \partial_r^2 f^{ap}\right] + (\alpha + \beta r)\partial_r f^{ap} - r = h(\tau, r). \quad (14.11)$$

If we insert the approximate solution into (14.2) then, after long but straightforward calculations based on expansion of all terms into a Taylor series expansion in τ we obtain:

$$h(\tau,r) = k_4(r)\tau^4 + k_5(r)\tau^5 + o(\tau^5), \qquad (14.12)$$

where the functions k_4 and k_5 are given by

$$\begin{aligned}k_4(r) &= \frac{1}{24}\gamma r^{2(\gamma-2)}\sigma^2 \left[2\alpha^2(-1+2\gamma)r^2 + 4\beta^2\gamma r^4 - 8r^{3+2\gamma}\sigma^2 \right.\\ &\quad + 2\beta(1-5\gamma+6\gamma^2)r^{2(1+\gamma)}\sigma^2 + \sigma^4 r^{4\gamma}(-3+16\gamma-28\gamma^2+16\gamma^3)\\ &\quad \left. + 2\alpha r\left(\beta(-1+4\gamma)r^2 + (2-7\gamma+6\gamma^2)r^{2\gamma}\sigma^2\right)\right],\end{aligned} \qquad (14.13)$$

$$\begin{aligned}k_5(r) &= \frac{\gamma\sigma^2}{120}r^{2(-2+\gamma)}\left[6\alpha^2\beta(-1+2\gamma)r^2 + 12\beta^3\gamma r^4 - 10(1-2\gamma)^2 r^{1+4\gamma}\sigma^4\right.\\ &\quad + 6\beta^2\sigma^2\left(1-5\gamma+6\gamma^2\right)r^{2(1+\gamma)}\\ &\quad + \beta r^{2\gamma}\sigma^2\left(-10(5+2\gamma)r^3 + 3(1-2\gamma)^2(-3+4\gamma)r^{2\gamma}\sigma^2\right)\\ &\quad + 2\alpha r\left(3\beta^2(-1+4\gamma)r^2 + 3\beta(2-7\gamma+6\gamma^2)r^{2\gamma}\sigma^2\right.\\ &\quad \left.\left. - 5(-1+2\gamma)r^{1+2\gamma}\sigma^2\right)\right].\end{aligned} \qquad (14.14)$$

Let us consider a function $g(\tau,r) = f^{ap} - f^{ex}$. As $(\partial_r g)^2 = (\partial_r f^{ap})^2 - (\partial_r f^{ex})^2 - 2\partial_r f^{ex}\partial_r g$ we have

$$\begin{aligned}&-\partial_\tau g + \frac{1}{2}\sigma^2 r^{2\gamma}\left[(\partial_r g)^2 + (\partial_r^2 g)\right] + (\alpha+\beta r)\partial_r g\\ &= \left\{-\partial_\tau f^{ap} + \frac{1}{2}\sigma^2 r^{2\gamma}\left[(\partial_r f^{ap})^2 + \partial_r^2 f^{ap}\right] + (\alpha+\beta r)\partial_r f^{ap}\right\}\\ &\quad - \left\{-\partial_\tau f^{ex} + \frac{1}{2}\sigma^2 r^{2\gamma}\left[(\partial_r f^{ex})^2 + (\partial_r^2 f^{ex})\right] + (\alpha+\beta r)\partial_r f^{ex}\right\}\\ &\quad - \sigma^2 r^{2\gamma}\partial_r f^{ex}\partial_r g.\end{aligned}$$

It follows from (14.10) and (14.11) that the function g satisfies the following PDE:

$$-\partial_\tau g + \frac{1}{2}\sigma^2 r^{2\gamma}\left[(\partial_r g)^2 + \partial_r^2 g\right] + (\alpha+\beta r)\partial_r g$$
$$= h(\tau,r) - \sigma^2 r^{2\gamma}(\partial_r f^{ex})(\partial_r g), \qquad (14.15)$$

where $h(\tau,r)$ satisfies (14.12). Let us expand the solution of (14.15) into a Taylor series with respect to τ with coefficients depending on r. We obtain $g(\tau,r) = \sum_{i=0}^\infty c_i(r)\tau^i = \sum_{i=\omega}^\infty c_i(r)\tau^i$, i.e., the first nonzero term in the expansion is $c_\omega(r)\tau^\omega$. Then $\partial_\tau g = \omega c_\omega(r)\tau^{\omega-1} + o(\tau^{\omega-1})$ and $h(\tau,r) = k_4(r)\tau^4 + o(\tau^4)$ as $\tau \to 0^+$. Here the term $k_4(r)$ is given by (14.13). The remaining terms in (14.12) are of the order $o(\tau^{\omega-1})$ as $\tau \to 0^+$. Hence $-\omega c_\omega(\tau) = k_4(r)\tau^4$ from which we deduce $\omega = 5$ and $c_5(r) = -\frac{1}{5}k_4(r)$. It means that $g(\tau,r) = \ln P^{ap}(\tau,r) - \ln P^{ex}(\tau,r) = -\frac{1}{5}k_4(r)\tau^5 + o(\tau^5)$ which completes the proof. $\square$

Remark 14.1. *The function $c_5(r)$ remains bounded as $r \to 0^+$ for the case of the CIR model in which $\gamma = 1/2$ or for the case when $\gamma \geq 1$. More precisely, $\lim_{r \to 0^+} c_5(r) = -\frac{\sigma^2}{120}$ for $\gamma = 1/2$. On the other hand, if $1/2 < \gamma < 1$, then $c_5(r)$ becomes singular, $c_5(r) = O(r^{2(\gamma-1)})$ as $r \to 0^+$.*

Corollary 14.1. *It follows from Theorem 14.3 that*

1. *the error in yield curves can be expressed as*

$$R^{ap}(\tau, r) - R^{ex}(\tau, r) = -c_5(r)\tau^4 + o(\tau^4) \text{ as } \tau \to 0^+;$$

2. *the relative error[2] of P is given by*

$$\frac{P^{ap}(\tau, r) - P^{ex}(\tau, r)}{P^{ex}(\tau, r)} = -c_5(r)\tau^5 + o(\tau^5) \text{ as } \tau \to 0^+.$$

Convergence is uniform w. r. to r on compact subintervals $[r_1, r_2] \subset (0, \infty)$.

Proof. The first corollary follows from the formula $R(\tau, r) = -\frac{\ln P(\tau, r)}{\tau}$ for calculating yield curves. To prove the second statement we note that Theorem 14.3 gives $\ln P^{ap} - \ln P^{ex} = c_5(r)\tau^5 + o(\tau^5)$. Hence $P^{ap}/P^{ex} = e^{c_5(r)\tau^5 + o(\tau^5)} = 1 + c_5(r)\tau^5 + o(\tau^5)$ and therefore $\frac{P^{ap} - P^{ex}}{P^{ex}} = -c_5(r)\tau^5 + o(\tau^5)$. □

Remark 14.2. *For the CIR model with $\gamma = 1/2$ the term $k_4(r)$ defined in (14.13) can be simplified to $\frac{1}{24}\sigma^2 [\alpha\beta + r(\beta^2 - 4\sigma^2)]$ and hence*

$$\ln P_{CIR}^{ap}(\tau, r) - \ln P_{CIR}^{ex}(\tau, r) = -\frac{1}{120}\sigma^2 [\alpha\beta + r(\beta^2 - 4\sigma^2)]\tau^5 + o(\tau^5)$$

as $\tau \to 0^+$. Now convergence is uniform w. r. to r on compact subintervals $[r_1, r_2] \subset [0, \infty)$.

14.1.3. Improved Higher Order Approximation Formula

In this section we recall the main result of the paper [114] by Stehlíková and Ševčovič. It follows from (14.3) that the term $\ln P^{ap}(\tau, r) - c_5(r)\tau^5$ is the higher order accurate approximation of $\ln P^{ex}$ when compared to the original approximation $\ln P^{ap}(\tau, r)$ obtained by Choi and Wirjanto in [25]. Furthemore, we show, that it is even possible to compute $O(\tau^6)$ term and to obtain a new approximation $\ln P^{ap2}(\tau, r)$ such that the difference $\ln P^{ap2}(\tau, r) - \ln P^{ex}(\tau, r)$ is $o(\tau^6)$ for small values of $\tau > 0$.

Let P^{ex} be the exact bond price in the model (14.1). Let us define an improved approximation P^{ap2} by the formula

$$\ln P^{ap2}(\tau, r) = \ln P^{ap}(\tau, r) - c_5(r)\tau^5 - c_6(r)\tau^6, \qquad (14.16)$$

where $\ln P^{ap}$ is given by (14.3), $c_5(\tau)$ is given by (14.9) in Theorem 1 and

$$c_6(r) = \frac{1}{6}\left(\frac{1}{2}\sigma^2 r^{2\gamma} c_5''(r) + (\alpha + \beta r)c_5'(r) - k_5(r)\right),$$

where c_5' and c_5'' stand for the first and second derivative of $c_5(r)$ w. r. to r and k_5 is defined in (14.14).

---
[2] In [25], this error is referred to as the relative mispricing.

Theorem 14.4 ([114, Thm. 4]). *The difference between the higher order approximation* $\ln P^{ap2}$ *given by* (14.16) *and the exact solution* $\ln P^{ex}$ *satisfies* $\ln P^{ap2}(\tau,r) - \ln P^{ex}(\tau,r) = o(\tau^6)$ *as* $\tau \to 0^+$. *Convergence is uniform w. r. to r on compact subintervals* $[r_1, r_2] \subset (0, \infty)$.

Proof. We have to prove that $g(\tau, r) = c_5(r)\tau^5 + c_6(r)\tau^6 + o(\tau^6)$, where c_5 and c_6 are given above. We already know the form of the coefficient $c_5 = c_5(r)$. Consider the following Taylor series expansions:

$$g(\tau,r) = \sum_{i=5}^{\infty} c_i(r)\tau^i, \quad h(\tau,r) = \sum_{i=4}^{\infty} k_i(r)\tau^i, \quad f(\tau,r) = \sum_{i=1}^{\infty} l_i(r)\tau^i.$$

The absolute term l_0 is zero because $f^{ex}(0,r) = \ln P^{ex}(0,r) = \ln 1 = 0$ for all $r > 0$. Substituting power series into equation (14.15) and comparing coefficients of the order τ^5 enables us to derive the identity:

$$-6c_6(r) + \frac{1}{2}\sigma^2 r^{2\gamma} c_5''(r) + (\alpha + \beta r) c_5'(r) - k_5(r) = 0$$

and hence

$$c_6(r) = \frac{1}{6}\left(\frac{1}{2}\sigma^2 r^{2\gamma} c_5''(r) + (\alpha + \beta r) c_5'(r) - k_5(r)\right).$$

The term $k_5(r)$ given by (14.14) is obtained by computing the expansion of h. □

The order of relative error of bond prices and order of error of interest rates for the new higher order approximation can be derived similarly as in Corollary 14.1.

Remark 14.3. *It is not obvious how to obtain the next higher order terms of expansion because the equations contain unknown coefficients* $l_i(r)$, $i \geq 1$, *of the logarithm of the exact solution, which is not known explicitly.*

Remark 14.4. *In the case of the CIR model we have*

$$c_5^{CIR}(r) = -\frac{\sigma^2}{120}\left(\alpha\beta + r(\beta^2 - 4\sigma^2)\right), \, k_5^{CIR}(r) = \frac{\beta\sigma^2}{40}\left(\alpha\beta + (\beta^2 - 10\sigma^2)r\right)$$

and so

$$c_6^{CIR}(r) = \frac{\sigma^2}{360}\left(-2\alpha\beta^2 + 17\beta\sigma^2 r - 2\beta^3 r + 2\alpha\sigma^2\right).$$

Hence

$$\ln P_{CIR}^{ap2} = \ln P_{CIR}^{ap} + \frac{\sigma^2}{120}\left(\alpha\beta + r(\beta^2 - 4\sigma^2)\right)\tau^5$$

$$-\frac{\sigma^2}{360}\left(-2\alpha\beta^2 + 17\beta\sigma^2 r - 2\beta^3 r + 2\alpha\sigma^2\right)\tau^6$$

The theorem yields $\ln P_{CIR}^{ap2}(\tau,r) - \ln P_{CIR}^{ex}(\tau,r) = o(\tau^6)$. *By computing the expansions of both exact and this approximative solutions we finally obtain*

$$\ln P_{CIR}^{ap2}(\tau,r) = \ln P_{CIR}^{ex}(\tau,r) - \frac{\sigma^2}{5040}\Big(11\alpha\beta^3 + 11\beta^4 r - 34\alpha\beta\sigma^2$$

$$- 180\beta^2 r\sigma^2 + 34r\sigma^4\Big)\tau^7 + o(\tau^7) \, \text{ as } \, \tau \to 0^+.$$

Convergence is uniform w. r. to r on compact subintervals $[r_1, r_2] \subset [0, \infty)$.

Table 14.1. The L_∞ and L_2 – errors for the original $\ln P_{CIR}^{ap}$ and improved $\ln P_{CIR}^{ap2}$ approximations. Source: Stehlíková and Ševčovič [114].

τ	$\|\ln P^{ap} - \ln P^{ex}\|_\infty$	EOC	$\|\ln P^{ap2} - \ln P^{ex}\|_\infty$	EOC
1	2.774×10^{-7}	4.930	4.682×10^{-10}	7.039
0.75	6.717×10^{-8}	4.951	6.181×10^{-11}	7.029
0.5	9.023×10^{-9}	4.972	3.576×10^{-12}	7.004
0.25	2.876×10^{-10}	–	2.786×10^{-14}	–

τ	$\|\ln P^{ap} - \ln P^{ex}\|_2$	EOC	$\|\ln P^{ap2} - \ln P^{ex}\|_2$	EOC
1	6.345×10^{-8}	4.933	9.828×10^{-11}	7.042
0.75	1.535×10^{-8}	4.953	1.296×10^{-11}	7.031
0.5	2.061×10^{-9}	4.973	7.492×10^{-13}	7.012
0.25	6.563×10^{-11}	–	5.805×10^{-15}	–

14.1.4. Comparison of Approximations to the Exact Solution for the CIR Model

In this section we present a comparison of the original and improved approximations in the case of the CIR model where the exact analytic solution is known. We use the parameter values from [25] and [114], i.e., $\alpha = 0.00315$, $\beta = -0.0555$ and $\sigma = 0.0894$.

In Table 14.1 we show L_∞ and L_2 – norms[3] with respect to r of the difference $\ln P^{ap} - \ln P^{ex}$ and $\ln P^{ap2} - \ln P^{ex}$ where we considered $r \in [0, 0.15]$. We also compute the experimental order of convergence (EOC) in these norms. Recall that the experimental order of convergence gives an approximation of the exponent α of expected power law estimate for the error $\|\ln P^{ap}(\tau, .) - \ln P^{ex}(\tau, .)\| = O(\tau^\alpha)$ as $\tau \to 0^+$. The EOC_i is given by a ratio

$$EOC_i = \frac{\ln(err_i/err_{i+1})}{\ln(\tau_i/\tau_{i+1})}, \quad \text{where} \quad err_i = \|\ln P^{ap}(\tau_i, .) - \ln P^{ex}(\tau_i, .)\|_p.$$

In Table 14.2 and Fig. 14.1 we show the L_2 – error of the difference between the original and improved approximations for larger values of τ. It turned out that the higher order approximation P^{ap2} gives about twice better approximation of bond prices in the long time horizon up to 10 years.

14.2. Mathematical Analysis of the Two-Factor Vasicek Model

The remaining three sections of this chapter deal with qualitative and quantitative analysis of bond prices and term structures in three popular two-factor models. First we analyze the two-factor Vasicek model. Then we present qualitative analysis of the two-factor Cox-Ingersoll-Ross model and, finally, the Fong-Vasicek model. In each of these particular

---

[3] L_p and L_∞ norms of a function f defined on a grid with step h are given by $\|f\|_p = (h \sum |f(x_i)|^p)^{1/p}$ and $\|f\|_\infty = \max |f(x_i)|$.

Table 14.2. The L_2 – error with respect to r for large values of τ. Source: Stehlíková and Ševčovič [114].

τ	1	2	5	10
$\|\ln P^{ap} - \ln P^{ex}\|_2$	6.345×10^{-8}	1.877×10^{-6}	1.427×10^{-4}	2.921×10^{-3}
$\|\ln P^{ap2} - \ln P^{ex}\|_2$	9.828×10^{-11}	1.314×10^{-8}	8.798×10^{-6}	1.200×10^{-3}

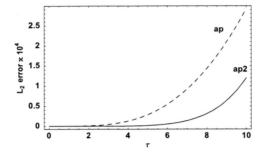

Figure 14.1. The error $\|\ln P^{ap}(\tau,.) - \ln P^{ex}(\tau,.)\|_2$ for the original approximation P^{ap} (dashed line) and the new approximation P^{ap2} (solid line). The horizontal axis is time τ to maturity. Source: Stehlíková and Ševčovič [114].

models we focus our attention to the analysis of averaged bond prices with respect to unobservable quantities of the models (e.g., the stochastic volatility).

We remind ourselves (see Chapters 7, 8) that in the two-factor Vasicek model, the short rate is a modelled as a sum of two independent Ornstein-Uhlenbeck processes r_1 and r_2, where

$$dr_1 = \kappa_1(\theta_1 - r_1)dt + \sigma_1 dw_1,$$
$$dr_2 = \kappa_2(\theta_2 - r_2)dt + \sigma_2 dw_2.$$

According to the general result for two-factor interest rate models derived in Chapter 8, the bond price $P(\tau, r_1, r_2)$ is then a solution to the following partial differential equation of parabolic type:

$$-\frac{\partial P}{\partial \tau} + \left(\kappa_1(\theta_1 - r_1) - \tilde{\lambda}_1 \sigma_1\right)\frac{\partial P}{\partial r_1} + \left(\kappa_2(\theta_2 - r_2) - \tilde{\lambda}_2 \sigma_2\right)\frac{\partial P}{\partial r_2} + \frac{\sigma_1^2}{2}\frac{\partial^2 P}{\partial r_1^2} + \frac{\sigma_2^2}{2}\frac{\partial^2 P}{\partial r_2^2} - (r_1 + r_2)P = 0, \quad (14.17)$$

which holds for any $r_1, r_2 \in (-\infty, \infty)$ and $\tau \in (0, \infty)$. A solution P satisfies the initial condition $P(0, r_1, r_2) = 1$ for each $r_1, r_2 \in (-\infty, \infty)$. Here, $\tilde{\lambda}_1$ and $\tilde{\lambda}_2$ are market prices of risk, corresponding to the factors r_1 and r_2. If these functions are chosen to be constant, λ_1 and λ_2 respectively, the solution of the resulting PDE has the form

$$P(\tau, r_1, r_2) = P_1(\tau, r_1) P_2(\tau, r_2), \quad (14.18)$$

where $P_i(\tau) = A_i(\tau)e^{-B_i(\tau)r}$, $i = 1,2$, are bond prices in the Vasicek model with respective parameters. This property is derived directly by inserting (14.18) into (14.17).

It follows from the form of a solution for bond prices (14.18) and the form of the bond price in the one-factor Vasicek model that the term structure in the two-factor Vasicek model is given by

$$R(\tau, r_1, r_2) = -\left(\frac{\ln A_1}{\tau} + \frac{\ln A_2}{\tau}\right) + \frac{B_1}{\tau}r_1 + \frac{B_2}{\tau}r_2. \tag{14.19}$$

Figures 14.2 and 14.3 show an example of a short rate process generated by the two-factor Vasicek model and term structures in this model.

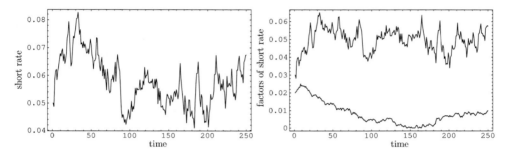

Figure 14.2. The short rate process r generated by the two-factor Vasicek model (left) and its components r_1 and r_2 (right).

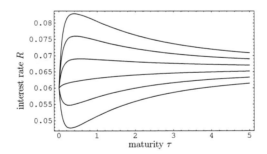

Figure 14.3. Term structures in the two-factor Vasicek model.

14.2.1. Statistical Properties of Bond Prices and Interest Rates

In practice, the components r_1 and r_2 of the short rate are not observable. An observable quantity is just the short rate expressed as their sum $r = r_1 + r_2$. Hence, we are interested in the conditional distribution of $P(\tau, r_1, r_2)$ and $R(\tau, r_1, r_2)$ under the constraint $r_1 + r_2 = r$, where the distributions of r_1 and r_2 are assumed to be their limiting distributions.

The limiting distributions of both Ornstein-Uhlenbeck processes forming the two-factor

Vasicek model, are known to be normal distributions with density

$$f_i(x) = \frac{1}{\sqrt{2\pi\tilde{\sigma}_i^2}} e^{-\frac{(x-\theta_i)^2}{2\tilde{\sigma}_i^2}}, \quad i = 1, 2, \tag{14.20}$$

with $\tilde{\sigma}_i^2 = \frac{\sigma_i^2}{2\kappa_i}$ (cf. [75]). Before deriving distributions of bond prices and interest rates, we formulate a theorem about conditional distribution of normal distributions. The proof can be found for example in [6].

Theorem 14.5 ([6, Thm. 4.12]). *Let X and Y be random variables with normal distributions $N(\mu_x, \sigma_x^2)$ and $N(\mu_y, \sigma_y^2)$ and let ρ be a correlation of X and Y. Then the conditional distribution of Y subject to $X = x$ is $N\left(\mu_y + \rho\frac{\sigma_y}{\sigma_x}(x - \mu_x), \sigma_y^2(1 - \rho^2)\right)$.*

Theorem 14.6. *Consider the two-factor Vasicek model and the limiting distribution of the factors r_1 and r_2. Then:*

1. *The conditional density of the interest rate $R(\tau, r_1, r_2)$, subject to the constraint $r_1 + r_2 = r$, is given by*

$$f_R(x; \tau, r) = \frac{1}{\sqrt{2\pi\sigma_R^2}} e^{-\frac{(x-\mu_R)^2}{2\sigma_R^2}},$$

where

$$\mu_R = -\left(\frac{\ln A_1}{\tau} + \frac{\ln A_2}{\tau}\right) + \frac{1}{\tau}\left[B_1\theta_1 + B_2\theta_2 + \frac{B_1\tilde{\sigma}_1^2 + B_2\tilde{\sigma}_2^2}{\tilde{\sigma}_1^2 + \tilde{\sigma}_2^2}(r - (\theta_1 + \theta_2))\right],$$

$$\sigma_R^2 = \frac{1}{\tau^2}(B_1^2\tilde{\sigma}_1^2 + B_2^2\tilde{\sigma}_2^2)\left(1 - \frac{(B_1\tilde{\sigma}_1^2 + B_2\tilde{\sigma}_2^2)^2}{(\tilde{\sigma}_1^2 + \tilde{\sigma}_2^2)(B_1^2\tilde{\sigma}_1^2 + B_2^2\tilde{\sigma}_2^2)}\right).$$

2. *The conditional density of the bond price $P(\tau, r_1, r_2)$, subject to the constraint condition $r_1 + r_2 = r$, is given by*

$$f_P(x) = \frac{1}{x}\frac{1}{\sqrt{2\pi\sigma_P^2}} e^{-\frac{(\ln x - \mu_P)^2}{2\sigma_P^2}}$$

for $x > 0$ and $f_P(x; \tau, r) = 0$ otherwise, where

$$\mu_P = \ln A_1 + \ln A_2 - \left((B_1\theta_1 + B_2\theta_2) + \frac{B_1\tilde{\sigma}_1^2 + B_2\tilde{\sigma}_2^2}{\tilde{\sigma}_1^2 + \tilde{\sigma}_2}(r - (\theta_1 + \theta_2))\right),$$

$$\sigma_P^2 = (B_1^2\tilde{\sigma}_1^2 + B_2^2\tilde{\sigma}_2^2)\left(1 - \frac{(B_1\tilde{\sigma}_1^2 + B_2\tilde{\sigma}_2^2)^2}{(\tilde{\sigma}_1^2 + \tilde{\sigma}_2^2)(B_1^2\tilde{\sigma}_1^2 + B_2^2\tilde{\sigma}_2^2)}\right).$$

It means that the distribution of interest rates is a normal distribution $N(\mu_R, \sigma_R^2)$ and the distribution of bond prices is lognormal with the logarithm of a bond price having a normal distribution $N(\mu_P, \sigma_P^2)$.

Proof.

1. Since the term $-\left(\frac{\ln A_1}{\tau} + \frac{\ln A_2}{\tau}\right)$ in (14.19) is constant with respect to r_1, r_2, we will consider distribution of $\frac{B_1}{\tau}r_1 + \frac{B_2}{\tau}r_2$, subject to the condition $r_1 + r_2 = r$. Define

$$X = r_1 + r_2, \quad Y = \frac{B_1}{\tau}r_1 + \frac{B_2}{\tau}r_2,$$

then

$$X \sim N\left(\theta_1 + \theta_2, \tilde{\sigma}_1^2 + \tilde{\sigma}_2^2\right), \quad Y \sim N\left(\frac{B_1}{\tau}\theta_1 + \frac{B_2}{\tau}\theta_2, \left(\frac{B_1}{\tau}\right)^2 \tilde{\sigma}_1^2 + \left(\frac{B_2}{\tau}\right)^2 \tilde{\sigma}_2^2\right),$$

$$Corr(X,Y) = \frac{\frac{B_1}{\tau}\tilde{\sigma}_1^2 + \frac{B_2}{\tau}\tilde{\sigma}_2^2}{\sqrt{\tilde{\sigma}_1^2 + \tilde{\sigma}_2^2}\sqrt{\left(\frac{B_1}{\tau}\right)^2\tilde{\sigma}_1^2 + \left(\frac{B_2}{\tau}\right)^2\tilde{\sigma}_2^2}}$$

and the claim follows from the previous theorem 14.5.

2. We have shown that $R \sim N(\mu_r, \sigma_R^2)$. Hence $-R\tau \sim N(-\mu_R\tau, \sigma_R^2\tau^2)$ and $P = e^{-R\tau}$ has a lognormal distribution with parameters given as in the theorem. For a density of a lognormal variable we refer to [6]. □

Fig. 14.4 shows examples of the distributions. With regard to the shape of distribution function one can expect that the variance of interest rates decreases for large maturities. In what follows, we will prove this property rigorously. Furthermore, we will give a condition guaranteeing a similar property for the variance of bond prices.

Theorem 14.7. *Consider the limiting distribution of factors r_1 and r_2 given by (14.20). Then:*

1. *The conditional variance $Var(R(\tau, r_1, r_2)|r_1 + r_2 = r)$ of interest rates (for a fixed r) converges to zero as time to maturity converges to infinity.*

2. *If*

$$\left(\theta_1 - \frac{\sigma_1\lambda_1}{\kappa_1} - \frac{\sigma_1^2}{2\kappa_1^2}\right) + \left(\theta_2 - \frac{\sigma_2\lambda_2}{\kappa_2} - \frac{\sigma_2^2}{2\kappa_2^2}\right) > 0 \qquad (14.21)$$

then the conditional variance $Var(P(\tau, r_1, r_2)|r_1 + r_2 = r)$ of bond prices (for a fixed r) converges to zero as time to maturity converges to infinity.

Remark 14.5. *Recall that in the one-factor Vasicek model, in which $R(\tau, r) = -\frac{\ln A(\tau)}{\tau} + \frac{B(\tau)}{\tau}r$, we have*

$$\lim_{\tau \to \infty} R(\tau, r) = \lim_{\tau \to \infty} -\frac{\ln A(\tau)}{\tau} + \frac{B(\tau)}{\tau}r = \theta - \frac{\sigma\lambda}{\kappa} - \frac{\sigma^2}{2\kappa^2}$$

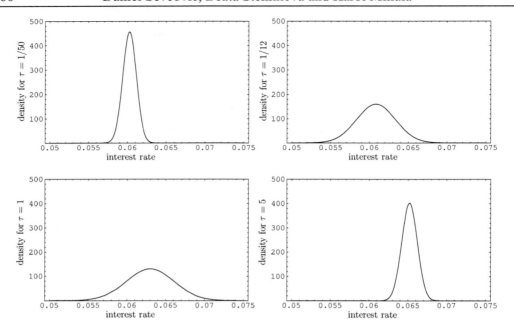

Figure 14.4. Distribution of interest rates in the two-factor Vasicek model.

(see Chapter 8). In the two-factor Vasicek model the limit of term structures can be expressed as:

$$\lim_{\tau \to \infty} R(\tau, r_1, r_2) = \lim_{\tau \to \infty} -\frac{\ln A_1(\tau)}{\tau} - \frac{\ln A_2(\tau)}{\tau} + \frac{B_1(\tau)}{\tau} r_1 + \frac{B_2(\tau)}{\tau} r_2 =$$
$$= \left(\theta_1 - \frac{\sigma_1 \lambda_1}{\kappa_1} - \frac{\sigma_1^2}{2\kappa_1^2} \right) + \left(\theta_2 - \frac{\sigma_2 \lambda_2}{\kappa_2} - \frac{\sigma_2^2}{2\kappa_2^2} \right).$$

Hence the condition (14.21), which is needed in the proof of Theorem 14.7, means that the limit of the term structures is positive.

Proof. We have already computed the variances. Therefore we only need to compute their limits.

1. In the previous section we derived variance of the interest rate R. Functions B_1 and B_2 have positive limits $1/\kappa_1$ and $1/\kappa_2$. Hence we obtain the expression

$$Var(R(\tau, r_1, r_2) | r_1 + r_2 = r)$$
$$= \frac{1}{\tau^2} \left(B_1^2 \tilde{\sigma}_1^2 + B_2^2 \tilde{\sigma}_2^2 \right) \left(1 - \frac{(B_1 \tilde{\sigma}_1^2 + B_2 \tilde{\sigma}_2^2)^2}{(\tilde{\sigma}_1^2 + \tilde{\sigma}_2^2)(B_1^2 \tilde{\sigma}_1^2 + B_2^2 \tilde{\sigma}_2^2)} \right),$$

which converges to zero as $\tau \to \infty$.

2. Since the bond price has a lognormal distribution with $\ln P \sim N(\mu_P, \sigma_P^2)$, its conditional variance is (cf. [6])

$$Var(P(\tau, r_1, r_2) | r_1 + r_2 = r) = e^{2\mu_P + \sigma_P^2} \left(e^{\sigma_P^2} - 1 \right).$$

Notice that A_1, A_2 converge to zero if (14.21) is satisfied and B_1, B_2 have finite limits as $\tau \to \infty$. Therefore we obtain $\mu_P \to -\infty$ and σ_P^2 has a finite limit as $\tau \to \infty$. We conclude that the variance of P converges to zero. □

14.2.2. Averaged Bond Price, Term Structures and Their Confidence Intervals

In this section we focus our attention on averaged values of bond prices and interest rates for given short rate r, with respect to the conditional distribution of factor components r_1, r_2 of the short rate. Moreover, we will analyze their confidence intervals.

Henceforth, we will use the following notation for averaged bond prices and interest rates:

$$\tilde{P}(\tau, r) = \langle P(\tau, r_1, r_2) | r_1 + r_2 = r \rangle,$$
$$\tilde{R}(\tau, r) = \langle R(\tau, r_1, r_2) | r_1 + r_2 = r \rangle.$$

Theorem 14.8 ([112]). *Averaged values of bond prices and interest rates, with respect to limit distributions of r_1, r_2, given that $r_1 + r_2 = r$, are*

1. $\tilde{R}(\tau, r) = -\left(\frac{\ln A_1}{\tau} + \frac{\ln A_2}{\tau}\right) + \frac{1}{\tau}\left[B_1\theta_1 + B_2\theta_2 + \frac{B_1\tilde{\sigma}_1^2 + B_2\tilde{\sigma}_2^2}{\tilde{\sigma}_1^2 + \tilde{\sigma}_2^2}(r - (\theta_1 + \theta_2))\right]$,

2. $\tilde{P}(\tau, r) = \tilde{A}(\tau) e^{-\tilde{B}(\tau)r}$, where

$$\tilde{A}(\tau) = A_1 A_2 \exp\left(-(B_1 - B_2)\left(\theta_1 - (\theta_1 + \theta_2)\frac{\tilde{\sigma}_1^2}{\tilde{\sigma}_1^2 + \tilde{\sigma}_2^2}\right)\right.$$
$$\left. + \frac{1}{2}\frac{\tilde{\sigma}_1^2 \tilde{\sigma}_2^2}{\tilde{\sigma}_1^2 + \tilde{\sigma}_2^2}(B_1 - B_2)^2\right),$$
$$\tilde{B}(\tau) = \frac{\tilde{\sigma}_1^2}{\tilde{\sigma}_1^2 + \tilde{\sigma}_2^2}B_1 + \frac{\tilde{\sigma}_2^2}{\tilde{\sigma}_1^2 + \tilde{\sigma}_2^2}B_2 \qquad (14.22)$$

and $A_i = A_i(\tau)$, $B_i = B_i(\tau)$ are given by solutions $A_i e^{-B_i r}$ of the respective one-factor Vasicek model, which were derived in Chapter 8.

Proof. We have already computed the averaged interest rate by means of the expected value of the interest rate distribution. The formula for the averaged bond price follows from the lognormal distribution for bond prices. For the expected value of a lognormal variable we again refer the reader to the book [6]. □

It follows from Theorem 14.5 that the conditional distribution of r_1, subject to $r_1 + r_2 = r$, is normal $N(\mu_c, \sigma_c^2)$ with parameters

$$\mu_c = \theta_1 + \frac{\tilde{\sigma}_1^2}{\tilde{\sigma}_1^2 + \tilde{\sigma}_2^2}(r - (\theta_1 + \theta_2)), \quad \sigma_c^2 = \frac{\tilde{\sigma}_1^2 \tilde{\sigma}_2^2}{\tilde{\sigma}_1^2 + \tilde{\sigma}_2^2}.$$

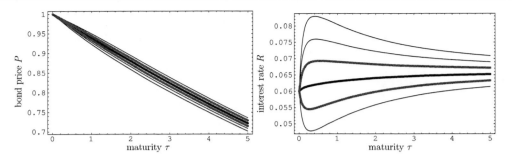

Figure 14.5. Averaged values (blue) and confidence intervals (red) for bond prices and interest rates in the two-factor Vasicek model.

Then the $p \times 100\%$ confidence interval (r_1^l, r_1^u) for r_1 can be constructed. We conclude that

$$P(\tau, r_1, r - r_1) = A_1 A_2 e^{-B_2 r - (B_1 - B_2) r_1},$$
$$R(\tau, r_1, r - r_1) = -\left(\frac{\ln A_1}{\tau} + \frac{\ln A_2}{\tau}\right) + \left(\frac{B_1}{\tau} - \frac{B_2}{\tau}\right) r_1 + \frac{B_2}{\tau} r,$$

are monotone functions of r_1 for fixed values of τ and r. Hence $P(\tau, r_1^l, r - r_1^l)$ and $P(\tau, r_1^u, r - r_1^u)$ are boundaries of the region, where the real bond price curve belongs to with a probability p. Similarly, $R(\tau, r_1^l, r - r_1^l)$ and $R(\tau, r_1^u, r - r_1^u)$ are boundaries of the confidence interval for term structures. Fig. 14.5 shows averaged values and confidence intervals constructed in this way.

14.2.3. Relation of Averaged Bond Prices to Solutions of One-Factor Models

Averaged values, computed in the previous section, are functions of the time to maturity τ and the short rate r. It is a similar dependence as that of one-factor models. Therefore it is natural question to study whether there exists a one-factor interest rate model such that the averaged value $\tilde{P}(\tau, r)$ satisfies the corresponding PDE for bond prices. We restrict ourselves to interest rate models having the short rate r driven by the SDE:

$$dr = \mu(r) dt + \sigma(r) dw, \qquad (14.23)$$

such that the drift μ, volatility σ and market price of risk λ are time independent functions.

Theorem 14.9 ([112, Thm. 3.1]). *Let us consider a class of one-factor models* (14.23) *where functions μ, σ, λ depend only on r and not on time t. Then there is no such a one-factor interest rate model for which the averaged bond prices from the two-factor Vasicek model $\tilde{P}(\tau, r)$ satisfy the PDE*

$$-\frac{\partial P}{\partial \tau} + (\mu(r) - \lambda(r)\sigma(r))\frac{\partial P}{\partial r} + \frac{1}{2}\sigma(r)^2 \frac{\partial^2 P}{\partial r^2} - rP = 0 \qquad (14.24)$$

for bond prices.

Proof. Suppose that the averaged bond price $\tilde{P}(\tau,r)$ is a solution of a one-factor model bond valuation PDE (14.24). Substituting it to this PDE yields

$$-\frac{\dot{\tilde{A}}(\tau)}{\tilde{A}(\tau)}+\dot{\tilde{B}}(\tau)r-(\mu(r)-\lambda(r)\sigma(r))\tilde{B}(\tau)+\frac{1}{2}\sigma^2(r)\tilde{B}^2(\tau)-r=0. \qquad (14.25)$$

It follows that $(\mu(r)-\lambda(r)\sigma(r))\tilde{B}(\tau)-\frac{1}{2}\sigma^2(r)\tilde{B}(\tau)^2$ is a linear function of r of the form

$$(\mu-\lambda\sigma)(r)\tilde{B}(\tau)-\frac{1}{2}\sigma^2(r)\tilde{B}(\tau)^2 = k_1(\tau)+k_2(\tau)r. \qquad (14.26)$$

Moreover, we show that the following stronger condition has to be satisfied:

$$\sigma^2(r) = l_1+l_2 r, \quad \text{where } l_2 \neq 0, \qquad (14.27)$$
$$\mu(r)-\lambda(r)\sigma(r) = l_3+l_4 r, \quad \text{where } l_4 \neq 0. \qquad (14.28)$$

It means that the terms $\mu(r)-\lambda(r)\sigma(r)$ and $\sigma^2(r)$ do not contain nonlinear terms that could eventually vanish in $(\mu(r)-\lambda(r)\sigma(r))\tilde{B}(\tau)-\frac{1}{2}\sigma^2(r)\tilde{B}(\tau)^2$. Then we obtain the equation

$$\left(-\frac{\dot{\tilde{A}}(\tau)}{\tilde{A}(\tau)}-l_3\tilde{B}(\tau)+\frac{1}{2}l_1\tilde{B}^2(\tau)\right)+r\left(\dot{\tilde{B}}(\tau)-l_4\tilde{B}(\tau)+\frac{1}{2}l_2\tilde{B}(\tau)-1\right)=0.$$

Thus, the equation for $\tilde{B}$ reads as follows:

$$\dot{\tilde{B}}(\tau) = 1-\left(\frac{1}{2}l_2-l_4\right)\tilde{B}(\tau),$$
$$\tilde{B}(0) = 0$$

with $l_2, l_4 \neq 0$. This is an equation of the same form as the one appearing in the Cox-Ingersoll-Ross model and its solution is known in a closed form and it is given in Chapter 8. However, the function

$$\tilde{B}(\tau) = \frac{\tilde{\sigma}_1^2}{\tilde{\sigma}_1^2+\tilde{\sigma}_2^2}B_1(\tau)+\frac{\tilde{\sigma}_2^2}{\tilde{\sigma}_1^2+\tilde{\sigma}_2^2}B_2(\tau) = c_0+c_1 e^{-\kappa_1\tau}+c_2 e^{-\kappa_2\tau}$$

for some constants c_0, c_1 and c_2, is not a function of this type.

To finish the proof, we prove (14.27) and (14.28). First, we write the PDE in terms of $B_1(\tau)$ and $B_2(\tau)$ only, i.e.

$$-B_1(\tau)\left(\lambda_1\sigma_1-\kappa_1(\theta_1+\theta_2)\frac{\tilde{\sigma}_1^2}{\tilde{\sigma}_1^2+\tilde{\sigma}_2^2}\right)-B_1(\tau)^2\left(\frac{1}{2}\sigma_1^2-\kappa_1\sigma_c^2\right)$$
$$-B_2(\tau)\left(\lambda_2\sigma_2-\kappa_2\theta_2-\kappa_2\left(\theta_1-(\theta_1+\theta_2)\frac{\sigma_1^2}{\tilde{\sigma}_1^2+\tilde{\sigma}_2^2}\right)\right)-B_2^2(\tau)\left(\frac{1}{2}\sigma_2^2-\kappa_2\sigma_c^2\right)$$
$$-(\sigma_c^2(\kappa_1+\kappa_2)B_1(\tau)B_2(\tau))+\left(-\kappa_1\frac{\tilde{\sigma}_1^2}{\tilde{\sigma}_1^2+\tilde{\sigma}_2^2}B_1(\tau)-\kappa_2\frac{\tilde{\sigma}_2^2}{\tilde{\sigma}_1^2+\tilde{\sigma}_2^2}B_2(\tau)\right)r$$
$$+\frac{1}{2}\sigma^2(r)\left(\frac{\tilde{\sigma}_1^2}{\tilde{\sigma}_1^2+\tilde{\sigma}_2^2}B_1(\tau)+\frac{\tilde{\sigma}_2^2}{\tilde{\sigma}_1^2+\tilde{\sigma}_2^2}B_2(\tau)\right)^2$$
$$-(\mu-\lambda\sigma)(r)\left(\frac{\tilde{\sigma}_1^2}{\tilde{\sigma}_1^2+\tilde{\sigma}_2^2}B_1(\tau)+\frac{\tilde{\sigma}_2^2}{\tilde{\sigma}_1^2+\tilde{\sigma}_2^2}B_2(\tau)\right)=0.$$

The equality holds for all r and $\tau > 0$. Consequently, the derivative of the left hand side with respect to τ is identically zero and its limit as $\tau \to 0^+$ is zero, too. It yields

$$-\left[\left(\lambda_1\sigma_1 - \kappa_1(\theta_1+\theta_2)\frac{\tilde{\sigma}_1^2}{\tilde{\sigma}_1^2+\tilde{\sigma}_2^2}\right) + \left(\lambda_2\sigma_2 - \kappa_2\theta_2 - \kappa_2\left(\theta_1 - (\theta_1+\theta_2)\frac{\tilde{\sigma}_1^2}{\tilde{\sigma}_1^2+\tilde{\sigma}_2^2}\right)\right)\right]$$

$$+\left[-\kappa_1\frac{\tilde{\sigma}_1^2}{\tilde{\sigma}_1^2+\tilde{\sigma}_2^2} - \kappa_2\frac{\tilde{\sigma}_2^2}{\tilde{\sigma}_1^2+\tilde{\sigma}_2^2}\right]r - (\mu(r) - \lambda(r)\sigma(r)) = 0.$$

The proof of proposition (14.27) now follows, with

$$l_4 = -\kappa_1\frac{\tilde{\sigma}_1^2}{\tilde{\sigma}_1^2+\tilde{\sigma}_2^2} - \kappa_2\frac{\tilde{\sigma}_2^2}{\tilde{\sigma}_1^2+\tilde{\sigma}_2^2}. \tag{14.29}$$

Hence $\sigma^2(r)$ is also a linear function of the form $l_1 + l_2 r$, as claimed in (14.28). What remains to show is that $l_2 \neq 0$. From (14.25) we can see that the linear coefficient $k_2(\tau)$ of $(\mu(r) - \lambda(r)\sigma(r))\tilde{B}(\tau) - \frac{1}{2}\sigma^2(r)\tilde{B}(\tau)^2$ in (14.26) is given by

$$k_2(\tau) = \dot{\tilde{B}}(\tau) - 1 = -\kappa_1\frac{\tilde{\sigma}_1^2}{\tilde{\sigma}_1^2+\tilde{\sigma}_2^2} - \kappa_2\frac{\tilde{\sigma}_2^2}{\tilde{\sigma}_1^2+\tilde{\sigma}_2^2}. \tag{14.30}$$

From (14.29) we obtain that the linear coefficient in $(\mu(r) - \lambda(r)\sigma(r))\tilde{B}(\tau)$ is equal to

$$\left(-\kappa_1\frac{\tilde{\sigma}_1^2}{\tilde{\sigma}_1^2+\tilde{\sigma}_2^2} - \kappa_2\frac{\tilde{\sigma}_2^2}{\tilde{\sigma}_1^2+\tilde{\sigma}_2^2}\right)\left(\frac{\tilde{\sigma}_1^2}{\tilde{\sigma}_1^2+\tilde{\sigma}_2^2}B_1(\tau) + \frac{\tilde{\sigma}_2^2}{\tilde{\sigma}_1^2+\tilde{\sigma}_2^2}B_2(\tau)\right).$$

But it has a different form when compared to (14.30). Hence, the linear coefficient in $\sigma^2(r)$ is not zero, which finishes the proof. $\square$

14.3. The Two-Factor Cox Ingersoll Ross Model

In the two-factor CIR model, the short rate is assumed to be a sum of two independent Bessel square root processes r_1 and r_2:

$$\begin{aligned} dr_1 &= \kappa_1(\theta_1 - r_1)dt + \sigma_1\sqrt{r_1}dw_1, \\ dr_2 &= \kappa_2(\theta_2 - r_2)dt + \sigma_2\sqrt{r_2}dw_2. \end{aligned}$$

According to a general form of the bond pricing PDE derived in Chapter 8, the bond price $P(\tau, r_1, r_2)$ is a solution to the following partial differential equation

$$-\frac{\partial P}{\partial \tau} + \left(\kappa_1(\theta_1 - r_1) - \tilde{\lambda}_1\sigma_1\sqrt{r_1}\right)\frac{\partial P}{\partial r_1} + \left(\kappa_2(\theta_2 - r_2) - \tilde{\lambda}_2\sigma_2\sqrt{r_2}\right)\frac{\partial P}{\partial r_2}$$
$$+\frac{\sigma_1^2}{2}r_1\frac{\partial^2 P}{\partial r_1^2} + \frac{\sigma_2^2}{2}r_2\frac{\partial^2 P}{\partial r_2^2} - (r_1+r_2)P = 0, \tag{14.31}$$

which holds for $r_1, r_2 \in (0, \infty)$ and $\tau \in (0, \infty)$. A solution P is subject to the initial condition $P(0, r_1, r_2) = 1$ for all $r_1, r_2 > 0$. Here, $\tilde{\lambda}_1$ and $\tilde{\lambda}_2$ are market prices of risk, corresponding to each of the factors r_1 and r_2. If these functions are chosen to be proportional to $\sqrt{r_1}$ and $\sqrt{r_2}$, i.e., $\tilde{\lambda}_1 = \lambda_1 \sqrt{r_1}$ and $\tilde{\lambda}_2 = \lambda_2 \sqrt{r_2}$ for some constants λ_1 and λ_2, the solution of the resulting PDE has the form

$$P(\tau, r_1, r_2) = P_1(\tau, r_1) P_2(\tau, r_2), \tag{14.32}$$

where $P_i(\tau, r_i) = A_i(\tau) e^{-B_i(\tau) r_i}$, $i = 1, 2$, are bond prices in the CIR model with corresponding parameters (indexed by 1 and 2) derived in Chapter 8. This can be easily shown by inserting (14.32) into (14.31).

To simplify further notation we define $A = A_1 A_2$. Then the bond price is given by

$$P(\tau, r_1, r_2) = A(\tau) e^{-B_1(\tau) r_1 - B_2(\tau) r_2} \tag{14.33}$$

and interest rate by

$$R(\tau, r_1, r_2) = -\frac{\ln A(\tau)}{\tau} + \frac{B_1(\tau)}{\tau} r_1 + \frac{B_2(\tau)}{\tau} r_2. \tag{14.34}$$

A typical decomposition of the short rate is similar as in the case of the two-factor Vasicek model discussed in the previous section. Let us note that a simulation of trajectories can be performed using the exact transition density, which is a multiple of the noncentral χ^2 distribution. An algorithm for generating random numbers from this distribution can be found for example in [56].

14.3.1. Distribution of Bond Prices and Interest Rates

Similarly, as in the case of the two-factor Vasicek model, we consider the limiting distribution of r_1 and r_2. Distributions of the variables will then be considered with respect to these limit distributions and the constraint condition on the observable short rate given by their sum. The limiting distribution for CIR processes

$$dr_i = \kappa_i(\theta_i - r_i)dt + \sigma_i \sqrt{r_i} dw_i, \ i = 1, 2,$$

is known to be (see e.g., [75]) the gamma distribution $\Gamma(a_i, b_i)$ with parameters

$$a_i = \frac{2\kappa_i}{\sigma_i}, \ b_i = \frac{2\kappa_i}{\sigma_i} \theta_i$$

and the corresponding densities

$$f_i(x) = \frac{a_i^{b_i}}{\Gamma(b_i)} e^{-a_i x} x^{b_i - 1}, \quad \text{for } x > 0 \tag{14.35}$$

and $f_i(x) = 0$ otherwise.

Theorem 14.10. *Let us consider the limiting gamma distribution of two factors r_1 and r_2 given as in* (14.35). *Then:*

1. The density function of the interest rate with maturity τ subject to a given level of short rate $r = r_1 + r_2$ is

$$f_R(x) = \frac{g_1(\tilde{r})g_2(r-\tilde{r})}{\int_0^r g_1(r_1)g_2(r-r_1)dr_1} \frac{1}{|B_2(\tau) - B_1(\tau)|}, \quad (14.36)$$

with

$$\tilde{r} = \frac{\tau x - (-\ln A(\tau) + B_2(\tau)r)}{B_1(\tau) - B_2(\tau)}$$

for x between values $-\frac{1}{\tau}\ln A(\tau) + \frac{1}{\tau}B_1(\tau)r$ and $-\frac{1}{\tau}\ln A(\tau) + \frac{1}{\tau}B_2(\tau)r$ and $f_R(x) = 0$ otherwise.

2. The density function of the bond price with a maturity τ subject to the given level of short rate $r = r_1 + r_2$ is

$$f_P(x) = \frac{1}{\tau x} f_R\left(-\frac{1}{\tau}\ln x\right), \quad (14.37)$$

where f_R is the density function of the interest rate given by (14.36).

Proof. The conditional density of r_1, subject to $r_1 + r_2 = r$, is

$$f(r_1, r) = \frac{f_1(r_1)f_2(r-r_1)}{\int_0^r f_1(s)f_2(r-s)ds}. \quad (14.38)$$

Now, we recall that if $r_1 + r_2 = r$, then the interest rate $R(\tau, r_1, r_2)$ can be written as $-\frac{1}{\tau}\ln A(\tau) + \frac{1}{\tau}B_2(\tau)r + \frac{1}{\tau}(B_1(\tau) - B_2(\tau))r_1$. Furthemore, we know that r_1 belongs to the interval $(0, r_1)$. Now, we consider two cases, depending on the sign of the term $B_1(\tau) - B_2(\tau)$.

Case 1: $B_1(\tau) - B_2(\tau) > 0$.
In this case, the minimal possible value of R is $-\frac{1}{\tau}\ln A(\tau) + \frac{1}{\tau}B_2(\tau)r$ and the maximal possible value is $-\frac{1}{\tau}\ln A(\tau) + \frac{1}{\tau}B_1(\tau)r$. For x from the interval $(-\frac{1}{\tau}\ln A(\tau) + \frac{1}{\tau}B_2(\tau)r, -\frac{1}{\tau}\ln A(\tau) + \frac{1}{\tau}B_1(\tau)r)$, the distribution function of R is given by

$$F_R(x) = \text{Prob}(R < x) = \frac{\int_{\substack{r_1 \in (0,r) \\ R(\tau, r_1, r-r_1) < x}} g_1(r_1)g_2(r-r_1)dr_1}{\int_{r_1 \in (0,r)} g_1(r_1)g_2(r-r_1)dr_1}.$$

The condition $R(\tau, r_1, r - r_1) < x$ can be rewritten in the form $r_1 < \tilde{r}$, where

$$\tilde{r} = \frac{\tau x - (-\ln A(\tau) + B_2(\tau)r)}{B_1(\tau) - B_2(\tau)}. \quad (14.39)$$

Because $x \in (-\frac{1}{\tau}\ln A(\tau) + \frac{1}{\tau}B_2(\tau)r, -\frac{1}{\tau}\ln A(\tau) + \frac{1}{\tau}B_1(\tau)r)$, this quantity $\tilde{r}$ belongs to the interval $(0, r)$. Hence the distribution function F can be written as

$$F_R(x) = \frac{\int_0^{\tilde{r}} g_1(r_1)g_2(r-r_1)dr_1}{\int_0^r g_1(r_1)g_2(r-r_1)dr_1}$$

and we obtain the density by taking the derivative

$$\begin{aligned} f_R(x) &= F_R'(x) = \frac{g_1(\tilde{r})g_2(r-\tilde{r})}{\int_0^r g_1(r_1)g_2(r-r_1)dr_1} \frac{d\tilde{r}}{dx} \\ &= \frac{g_1(\tilde{r})g_2(r-\tilde{r})}{\int_0^r g_1(r_1)g_2(r-r_1)dr_1} \frac{\tau}{B_1(\tau) - B_2(\tau)}. \end{aligned}$$

Case 2: $B_1(\tau) - B_2(\tau) < 0$.

In this case, the minimal possible value of R is equal to $-\frac{1}{\tau}\ln A(\tau) + \frac{1}{\tau}B_1(\tau)r$ and the maximal possible value is equal to $-\frac{1}{\tau}\ln A(\tau) + \frac{1}{\tau}B_2(\tau)r$. For x from the interval $(-\frac{1}{\tau}\ln A(\tau) + \frac{1}{\tau}B_1(\tau)r, -\frac{1}{\tau}\ln A(\tau) + \frac{1}{\tau}B_2(\tau)r)$ we compute the distribution function F in the same way as before:

$$F_R(x) = \frac{\int_0^{\tilde{r}} g_1(r_1)g_2(r-r_1)dr_1}{\int_0^r g_1(r_1)g_2(r-r_1)dr_1} = \frac{\int_{\tilde{r}}^r g_1(r_1)g_2(r-r_1)dr_1}{\int_0^r g_1(r_1)g_2(r-r_1)dr_1},$$

where

$$\tilde{r} = \frac{x\tau - (-\ln A(\tau) + B_2(\tau)r)}{B_1(\tau) - B_2(\tau)} \qquad (14.40)$$

belongs to the interval $(0, r)$. Taking the derivative with respect to x we obtain the density function:

$$f_R(x) = F'_R(x) = -\frac{g_1(\tilde{r})g_2(r-\tilde{r})}{\int_0^r g_1(r_1)g_2(r-r_1)dr_1} \frac{d\tilde{r}}{dx}$$

$$= \frac{g_1(\tilde{r})g_2(r-\tilde{r})}{\int_0^r g_1(r_1)g_2(r-r_1)dr_1} \frac{\tau}{B_2(\tau) - B_1(\tau)}. \qquad (14.41)$$

Comparing (14.39) and (14.40) we see that, in the both cases, the quantity $\tilde{r}$ entering the formula for the density function is the same. □

Similarly as in the case of the two-factor Vasicek model, we are able to prove a theorem on the limit of variance for τ approaching infinity. For the proof of the next theorem we need the following lemma.

Lemma 14.1. *Consider a family of the random variables X_τ, $\tau > 0$, with densities f_τ, means μ_τ and variances σ_τ^2. Suppose that X_τ takes values from the interval (a_τ, b_τ), where $(b_\tau - a_\tau) \to 0$ as $\tau \to \infty$. Then $\sigma_\tau^2 \to 0$ as $\tau \to \infty$.*

Proof. Since X_τ takes values only in the interval (a_τ, b_τ), so does the mean value $\mu_\tau \in (a_\tau, b_\tau)$. The variance is given by

$$\sigma_\tau^2 = E\left((X_\tau - \mu_\tau)^2\right) = \int_{a_\tau}^{b_\tau} (x - \mu_\tau)^2 f_\tau(x)dx.$$

By the mean value theorem in the integral form, we have

$$\sigma_\tau^2 = (\xi_\tau - \mu_\tau)^2 \int_{a_\tau}^{b_\tau} f_\tau(x)dx = (\xi_\tau - \mu_\tau)^2$$

for some $\xi_\tau \in (a_\tau, b_\tau)$. Since both ξ_τ and μ_τ belong to (a_τ, b_τ), we have

$$\sigma_\tau^2 = (\xi_\tau - \mu_\tau)^2 < (b_\tau - a_\tau)^2.$$

If $b_\tau - a_\tau$ converges to zero as $\tau \to \infty$, then the variance σ_τ^2 also converges to zero as $\tau \to \infty$. □

Now we can state our next result regarding convergence the conditional variances.

Theorem 14.11. *The conditional variance of both interest rates $R(\tau, r_1, r_2 | r_1 + r_2 = r)$ and bond prices $P(\tau, r_1, r_2 | r_1 + r_2 = r)$ converges to zero for a fixed r as $\tau \to \infty$.*

Proof. The proof is based on the intervals of possible values of interest rates and bond prices for different maturities. According to the previous lemma, it suffices to show that the lengths in the intervals of interest rate and bonds values converge to zero. Let us denote the interval of interest rates by (a_τ, b_τ). Then the interval of bond values is $(e^{-b_\tau \tau}, e^{-a_\tau \tau})$. We show that a_τ and b_τ have the same positive limit, from what it follows that both $b_\tau - a_\tau$ and $e^{-b_\tau \tau} - e^{a_\tau \tau}$ converge to zero.

First, we note the that for functions $A_{cir}(\tau)$ and $B_{cir}(\tau)$ from the one-factor CIR model we have:
$$\lim_{\tau \to \infty} B_{cir}(\tau) = \lim_{\tau \to \infty} \frac{2(e^{\gamma \tau} - 1)}{2\gamma + (\kappa + \lambda + \gamma)(e^{\gamma \tau} - 1)} = \frac{2}{\kappa + \lambda + \gamma},$$
because $\kappa + \lambda + \gamma = (\kappa + \lambda) + \sqrt{(\kappa + \lambda)^2 + 2\sigma^2} \neq 0$, for $\sigma \neq 0$. Using l'Hospital's rule we compute the limit
$$\lim_{\tau \to \infty} \frac{1}{\tau} \log A_{cir}(\tau) = \lim_{\tau \to \infty} \frac{2\kappa\theta}{\sigma^2} \frac{1}{\tau} \log \left(\frac{2\gamma e^{\frac{1}{2}(\kappa + \lambda + \gamma)\tau}}{2\gamma + (\kappa + \lambda + \gamma)(e^{\gamma \theta} - 1)} \right) = \frac{\kappa\theta}{\sigma^2}(\kappa + \lambda - \gamma).$$

Now, it follows that for the two-factor CIR model we have
$$\lim_{\tau \to \infty} -\frac{1}{\tau} \ln A(\tau) = \lim_{\tau \to \infty} -\frac{1}{\tau} \ln A_1(\tau) - \frac{1}{\tau} \ln A_2(\tau)$$
$$= -\frac{\kappa_1 \theta_1}{\sigma_1^2}(\kappa_1 + \lambda_1 - \gamma_1) - \frac{\kappa_2 \theta_2}{\sigma_2^2}(\kappa_2 + \lambda_2 - \gamma_2) > 0$$

and
$$\lim_{\tau \to \infty} \frac{1}{\tau} B_i(\tau) = 0, \quad \text{for } i = 1, 2.$$

Hence both $-\frac{1}{\tau} \ln A(\tau) + \frac{1}{\tau} B_1(\tau)$ and $-\frac{1}{\tau} \ln A(\tau) + \frac{1}{\tau} B_2(\tau)$ converge to same positive limit. Since a_τ and b_τ take values of $-\frac{1}{\tau} \ln A(\tau) + \frac{1}{\tau} B_1(\tau) r$ and $-\frac{1}{\tau} \ln A(\tau) + \frac{1}{\tau} B_2(\tau) r$, our claim is proved. □

14.3.2. Averaged Bond Prices, Term Structures and Their Confidence Intervals

In this section we will show that the averaged bond price for the two-factor CIR model can be expressed in terms of the so-called Kummer hypergeometric function. Therefore, before stating our results regarding averaged bond prices and term structures, we need to recall the definition and basic properties of the Kummer hypergeometric function $_1F_1$ (cf. Abramowitz and Stegun [1]). The function $_1F_1$ defined as a parametric integral:
$$_1F_1(a, b, z) = \frac{\Gamma(b)}{\Gamma(b-a)\Gamma(a)} \int_0^1 e^{zt} t^{a-1} (1-t)^{b-a-1} dt.$$

Next we remind ourselves two useful properties of the Kummer confluent hypergeometric function $_1F_1$ (cf. [1, (13.2.1-2)]). They will be used in the proof of Theorem 14.13.

The following equality holds:

$$\int_0^r e^{-ax} x^{b-1} (r-x)^c dx = r^{b+c} \frac{\Gamma(b)\Gamma(1+c)}{\Gamma(1+b+c)} {}_1F_1(b, 1+b+c, -ar).$$

The first terms in power series expansion of $_1F_1(a,b,z)$ are given by:

$${}_1F_1(a,b,z) = 1 + \frac{a}{b}z + \frac{a(a+1)}{b(b+1)}z^2 + \frac{a(a+1)(a+2))}{b(b+1)(b+2)}z^3 + \ldots.$$

Now we can turn back to the problem of averaging the bond prices and term structures in the two-factor CIR model. Similarly, as in the case of the two-factor Vasicek model, we introduce the notation for the averaged bond price and interest rate:

$$\begin{aligned} \tilde{P}(\tau, r) &= \langle P(\tau, r_1, r_2) | r_1 + r_2 = r \rangle, \\ \tilde{R}(\tau, r) &= \langle R(\tau, r_1, r_2) | r_1 + r_2 = r \rangle. \end{aligned}$$

In the next theorem we explicitly compute these averaged value.

Theorem 14.12 ([113]). *The averaged bond price with respect to the limiting distributions of the processes r_1, r_2 given by (14.35) subject to the constraint $r_1 + r_2 = r$ is given by*

$$\tilde{P}(\tau, r) = A e^{-Br} \frac{{}_1F_1(b_1, b_1+b_2, -((B_1-B_2)+(a_1-a_2)r))}{{}_1F_1(b_1, b_1+b_2, -(a_1-a_2)r)}, \tag{14.42}$$

where $A = A_1(\tau) A_2(\tau)$, $B_i = B_i(\tau)$, $i=1,2$, are given by bond prices $P_i(\tau) = A_i(\tau) e^{-B_i(\tau) r}$, $i=1,2$, in one-factor CIR model, derived in Chapter 8.

The averaged interest rate with respect to limiting distributions of the processes r_1, r_2 given by (14.35) subject to $r_1 + r_2 = r$ is given by

$$\tilde{R}(\tau, r) = -\frac{\ln A}{\tau} + \frac{B_2}{\tau} r + \left(\frac{B_1}{\tau} - \frac{B_2}{\tau}\right) r \frac{b_1}{b_1+b_2} \frac{{}_1F_1(b_1, b_1+b_2+1, -(a_1+a_2)r)}{{}_1F_1(b_1, b_1+b_2, -(a_1+a_2)r)},$$

where $A = A_1(\tau) A_2(\tau)$, $B_i = B_i(\tau)$, $i=1,2$, are given by bond prices $P_i(\tau) = A_i(\tau) e^{-B_i(\tau) r}$, $i=1,2$, in one-factor CIR model, derived in Chapter 8.

Proof. First, we write the denominator appearing in the expression (14.38) for the density function $f(r_1, r)$ and the density itself in a form which will be useful later. Using the definition of the Kummer hypergeometric function and the previous lemma we obtain:

$$M(r) := \int_0^r f_1(r_1) f_2(r-r_1) dr_1 = \frac{a_1^{b_1} a_2^{b_2}}{\Gamma(b_1+b_2)} e^{-a_2 r} r^{b_1+b_2-1} {}_1F_1(b_1, b_1+b_2, -(a_1-a_2)r).$$

Substituting it into the density function yields

$$\begin{aligned} f(r_1, r) &= \frac{1}{M(r)} f_1(r_1) f_2(r - r_1) \\ &= \frac{1}{{}_1F_1(b_1, b_1+b_2, -(a_1-a_2)r)} \frac{\Gamma(b_1+b_2)}{\Gamma(b_1)\Gamma(b_2)} \frac{e^{-(a_1-a_2)r_1} r_1^{b_1-1} (r-r_1)^{b_2-1}}{r^{b_1+b_2-1}}. \end{aligned}$$

Now, we can compute the expected values of bond prices and interest rates. Substituting (14.43) into the expression for the averaged bond price gives

$$\tilde{P}(\tau,r) = \int_0^r P(\tau,r_1,r-r_1)f(r_1,r)dr_1$$
$$= Ae^{-Br}\frac{{}_1F_1(b_1,b_1+b_2,-((B_1-B_2)+(a_1-a_2)r))}{{}_1F_1(b_1,b_1+b_2,-(a_1-a_2)r)}. \quad (14.43)$$

As far as the term structure R is concerned, we obtain

$$R(\tau,r_1,r_2|r_1+r_2=r) = -\frac{\ln A}{\tau} + \frac{B_2}{\tau}r + \left(\frac{B_1}{\tau} - \frac{B_2}{\tau}\right)r_1,$$

we need to compute the expected value of r_1. Substituting the density $f(r_1,r)$ yields

$$\langle r_1 \rangle = \int_0^r r_1 \frac{f_1(r_1)f_2(r-r_1)}{\int_0^r f_1(s)f_2(r-s)ds} dr_1$$
$$= r\frac{b_1}{b_1+b_2}\frac{{}_1F_1(b_1,b_1+b_2+1,-(a_1+a_2)r)}{{}_1F_1(b_1,b_1+b_2,-(a_1+a_2)r)}. \quad \square$$

Since for a given r, the bond prices and interest rates are monotone functions of r_1, we can construct confidence intervals following the same methodology as in the case of the two-factor Vasicek model.

14.3.3. Relation of Averaged Bond Prices to Solutions of One-Factor Interest Rate Models

Before stating the main theorem concerning the relationship between averaged bond prices from the two-factor CIR model and solutions to general one-factor models, we derive several usefull properties of the averaging operator.

Theorem 14.13 ([113, Theorem 3.1]). *Consider the averaged bond prices $\tilde{P}(\tau,r)$ from the previous section. They have the following properties:*

1. $\tilde{P}(\tau,r) \to A(\tau)$ as $r \to 0$,

2. $\frac{\partial \tilde{P}}{\partial \tau}(\tau,r) \to \dot{A}(\tau)$ as $r \to 0$,

3. $\frac{\partial \tilde{P}}{\partial r}(\tau,r) \to -A(\tau)\left(\frac{b_1}{b_1+b_2}B_1(\tau) + \frac{b_2}{b_1+b_2}B_2(\tau)\right)$ as $r \to 0$,

4. $\frac{\partial^2 \tilde{P}}{\partial r^2}(\tau,r)$ *is bounded on the neighborhood of $r=0$.*

Proof.
1) Since both denominator and numerator of the fraction in (14.43) converge to the unity as $r \to 0$, we have
$$\lim_{r \to 0} \tilde{P}(\tau,r) = A(\tau).$$

Advanced Topics in the Term Structure Modeling

2) We compute the derivative of $\tilde{P}$ with respect to τ:

$$\frac{\partial \tilde{P}}{\partial \tau} = \int_0^r \frac{\partial P}{\partial \tau}(\tau, r_1, r - r_1) f(r_1, r) dr_1 =$$

$$= \left[\left(\frac{\dot{A}}{A} - \dot{B}_2 r \right) - (\dot{B}_1 - \dot{B}_2) \frac{\int_0^r r_1 P(\tau, r_1, r - r_1) f(r_1, r) dr_1}{\int_0^r P(\tau, r_1, r - r_1) f(r_1, r) dr_1} \right] \tilde{P}. \quad (14.44)$$

The numerator of the last fraction in (14.44) is positive for all $r > 0$ and can be bounded from above by $r \int_0^r P(\tau, r_1, r - r_1) f(r_1, r) dr_1$. Hence the fraction is positive and bounded from above by r, which implies that it converges to zero as $r \to 0$. Since we already know that $\tilde{P}(\tau, r) \to A(\tau)$ for $r \to 0$, we obtain from (14.44) that

$$\lim_{r \to 0} \frac{\partial \tilde{P}}{\partial \tau}(\tau, r) = \dot{A}(\tau).$$

3) By computing the derivative $\frac{\partial \tilde{P}}{\partial r}$ we obtain

$$\frac{\partial \tilde{P}}{\partial r} = \int_0^r \frac{\partial P}{\partial r}(\tau, r_1, r - r_1) f(r_1, r) + P(\tau, r_1, r - r_1) \frac{\partial f}{\partial r}(r_1, r) dr_1. \quad (14.45)$$

There are two derivatives that have to be computed: $\frac{\partial P}{\partial r}$ and $\frac{\partial f}{\partial r}$. Now, we evaluate these expressions. First,

$$\frac{\partial P}{\partial r}(\tau, r_1, r - r_1) = -B_2(\tau) P(\tau, r_1, r - r_1). \quad (14.46)$$

Secondly,

$$\frac{\partial f}{\partial r}(r_1, r) = \frac{f_1(r_1) f_2'(r - r_1)}{M(r)} - \frac{f_1(r_1) f_2(r - r_1)}{M^2(r)} M'(r)$$

$$= f(r_1, r) \left[\frac{f_2'(r - r_1)}{f_2(r - r_1)} - \frac{\int_0^r f_1(s) f_2'(r - s) ds}{\int_0^r f_1(s) f_2(r - s) ds} \right]. \quad (14.47)$$

Notice that

$$\frac{f_2'(x)}{f_2(x)} = -a_2 + (b_2 - 1) \frac{1}{x}$$

and using it in (14.47) enables us to conclude

$$\frac{\partial f}{\partial r}(r_1, r) = f(r_1, r)(b_2 - 1) \left[\frac{1}{r - r_1} - \frac{\int_0^r \frac{1}{r-s} f_1(s) f_2(r - s) ds}{\int_0^r f_1(s) f_2(r - s) ds} \right]. \quad (14.48)$$

Substituting (14.46) and (14.48) into (14.45) we obtain, after some rearrangements,

$$\frac{\partial \tilde{P}}{\partial r} = \left[-B_2 + (b_2 - 1) \left(\frac{\int_0^r \frac{1}{r-r_1} \pi(\tau, r_1, r - r_1) f(r_1, r) dr_1}{\int_0^r \pi(\tau, r_1, r - r_1) f(r_1, r) dr_1} \right. \right.$$

$$\left. \left. - \frac{\int_0^r \frac{1}{r-r_1} f_1(r_1) f_2(r - r_1) dr_1}{\int_0^r f_1(r_1) f_2(r - r_1) dr_1} \right) \right] \tilde{P}. \quad (14.49)$$

Let us denote

$$X_1 = \frac{\int_0^r \frac{1}{r-r_1}\pi(\tau,r_1,r-r_1)f(r_1,r)dr_1}{\int_0^r \pi(\tau,r_1,r-r_1)f(r_1,r)dr_1}, \quad X_2 = \frac{\int_0^r \frac{1}{r-r_1}f_1(r_1)f_2(r-r_1)dr_1}{\int_0^r f_1(r_1)f_2(r-r_1)dr_1}.$$

With this notation we have

$$\frac{\partial \tilde{P}}{\partial r} = [-B_2 + (b_2 - 1)(X_1 - X_2)]\tilde{P}. \tag{14.50}$$

Next we rewrite each of the expressions X_1 and X_2 in terms of the Kummer hypergeometric functions ${}_1F_1$:

$$X_1 = \frac{1}{r}\frac{b_1 + b_2 - 1}{b_2 - 1}\frac{{}_1F_1(b_1, b_1 + b_2 - 1, -((B_1 - B_2) + (a_1 - a_2)r))}{{}_1F_1(b_1, b_1 + b_2, -((B_1 - B_2) + (a_1 - a_2)r))} \tag{14.51}$$

and, in a similar way,

$$X_2 = \frac{1}{r}\frac{b_1 + b_2 - 1}{b_2 - 1}\frac{{}_1F_1(b_1, b_1 + b_2 - 1, -(a_1 - a_2)r)}{{}_1F_1(b_1, b_1 + b_2, -(a_1 - a_2)r)}. \tag{14.52}$$

Hence

$$X_1 - X_2 = \frac{1}{r}\frac{b_1 + b_2 - 1}{b_2 - 1}\left[\frac{G_1}{G_2} - \frac{G_3}{G_4}\right],$$

where we have denoted

$$\begin{aligned} G_1 &= {}_1F_1(b_1, b_1 + b_2 - 1, -((B_1 - B_2) + (a_1 - a_2))r), \\ G_2 &= {}_1F_1(b_1, b_1 + b_2, -((B_1 - B_2) + (a_1 - a_2))r), \\ G_3 &= {}_1F_1(b_1, b_1 + b_2 - 1, -(a_1 - a_2)r), \\ G_4 &= {}_1F_1(b_1, b_1 + b_2, -(a_1 - a_2)r). \end{aligned} \tag{14.53}$$

Because $G_2 G_4 \to 1$ as $r \to 0$, we need to compute $G_1 G_4 - G_2 G_3$ in order to be able to compute the limit of (14.49). Since

$$\begin{aligned} G_1 &= 1 - \frac{b_1}{b_1 + b_2 - 1}((B_1 - B_2) + (a_1 - a_2))r + o(r), \\ G_2 &= 1 - \frac{b_1}{b_1 + b_2}((B_1 - B_2) + (a_1 - a_2))r + o(r), \\ G_3 &= 1 - \frac{b_1}{b_1 + b_2 - 1}(a_1 - a_2)r + o(r), \\ G_4 &= 1 - \frac{b_1}{b_1 + b_2}(a_1 - a_2)r + o(r), \end{aligned} \tag{14.54}$$

as $r \to 0$, we have

$$G_1 G_4 - G_2 G_3 = r\left(-\frac{b_1}{b_1 + b_2 - 1} + \frac{b_1}{b_1 + b_2}\right) + o(r) \tag{14.55}$$

as $r \to 0$. Hence

$$X_1 - X_2 = \frac{b_1 + b_2 - 1}{b_2 - 1}\frac{1}{G_2 G_4}\left[(B_1 - B_2)\left(-\frac{b_1}{b_1 + b_2 - 1} + \frac{b_1}{b_1 + b_2}\right) + \frac{o(r)}{r}\right]$$

and
$$\lim_{r \to 0} X_1 - X_2 = \frac{b_1 + b_2 - 1}{b_2 - 1}(B_1 - B_2)\left(-\frac{b_1}{b_1 + b_2 - 1} + \frac{b_1}{b_1 + b_2}\right).$$

Finally, we can compute the limit of (14.49)

$$\begin{aligned}\lim_{r \to 0} \frac{\partial \tilde{P}}{\partial r}(\tau, r) &= \lim_{r \to 0}\left[-B_2 + (b_2 - 1)(X_1 - X_2)\right]\tilde{P} \\ &= A\left[-B_2 + (b_1 + b_2 - 1)(B_1 - B_2)\left(-\frac{b_1}{b_1 + b_2 - 1} + \frac{b_1}{b_1 + b_2}\right)\right] \\ &= -A\left[\frac{b_1}{b_1 + b_2}B_1 + \frac{b_2}{b_1 + b_2}B_2\right].\end{aligned}$$

4) We show that there is a finite limit of $\frac{\partial^2 \tilde{P}}{\partial r^2}(\tau, r)$ as $r \to 0$, from which the boundedness of $\frac{\partial^2 \tilde{P}}{\partial r^2}$ follows.

According to (14.49) we have

$$\frac{\partial^2 \tilde{P}}{\partial r^2} = \frac{\partial \tilde{P}}{\partial r}\left[-B_2 + (b_2 - 1)(X_1 - X_2)\right] + \tilde{P}\frac{\partial\left[-B_2 + (b_2 - 1)(X_1 - X_2)\right]}{\partial r}.$$

From the definition of X_1 and X_2 and already computed limits it follows that it suffices to show the existence of the finite limit of $\frac{\partial}{\partial r}\left(\frac{1}{r}F(r)\right)$ for $r \to 0+$, where

$$F(r) = \frac{G_1(r)}{G_2(r)} - \frac{G_3(r)}{G_4(r)}. \tag{14.56}$$

Assuming $F(r)$ has the power series expansion $F(r) = \sum_{k=0}^{\infty} a_k r^k$, the condition $a_0 = 0$ is sufficient for boundedness of the term $\frac{\partial}{\partial r}\left(\frac{1}{r}F(r)\right)$ in the neighborhood of $r = 0$, which holds for (14.56). □

Now we state the main result on the nonexistence of a one-factor model describing the averaged bond price $\tilde{P}$.

Theorem 14.14 ([113, Theorem 3.3]). *Consider averaged bond prices $\tilde{P}(\tau, r)$ obtained from the two-factor CIR model and a class of one-factor short rate models with the underlying short rate satisfying the SDE:*

$$dr = \mu(t, r)dt + \sigma(t, r)dw,$$

with the drift and volatility functions such that:

1. *μ, σ and the market price of risk λ depend only on r and not on t,*

2. *functions μ, σ, λ are continuous at $r = 0$, $\sigma(0) = 0$,*

3. *volatility parameters of the factors from the two-factor CIR model are mutually different, i.e., $\sigma_1 \neq \sigma_2$.*

Then, there is no such a one-factor interest rate model, for which the averaged bond price $\tilde{P}(\tau,r)$ satisfies the PDE for bond prices

$$-\frac{\partial P}{\partial \tau} + (\mu(r) - \lambda(r)\sigma(r))\frac{\partial P}{\partial r} + \frac{1}{2}\sigma(r)^2\frac{\partial^2 P}{\partial r^2} - rP = 0 \qquad (14.57)$$

for all $r \geq 0$, $\tau > 0$.

Proof. By taking the limit $r \to 0$ in the PDE (14.57) and using the results from the previous theorem, we obtain for all $\tau > 0$:

$$-\dot{A}(\tau) + \mu(0^+)(-A(\tau))\left(\frac{b_1}{b_1+b_2}B_1(\tau) + \frac{b_2}{b_1+b_2}B_2(\tau)\right) = 0.$$

From this we can calculate the value of the function μ for $r = 0$:

$$\mu(0^+) = -\frac{\dot{A}(\tau)}{A(\tau)}\frac{1}{\frac{b_1 B_1(\tau)}{b_1+b_2} + \frac{b_2 B_2(\tau)}{b_1+b_2}} = -\frac{\dot{A}(\tau)}{A(\tau)}\frac{b_1+b_2}{b_1 B_1(\tau) + b_2 B_2(\tau)}.$$

It follows that

$$-\frac{\dot{A}(\tau)}{A(\tau)}\frac{b_1+b_2}{b_1 B_1(\tau) + b_2 B_2(\tau)} = K_1, \qquad (14.58)$$

for all $\tau > 0$ where K_1 is a constant independent of τ.

Now we recall that the function $A(\tau)$ from the two-factor CIR model can be written as $A(\tau) = A_1(\tau)A_2(\tau)$, where $A_1(\tau)$ and $A_2(\tau)$ are functions appearing in the original CIR model, corresponding to each of the equations for P_1 and P_2, respectively. Hence they satisfy ODEs:

$$\dot{A}_i(\tau) = -\kappa_i\theta_i A_i(\tau)B_i(\tau), \quad i = 1, 2.$$

Therefore

$$\frac{\dot{A}(\tau)}{A(\tau)} = \frac{\dot{A}_1(\tau)A_2(\tau) + A_1(\tau)\dot{A}_2(\tau)}{A_1(\tau)A_2(\tau)} = \frac{\dot{A}_1(\tau)}{A_1(\tau)} + \frac{\dot{A}_2(\tau)}{A_2(\tau)} = -\kappa_1\theta_1 B_1(\tau) - \kappa_2\theta_2 B_2(\tau).$$

Thus the equality (14.58) can be rewritten as

$$K_1 = -\frac{\dot{A}(\tau)}{A(\tau)}\frac{b_1+b_2}{b_1 B_1(\tau) + b_2 B_2(\tau)} = (\kappa_1\theta_1 B_1(\tau) + \kappa_2\theta_2 B_2(\tau))\frac{b_1+b_2}{b_1 B_1(\tau) + b_2 B_2(\tau)}.$$

Since $b_1 + b_2$ is constant, the only important part is the following fraction:

$$\frac{\kappa_1\theta_1 B_1(\tau) + \kappa_2\theta_2 B_2(\tau)}{b_1 B_1(\tau) + b_2 B_2(\tau)} = K,$$

which has to be equal to some constant K. It implies that

$$\kappa_1\theta_1 B_1(\tau) + \kappa_2\theta_2 B_2(\tau) = K(b_1 B_1(\tau) + b_2 B_2(\tau))$$

and so

$$(\kappa_1\theta_1 - Kb_1)B_1(\tau) = (Kb_2 - \kappa_2\theta_2)B_2(\tau)$$

for each $\tau > 0$. It is possible in two ways:

1. $\kappa_1\theta_1 - Kb_1 = 0$, $Kb_2 - \kappa_2\theta_2 = 0$,

2. $B_1(\tau) = cB_2(\tau)$, where c is a constant.

Now we look at each of these possibilities:

1. The same constant K appears in both equalities. From the first one (i.e., $\kappa_1\theta_1 - Kb_1 = 0$), we get $K = \frac{\kappa_1\theta_1}{b_1}$ and by substituting the value of $b_1 = \frac{2\kappa_1\theta_1}{\sigma_1^2}$, we obtain $K = \frac{\sigma_1^2}{2}$. In the same way, from the second equality (i.e., $Kb_2 - \kappa_2\theta_2 = 0$), we obtain $K = \frac{\sigma_2^2}{2}$. But by the hypothesis, $\sigma_1^2 \neq \sigma_2^2$, which is a contradiction.

2. We recall the equation for B_1 from the CIR model:

$$-\dot{B}_1(\tau) = (\kappa_1 + \lambda_1\sigma_1)B_1(\tau) + \frac{1}{2}\sigma_1^2 B_1(\tau)^2 - 1. \tag{14.59}$$

An analogous equation for $B_2(\tau)$ yields

$$-\dot{B}_2(\tau) = (\kappa_2 + \lambda_2\sigma_2)B_2(\tau) + \frac{1}{2}\sigma_2^2 B_2(\tau)^2 - 1. \tag{14.60}$$

Since $B_1(\tau) = cB_2(\tau)$, we obtain another expression for B_1:

$$-\dot{B}_1(\tau) = c\left[(\kappa_2 + \lambda_2\sigma_2)B_2(\tau) + \frac{1}{2}\sigma^2 B_2(\tau)^2 - 1\right]. \tag{14.61}$$

The right-hand sides of (14.59) and (14.61) have to be equal to:

$$c\left[(\kappa_2 + \lambda_2\sigma_2)B_2(\tau) + \frac{1}{2}\sigma^2 B_2(\tau)^2 - 1\right] = (\kappa_1 + \lambda_1\sigma_1)B_1(\tau) + \frac{1}{2}\sigma_1^2 B_1(\tau)^2 - 1$$

for all $\tau > 0$. By continuity, the equality holds also in the limit $\tau = 0^+$. From this, we obtain $c = 1$. Hence the functions $B_1(\tau)$ and $B_2(\tau)$ coincide. We denote this function by $B(\tau)$. By subtracting equations (14.59) and (14.60) we obtain:

$$[-(\kappa_1 + \lambda_1\sigma_1) + (\kappa_2 + \lambda_1\sigma_1)]B(\tau) + \left[-\frac{1}{2}\sigma_1^2 + \frac{1}{2}\sigma_2^2\right]B^2(\tau) = 0$$

and, dividing by a nonzero term $B(\tau)$ we obtain

$$[-(\kappa_1 + \lambda_1\sigma_1) + (\kappa_2 + \lambda_1\sigma_1)] - \frac{1}{2}\left[\sigma_2^2 - \sigma_1^2\right]B(\tau) = 0.$$

Since $\sigma_1 \neq \sigma_2$, it implies that $B(\tau)$ is a constant function, which is an obvious contradiction.

Since both possibilities lead to a contradiction, the theorem is proved. □

14.4. The Fong-Vasicek Model with a Stochastic Volatility

Another popular two-factor interest rate model for pricing zero copupon bonds is the Fong-Vasicek model introduced in Chapter 8. It belongs to the class of so-called stochastic volatility models in which the volatility of the short rate process follows another stochastic process. Recall that the stochastic process driving the short rate in the Fong-Vasicek model is given by the following system of stochastic differential equations:

$$\begin{aligned} dr &= \kappa_1(\theta_1 - r)dt + \sqrt{y}dw_1, \\ dy &= \kappa_2(\theta_2 - y)dt + v\sqrt{y}dw_2, \end{aligned} \quad (14.62)$$

where the correlation $E(dw_1 dw_2)$ of differentials dw_1 and dw_2 of Wiener processes is ρdt. The parameters $\theta_1, \kappa_1 > 0$ represent the long term interest rate and the rate of return, respectively. On the other hand, the parameters $\theta_2, \kappa_2, v > 0$ represent the long term volatility, its rate of reversion and the volatility of volatility process, respectively.

If the market prices of risk are given as $\lambda_1 \sqrt{y}$, respectively $\lambda_2 \sqrt{y}$, then the PDE for the bond prices reads as follows:

$$-\frac{\partial P}{\partial \tau} + (\kappa_1(\theta_1 - r) - \lambda_1 y)\frac{\partial P}{\partial r} + (\kappa_2(\theta_2 - y) - \lambda_2 v y)\frac{\partial P}{\partial y}$$
$$+ \frac{y}{2}\frac{\partial^2 P}{\partial r^2} + \frac{v^2 y}{2}\frac{\partial^2 P}{\partial y^2} + \rho v y \frac{\partial^2 P}{\partial r \partial y} - rP = 0. \quad (14.63)$$

It is known that a solution to (14.63) can be constructed in the form $P(\tau, r, y) = A(\tau)e^{-B(\tau)r - C(\tau)y}$ (see [48]). Although there are several possibilities how to characterize the functions A, B, C, we will use a description through a system of ordinary differential equations for these functions. It is useful approach in deriving the properties of solutions to the Fong-Vasicek model. Results presented in this section have been obtained by Stehlíková et al. in papers [115] and [111].

14.4.1. Qualitative Properties of Bond Prices and Term Structures

In the following theorem we give the characterization of the bond price by a system of ordinary differential equations. This form will be used later when proving the properties of the model.

Theorem 14.15 ([115]). *A solution of the PDE for bond prices* (14.63) *has the form*

$$P(\tau, r, y) = A(\tau)e^{-B(\tau)r - C(\tau)y}, \quad (14.64)$$

for $r > 0$, $y > 0$ and $\tau > 0$, where functions $A = A(\tau)$, $B = B(\tau)$, $C = C(\tau)$ satisfy the following system of ordinary differential equations:

$$\begin{aligned} \dot{A} &= -A(\kappa_1 \theta_1 B + \kappa_2 \theta_2 C), \\ \dot{B} &= -\kappa_1 B + 1, \\ \dot{C} &= -\lambda_1 B - \kappa_2 C - \lambda_2 v C - \frac{B^2}{2} - \frac{v^2 C^2}{2} - v\rho BC, \end{aligned} \quad (14.65)$$

with initial conditions $A(0) = 1$, $B(0) = 0$, $C(0) = 0$. This can be represented in the following form:

$$B = \frac{1}{\kappa_1}\left(1 - e^{-\kappa_1 \tau}\right), \tag{14.66}$$

$$\dot{C} = -\lambda_1 B - \frac{B^2}{2} - (\kappa_2 + \lambda_2 v + v\rho B)C - \frac{v^2}{2}C^2, \quad C(0) = 0, \tag{14.67}$$

$$A = \exp\left(-\theta_1 \tau + \theta_1 B - \kappa_2 \theta_2 \int_0^\tau C(s) ds\right). \tag{14.68}$$

Proof. By inserting the ansatz (14.64) on the solution into the PDE (14.63) the assertions of theorem easily follow. Furthermore, the ODE for $B(\tau)$ can be solved analytically. The result is then substituted into the equation for $C(\tau)$, which we can solve numerically by the Runge-Kutta method. Finally, we integrate the equation for $A(\tau)$ and use the results for functions B and C. □

It is important to emphasize that the term structure R depends linearly in both the short rate r as well as volatility y variables. Indeed, the term structure can be expressed as

$$R(\tau, r, y) = -\frac{\ln A(\tau)}{\tau} + \frac{B(\tau)}{\tau} r + \frac{C(\tau)}{\tau} y. \tag{14.69}$$

In the next theorem we prove some useful properties of the functions A, B, C forming the zero coupon bond price $P(\tau, r, y) = A(\tau) e^{-B(\tau)r - C(\tau)y}$.

Theorem 14.16 ([115]). *Suppose that the market price of risk λ_1 and the rate of reversion κ_1 satisfy*

$$\lambda_1 \leq -\frac{1}{2\kappa_1}. \tag{14.70}$$

Then the function A, B, C forming the zero coupon bond price fulfill the following conditions:

1. $\dot{C}(0) = 0$, $\ddot{C}(0) = -\lambda_1$,

2. *For every $\tau > 0$: $0 < A(\tau) < 1$, $B(\tau) > 0$, $C(\tau) > 0$,*

3. $A(\tau) \to 0$ *for* $\tau \to \infty$,

4. $C(\tau)$ *is bounded on* $[0, \infty)$.

Proof.
1) It follows from the differentiating the ODE for $C(\tau)$ and using the continuity of C and its derivatives at $\tau = 0$.
2) From the previous statement and the assumption $\lambda_1 \leq -1/2\kappa_1 < 0$ it follows that $C(\tau) > 0$ on some neighborhood of $\tau = 0$. Hence it suffices to show that $\dot{C}(\tau) > 0$ whenever $C(\tau) = 0$. To prove the above claim, we write $\dot{C}(\tau)$ in the following form:

$$\dot{C}(\tau) = -\lambda_1 B(\tau) - \frac{B^2(\tau)}{2} = -\frac{1 - e^{-\kappa_1 \tau}}{2\kappa_1^2}\left(2\lambda_1 \kappa_1 + 1 - e^{-\kappa_1 \tau}\right) > 0,$$

provided $C(\tau) = 0$ and $\lambda_1 \leq -1/2\kappa_1$.

Positiveness of $B(\tau)$ follows directly from the expression of this function. Furthermore, we notice that $B(\tau) < \tau$. The function $A(\tau)$ is positive. Its upper bound follows from the following estimate:

$$A(\tau) = \exp\left(-\theta_1(\tau - B(\tau)) - \kappa_2\theta_2 \int_0^\tau C(s)ds\right)$$
$$< \exp(-\theta_1(\tau - B(\tau))) < 1.$$

The first inequality follows from the positiveness of C and the second one from the positiveness of the difference $\tau - B(\tau)$.

3) We have already shown that

$$0 < A(\tau) < \exp(-\theta(\tau - B(\tau))).$$

Since $\tau - B(\tau)$ converges to infinity as $\tau \to \infty$, we obtain that $A(\tau) \to 0$ as $\tau \to \infty$.

4) It suffices to show that there exists a constant $K > 0$ such that $\dot{C}(\tau) < 0$ whenever $C(\tau) = K$. Notice that $B(\tau) < 1/\kappa_1$. Since $-\nu\rho B < 0$ for $\rho < 0$ and $-\nu\rho B < \frac{\rho\nu}{\kappa_1}$ for $\rho > 0$, we have $-\nu\rho B < \min\left(0, -\frac{\rho\nu}{\kappa_1}\right)$. Using this inequality and the assumption (14.70), from (14.67) we obtain the estimate

$$\dot{C}(\tau) < -\frac{\lambda_1}{\kappa_1} + \left(-\kappa_2 - \lambda_2 \nu + \min\left(0, -\frac{\rho\nu}{\kappa_1}\right)\right)K - \frac{\nu^2}{2}K^2,$$

which is satisfied for any τ such that $C(\tau) = K$. Taking K sufficiently large, the statement 4 follows.

□

Notice that the structural condition (14.70) is not very restrictive because one may expect negative values of the market risk premium λ_1 calibrated from real market term structure data (see Chapter 13). It should be also noted if $\lambda_1 > 0$, then $\ddot{C}(0) = -\lambda_1 < 0$. Since $C(0) = 0$ and $\dot{C}(0) = 0$, we thus have that $C(\tau) < 0$ for small $\tau > 0$. Hence, for small times to maturity τ, the interest rates $R(\tau, r, y)$ becomes negative for large volatility y. Such a undesirable behavior is avoided by assuming (14.70). It is a stronger condition, but we needed it in this form in order to prove the assertions of the theorem.

14.4.2. Distribution of Stochastic Bond Prices and Term Structures for the Fong–Vasicek Model

Let us recall that $P(\tau, r, y)$ is the price of a bond maturing at time τ for a given values of the short rate r and volatility y. Unlike the short rate r, the volatility y is not an observable variable in the real market. Similarly as in the case of previously investigated two-factor interest rate models, it is reasonable to study the bond price $P(\tau, r, y)$ for the given τ and r as a function of the random variable y.

In what follows, we will assume that the value of the short rate r at time to maturity τ is known from the spot market data. The hidden parameter in the model is the volatility y, which is supposed to be driven by a Bessel square root process

$$dy = \kappa_2(\theta_2 - y)dt + \nu\sqrt{y}dw_2.$$

We already know from Chapter 8 that its limiting density f_y is a density of the Gamma distribution $\Gamma(\beta,\alpha)$ i.e.

$$f_y(x) = \frac{\beta^\alpha}{\Gamma(\alpha)}e^{-\beta x}x^{\alpha-1}, \qquad \beta = \frac{2\kappa_2}{v^2}, \quad \alpha = \frac{2\kappa_2}{v^2}\theta_2, \tag{14.71}$$

for $x > 0$ and $f_y(x) = 0$ otherwise. It enables us to compute the distribution function of $P(\tau,r,y)$ and $R(\tau,r,y)$ now.

Theorem 14.17. *Under the assumption* (14.70) *the density functions of the bond prices $P(\tau,r,y)$ and term structures $R(\tau,r,y)$ with respect to the limiting distribution* (14.71) *of the process y are given by*

$$f_P(x) = f_y\left(-\frac{B(\tau)}{C(\tau)}r - \frac{1}{C(\tau)}\ln\frac{x}{A(\tau)}\right)\frac{1}{C(\tau)x} \tag{14.72}$$

and

$$f_R(x) = f_y\left(\frac{1}{C(\tau)}(\tau x + \ln A(\tau) - B(\tau))\right)\frac{\tau}{C(\tau)}, \tag{14.73}$$

where f_y is density of the limiting $\Gamma(\beta,\alpha)$ distribution with shape parameters $\beta = \frac{2\kappa_2}{v^2}$, $\alpha = \frac{2\kappa_2}{v^2}\theta_2$.

Proof. First, we compute distribution functions and then by differentiating them we obtain densities.

Since $P(\tau,r,y)$ is a decreasing function of y, the range of its possible values is the interval $\left(0, A(\tau)e^{-B(\tau)r}\right]$. Hence outside of this interval, the density is vanishing. For $x \in \left(0, A(\tau)e^{-B(\tau)r}\right]$ we have the following expression for the cumulative distribution function $F_P(x)$:

$$\begin{aligned}F_P(x) &= \text{Prob}\left[A(\tau)e^{-B(\tau)r-C(\tau)y} < x\right] = \text{Prob}\left[y > -\frac{B(\tau)}{C(\tau)}r - \frac{1}{C(\tau)}\ln\frac{x}{A(\tau)}\right]\\ &= 1 - F_y\left(-\frac{B(\tau)}{C(\tau)}r - \frac{1}{C(\tau)}\ln\frac{x}{A(\tau)}\right),\end{aligned}$$

where we have used the positiveness of $C(\tau)$, and so

$$f_P(x) = F_P'(x) = f_y\left(-\frac{B(\tau)}{C(\tau)}r - \frac{1}{C(\tau)}\ln\frac{x}{A(\tau)}\right)\frac{1}{C(\tau)x}.$$

Similarly, because of increasing dependence of $R(\tau,r,y)$ on y, the range of possible values for R is the interval $\left[-\frac{\ln A(\tau)}{\tau} + \frac{B(\tau)}{\tau}r, \infty\right)$. Hence its density is zero outside this interval. For $x \in \left[-\frac{\ln A(\tau)}{\tau} + \frac{B(\tau)}{\tau}r, \infty\right)$ we have

$$\begin{aligned}F_R(x) &= \text{Prob}\left[-\frac{\ln A(\tau)}{\tau} + \frac{B(\tau)}{\tau}r + \frac{C(\tau)}{\tau}y < x\right]\\ &= \text{Prob}\left[y < \frac{\tau x + \ln A(\tau) - B(\tau)}{C(\tau)}\right] = F_y\left(\frac{\tau x + \ln A(\tau) - B(\tau)}{C(\tau)}\right),\end{aligned}$$

where we have used the positiveness of $C(\tau)$ again. Hence

$$f_R(x) = F_R'(x) = f_y\left(\frac{1}{C(\tau)}(\tau x + \ln A(\tau) - B(\tau))\right)\frac{\tau}{C(\tau)}. \qquad \square$$

14.4.3. Averaged Bond Prices, Term Structures and Their Confidence Intervals

In analogy with previously discussed two-factor models, for the Fong-Vasicek model we can also prove that the variances of bond prices and interest rates tend to zero as $\tau \to \infty$ and other useful properties regarding the averaged bond prices and term structures.

As usual, we define the averaged bond prices and interest rates as follows:

$$\tilde{P}(\tau,r) = \langle P(\tau,r,y)\rangle_y,$$
$$\tilde{R}(\tau,r) = \langle R(\tau,r,y)\rangle_y,$$

where the expectations are taken with respect to the limiting distribution (14.71) of the process y describing the stochastic volatility.

Theorem 14.18 ([115]). *Averaged bond prices and interest rates, with respect to the limiting Gamma distribution (14.71) of the random variable y are given by:*

1. $\tilde{P}(\tau,r) = A(\tau)e^{-B(\tau)r}\left(1 + \frac{C(\tau)}{\beta}\right)^{-\alpha},$

2. $\tilde{R}(\tau,r) = -\frac{1}{\tau}\ln A(\tau) + \frac{B(\tau)}{\tau}r + \frac{C(\tau)}{\tau}\theta_2.$

Proof. 1) We compute the averaged bond price:

$$\tilde{P}(\tau,r) = \int_0^\infty P(\tau,r,y)g(y)dy = A(\tau)e^{-B(\tau)r}\frac{\beta^\alpha}{\Gamma(\alpha)}\int_0^\infty e^{-(C(\tau)+\beta)y}y^{\alpha-1}dy$$
$$= A(\tau)e^{-B(\tau)r}\frac{\beta^\alpha}{(C(\tau)+\beta)^\alpha} = A(\tau)e^{-B(\tau)r}\left(1+\frac{C(\tau)}{\beta}\right)^{-\alpha}.$$

2) The formula for the averaged interest rate follows from linearity of interest rate in the y variable and taking into account the expected value of the limiting gamma distribution $\Gamma(\beta,\alpha) = \Gamma\left(\frac{2\kappa_2}{v^2}, \frac{2\kappa_2}{v^2}\theta_2\right)$, which is equal to θ_2. □

We remark that the function $y \to P(\tau,r,y)$ is strictly convex because its second derivative ∂_y^2 is equal to $C^2(\tau)P(\tau,r,y) > 0$. Hence, by applying Jensen's inequality, we obtain

$$\tilde{P}(\tau,r) = \int_0^\infty P(\tau,r,y)g(y)dy > P\left(\tau,r,\int_0^\infty yg(y)dy\right) = P(\tau,r,\theta_2),$$

since $\int_0^\infty yg(y)dy = \theta_2$. It means that the averaged bond price $\tilde{P}(\tau,r) = \langle P(\tau,r,y)\rangle_y$ is always greater than the bond price of the two-factor CIR model corresponding to the limiting mean value $\langle y \rangle_y = \theta_2$ of the stochastic volatility y.

Now, we are able to prove the assertion on limiting behavior of variances.

Theorem 14.19 ([115]). *For fixed values of τ and r we have*

$$\lim_{\tau\to\infty} Var_y P(\tau,r,y) = 0, \quad \lim_{\tau\to\infty} Var_y R(\tau,r,y) = 0,$$

where variances are computed with respect to the limiting distribution of the stochastic volatility y.

Proof. We have already computed $\langle P(\tau,r,y)\rangle_y$. In the same way we can compute the expected value of P^2:

$$\begin{aligned}\langle P^2(\tau,r,y)\rangle_y &= \int_0^\infty \left(A(\tau)e^{-B(\tau)r-C(\tau)y}\right)^2 g(y)dy \\ &= A(\tau)^2 e^{-2B(\tau)r} \int_0^\infty e^{-(2C(\tau)+\beta)y} y^{\alpha-1} dy \\ &= A(\tau)^2 e^{-2B(\tau)r} \left(1+\frac{2C(\tau)}{\beta}\right)^{-\alpha}.\end{aligned}$$

Hence

$$\begin{aligned}Var_y(P(\tau,r,y)) &= \langle P^2(\tau,r,y)\rangle_y - \langle P(\tau,r,y)\rangle_y^2 \\ &= A(\tau)^2 e^{-2B(\tau)r} \left[\left(1+\frac{2C(\tau)}{\beta}\right)^{-\alpha} - \left(1+\frac{C(\tau)}{\beta}\right)^{-2\alpha}\right] \\ &= A(\tau)^2 e^{-2B(\tau)r} \left[\left(1+\frac{2C(\tau)}{\beta}\right)^{-\alpha} - \left(1+\frac{2C(\tau)}{\beta}+\frac{C(\tau)^2}{\beta^2}\right)^{-\alpha}\right].\end{aligned}$$

By the mean value theorem

$$\left(1+\frac{2C(\tau)}{\beta}\right)^{-\alpha} - \left(1+\frac{2C(\tau)}{\beta}+\frac{C(\tau)^2}{\beta^2}\right)^{-\alpha} = (-\alpha\xi^{-\alpha-1})\left(-\frac{C(\tau^2)}{\beta^2}\right)$$

for some real number ξ belonging to the interval $\left(1+\frac{2C(\tau)}{\beta}, 1+\frac{2C(\tau)}{\beta}+\frac{C(\tau)^2}{\beta^2}\right)$. Hence $\xi > 1$ and therefore

$$\begin{aligned}Var_y(P(\tau,r,y)) &= A(\tau)^2 e^{-2B(\tau)r} \alpha \xi^{-\alpha-1} \frac{C(\tau^2)}{\beta^2} \\ &< A(\tau)^2 e^{-2B(\tau)r} \frac{\alpha}{\beta^2} C(\tau^2).\end{aligned}$$

Since $C(\tau)$ and $B(\tau)$ are bounded on $[0,\infty)$ and $A(\tau) \to 0$ as $\tau \to \infty$, we conclude that

$$Var_y(P(\tau,r,y)) \to 0 \text{ as } \tau \to \infty.$$

Since R is linear in the y variable and the variance of y is $Var(y) = \frac{\alpha}{\lambda^2}$, we obtain

$$Var_y(R(\tau,r,y)) = \left(\frac{C(\tau)}{\tau}\right)^2 Var(y) = \frac{v^2\theta_2}{2\kappa_2} \frac{C^2(\tau)}{\tau^2}.$$

Because $C(\tau)$ is bounded and $\frac{1}{\tau^2} \to 0$ for $\tau \to \infty$ we obtain

$$Var_y(R(\tau,r,y)) \to 0 \text{ for } \tau \to \infty,$$

as claimed in the theorem. □

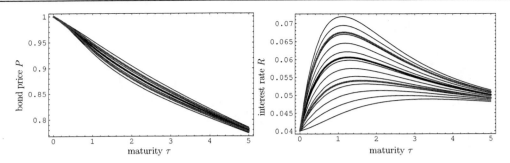

Figure 14.6. Bond prices (left) and term structures (right) in the Fong-Vasicek model (grey), their averaged values (blue) and confidence intervals (red). Source: Stehlíková and Ševčovič [115].

As we already know, $P(\tau,r,y)$ and $R(\tau,r,y)$ are monotone functions in the variable y. Hence the area containing a given percentile of bond prices and term structures can be bounded by $P(\tau,r,y_1)$ and $P(\tau,r,y_2)$, respectively $R(\tau,r,y_1)$ and $R(\tau,r,y_2)$, where $\int_{y_1}^{y_2} g(y)dy$ is equal to a given percentile.

In Fig. 14.6 we can observe examples of bond prices and term structures (corresponding to the same short rate but different volatilities) together with their averaged values and 95% confidence intervals.

14.4.4. Relation of Averaged Bond Prices to Solutions to One-Factor Interest Rate Models

For the Fong–Vasicek model we can again investigate the problem whether there are functions $\mu = \mu(r)$ and $\sigma = \sigma(r)$ such that the bond prices, given as a solution to a one-factor model with the short rate satisfying $dr = \mu dt + \sigma dw$, are the same as those obtained as averaged prices from the Fong-Vasicek two-factor model. We restrict ourselves to certain processes only. The drift and volatility of the process, as well as the market price of risk λ are assumed to be time-independent. For the vanishing short rate we require the volatility to be zero. This condition is needed for the proof of nonnegativity of short rate.

Theorem 14.20 ([112, Theorem 1]). *Assume the following conditions on the one-factor bond pricing model with the short rate*

$$dr = \mu dt + \sigma dw,$$

1. *functions μ, σ, λ depend only on the short rate r and not on time t,*
2. *functions μ, σ, λ are continuous in $r = 0$, $\sigma(0) = 0$.*

Consider the averaged bond prices $\tilde{P}(\tau,r)$ computed from the Fong-Vasicek model. Then there is no such one-factor interest rate model, for which the averaged bond prices satisfy the bond pricing PDE

$$-\frac{\partial P}{\partial \tau} + (\mu(r) - \lambda(r)\sigma(r))\frac{\partial P}{\partial r} + \frac{1}{2}\sigma(r)^2\frac{\partial^2 P}{\partial r^2} - rP = 0, \quad (14.74)$$

for $r \geq 0$ and $\tau > 0$.

Proof. Suppose to the contrary the existence of such a one-factor model.

In order to insert the averaged price $\tilde{P}(\tau,r)$ into the PDE (14.74) we begin with computing the necessary partial derivatives of $\tilde{P}$:

$$\frac{\partial \tilde{P}}{\partial \tau} = \left(\frac{\dot{A}}{A} - \dot{B}r - \frac{\alpha \dot{C}}{\beta+C}\right)\tilde{P},$$

$$\frac{\partial \tilde{P}}{\partial r} = -B\tilde{P},$$

$$\frac{\partial^2 \tilde{P}}{\partial r^2} = B^2 \tilde{P}.$$

Now we can use a similar idea as in the proof of Theorem 14.14 for the case of the two-factor CIR model. Indeed, suppose that the averaged bond price $\tilde{P}(\tau,r)$ satisfies a one-factor PDE for bond prices. Then, when taking the limit $r \to 0^+$, the terms involving the volatility vanish because $\sigma(0) = 0$. Terms involving the drift converge to $\mu(0)$, which has to be constant with respect to τ. In the case of the Fong-Vasicek model we obtain

$$\mu(0) = \left.\frac{\partial_\tau \tilde{P}}{\partial_r \tilde{P}}\right|_{r=0} = \left.\frac{\frac{\dot{A}}{A} - \dot{B}r - \frac{\alpha}{\beta+C}\dot{C}}{-B}\right|_{r=0} = \frac{\frac{\dot{A}}{A} - \frac{\alpha}{\beta+C}\dot{C}}{-B}.$$

Hence the necessary condition for $\tilde{P}$ to be a bond price in a one-factor model is that the following identity

$$\frac{-\kappa_1 \theta_1 B - \kappa_2 \theta_2 C - \dot{C}\frac{\alpha}{\beta+C}}{-B} = k$$

holds for all $\tau > 0$ and some constant $0 < k < \infty$. Then

$$-\kappa_1 \theta_1 B - \kappa_2 \theta_2 C - \alpha\frac{\dot{C}}{\beta+C} + kB = 0, \quad (14.75)$$

for all $\tau > 0$. Therefore the derivative with respect to τ of the left hand side is identically zero. Thus

$$-\kappa_1 \theta_1 \dot{B} - \kappa_2 \theta_2 \dot{C} - \alpha\frac{(\beta+C)\ddot{C} - (\dot{C})^2}{(\beta+C)^2} + k\dot{B} = 0.$$

Since this equality holds for all $\tau > 0$, also the limit of its left hand side for $\tau \to 0+$ equals zero. Using the initial condition for the function C and the values of its derivatives for $\tau = 0$ (see Theorem 14.16), we obtain

$$\kappa_1 \theta_1 - \lambda_1 \theta_2 = k.$$

Substituting this expression for k into (14.75) yields

$$-\kappa_2 \theta_2 C - \frac{\alpha}{\beta+C}\dot{C} - \lambda_2 \theta_2 B = 0,$$

from which we can express the term $\alpha \dot{C}$ as follows:

$$\alpha \dot{C} = (\beta+C)(\kappa_2 \theta_2 C - \lambda_1 \theta_2 B). \quad (14.76)$$

On the other hand, from (14.67) we know that

$$\alpha \dot{C} = \alpha \left(-\lambda_1 B - \frac{B^2}{2} - (\kappa_2 + \lambda_2 v + v\rho B)C - \frac{v^2}{2}C^2 \right). \tag{14.77}$$

Hence the right hand sides of (14.76) and (14.77) have to be equal. From this equality, using the relation between parameters α and β, we are able to express the function C explicitly as

$$C(\tau) = \frac{\alpha \left(\lambda_2 - \lambda_1 - \frac{\beta}{2} \right)}{v\alpha\rho - \lambda_2\theta_2 + \lambda_2 v \frac{1}{B(\tau)}},$$

where $B(\tau)$ is given by (14.66). The derivative $\dot{C}(\tau)$ can be computed to be

$$\dot{C}(\tau) = -\frac{2e^{\kappa_1\tau}\kappa_1^2\kappa_2\lambda_2\theta_2(\kappa_2 + (\lambda_1 - \lambda_2)v^2)}{v(2(e^{\kappa_1\tau} - 1)\kappa_2\rho\theta_2 + \lambda_2 v(\theta_2 - e^{\kappa_1\tau}\theta_2 + e^{\kappa_1\tau}\kappa_1 v))^2}$$

and so

$$\dot{C}(0) = \frac{2\kappa_2\theta_2(\kappa_2 + (\lambda_1 - \lambda_2)v^2)}{\lambda_2 v^5}.$$

We already know that $\dot{C}(0) = 0$. Since κ_2 and θ_2 are positive parameters, it implies that $\kappa_2 + (\lambda_1 - \lambda_2)v^2 = 0$. But then $\dot{C}(\tau)$ would be identically zero and hence $C(\tau)$ would be constant with respect to τ, which is a contradiction. It completes the proof of theorem. □

14.5. Stochastic CIR Model Describing Volatility Clustering

In this section we are interested in modeling and analysis of the two factor term structure model for pricing zero coupon bonds under the assumption of rapidly oscillating stochastic volatility. We analyze solutions to the generalized Cox–Ingersoll-Ross two factors model describing clustering of interest rate volatilities. The main goal is to derive an asymptotic expansion of the bond price with respect to a singular parameter representing the fast scale for the stochastic volatility process. We derive the second order asymptotic expansion of a solution to the two factors generalized CIR model and we show that the first two terms in the expansion are independent of the variable representing stochastic volatility. The results of this section have been obtained by Stehlíková and Ševčovič in the paper [116]. Analogous results

14.5.1. Empirical Evidence of Existence of Volatility Clusters and Their Modeling

The key feature of the CIR modeling consists of the assumption made on constant volatility of the stochastic process (7.4) driving the short rate r. However, in real financial markets we can observe a substantial deviation from this assumption. To provide an empirical evidence for such a volatility process, we computed maximum likelihood estimates of the dispersion for the CIR model for 20-day-long intervals using three months treasury bills data. Figure 14.7 (left) shows the estimated dispersion as a function of time. Higher and lower volatility

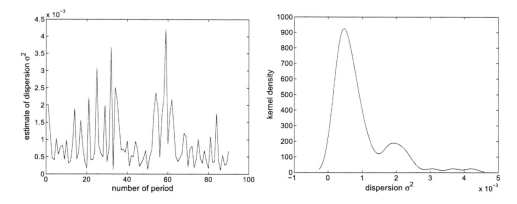

Figure 14.7. Left: Estimates of CIR model's dispersion σ^2 from 20-day intervals (3-months Treasury bills, 90 intervals starting in January 1990). Right: the density distribution of estimates of the dispersion σ^2. Source: Stehlíková and Ševčovič [116].

periods can be distinguished. They can be seen also on the kernel density estimates of the values in Figure 14.7 (right).

In order to capture such a behavior of the dispersion σ^2 we shall consider a model in which the limiting density of the dispersion (as $t \to \infty$) has two local maxima. It corresponds to the so-called volatility clustering phenomenon where the dispersion can be observed in the vicinity of two local maxima of the density distribution (see [110]). The desired behavior of the process and its limiting density are shown in the Figure 14.8. A natural candidate for such a volatility process is

$$dy = \alpha(y)dt + \omega(y)dw, \qquad (14.78)$$

having a drift function $\alpha(y)$ such that the differential equation $\frac{dy}{dt} = \alpha(y)$ has two stable stationary solutions. With added stochastic part $\omega(y)dw$ of the process, these stationary solutions become values, around which the volatility concentrates. Recall that the cumulative distribution function $\tilde{G} = \tilde{G}(y,t) = \text{Prob}(y(t) < y | y(0) = y_0)$ of the process $y = y(t)$ satisfying (14.78) and starting almost surely from the initial datum y_0 can be obtained from a solution $\tilde{g} = \partial \tilde{G}/\partial y$ to the so-called Fokker-Planck equation for the density function:

$$\frac{\partial \tilde{g}}{\partial t} = \frac{1}{2} \frac{\partial^2}{\partial y^2}(\omega(y)^2)\tilde{g}) - \frac{\partial}{\partial y}(\alpha(y)\tilde{g}), \ \tilde{g}(y,0) = \delta(y - y_0) \qquad (14.79)$$

(see Chapter 7). Here $\delta(y - y_0)$ denotes the Dirac delta function located at y_0. The limiting density $g(y) = \lim_{t \to \infty} \tilde{g}(y,t)$ of the process is therefore a stationary solution to the Fokker-Planck equation (14.79) and it forgets any information about the initial datum y_0, i.e

$$L_0^* g \equiv \frac{1}{2} \frac{\partial^2}{\partial y^2}(\omega(y)^2 g) - \frac{\partial}{\partial y}(\alpha(y)g) = 0. \qquad (14.80)$$

In [110] one of the authors proposed a model with a property that the limiting density is a combination of two Gamma densities. Indeed, let us consider the following two mean

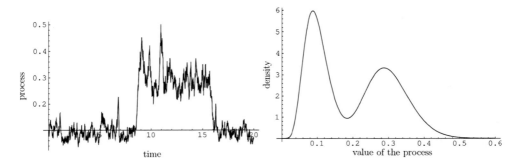

Figure 14.8. Simulation of a process (left) and its asymptotic distribution (right). Source: Stehlíková and Ševčovič [116].

reverting Bessel square root stochastic processes:

$$dy_i = \kappa_y(\theta_i - y_i)dt + v\sqrt{y_i}dw_i, \quad i = 1,2, \tag{14.81}$$

where $\theta_i > 0$, $2\kappa_y\theta_i > v^2 > 0$ for $i = 1,2$, and dw_1, dw_2 are uncorrelated differentials of the Wiener processes. Solving the stationary Fokker-Planck equation (14.80) it turns out that the limiting distributions of the processes y_1, y_2 are the Gamma distributions with shape parameters $2\kappa_y/v^2$ and $2\kappa_y\theta_i/v^2$. Denote their densities by g_1 and g_2. Then $g_i(y) = C_i y^{\frac{2\kappa_y\theta_i}{v^2}-1} \exp(-\frac{2\kappa_y}{v^2}y)$ for $y > 0$ and $g_i(y) = 0$ otherwise. Here $C_i > 0$ is a normalization constant such that $\int_R g_i(y)dy = 1$. Choose a parameter $k \in (0,1)$. Our aim is to construct a process with asymptotic density

$$g(y) = kg_1(y) + (1-k)g_2(y), \tag{14.82}$$

corresponding to a convex mixture of densities g_1 and g_2. In the following theorem we see that for the same square root volatility function of the form $v\sqrt{y}$ it is possible to achieve this goal. Drift of the process $\alpha(y)$ can be written as a weighted sum of drifts $\alpha_i(y) = \kappa(\theta_i - y), i = 1,2$, with the weights depending on y.

Theorem 14.21 ([110, Section 5]). *Suppose that the drift term α has the form: $\alpha(y) = w(y)\alpha_1(y) + (1-w(y))\alpha_2(y)$ where $w(y) = kg_1(y)/(kg_1(y) + (1-k)g_2(y))$ and $\alpha_i(y) = \kappa(\theta_i - y)$. Then the stochastic process driven by the SDE: $dy = \alpha(y)dt + v\sqrt{y}dw$ has the limiting distribution g given by the convex combination (14.82) of densities g_1, g_2.*

14.5.2. Generalized CIR Model with Rapidly Oscillating Stochastic Volatility and Its Asymptotic Analysis

The aim of this section is to provide a tool for modeling the effects of rapidly oscillating stochastic volatility that can be observed in real markets. If the length of the time scale for dispersion y is denoted by ε, the equation for the variable y reads as follows:

$$dy = \frac{\alpha(y)}{\varepsilon}dt + \frac{v\sqrt{y}}{\sqrt{\varepsilon}}dw_y. \tag{14.83}$$

In what follows we will assume that $0 < \varepsilon \ll 1$ is a small singular parameter. Notice that the limiting density function g of the stochastic process driven by SDE (14.83) is independent of the scaling parameter $\varepsilon > 0$. The statement follows directly from the stationary Fokker-Planck equation (14.80). Concerning structural assumption made on the drift function $\alpha : R \to R$ we shall henceforth assume the following hypothesis:

(A) $\qquad \alpha$ is a C^1 function on $[0, \infty)$, $\dfrac{2\alpha(0)}{v^2} > 1$, $\limsup_{y \to \infty} \dfrac{\alpha(y)}{y} < 0.$

Now it is straightforward computation to verify the following auxiliary lemma. The original proofs can be found in the paper [116].

Lemma 14.2. *Let the drift function $\alpha(y)$ be defined as a mixture of two Gamma limiting distributions as in Theorem 14.21. Then the function α satisfies the hypothesis (A) with $\alpha(0) = \kappa \min(\theta_1, \theta_2)$ and $\limsup_{y \to \infty} \frac{\alpha(y)}{y} = -\kappa < 0$.*

Next we shall show the limiting density g of the process driven by SDE (14.83) is uniquely given by the following lemma:

Lemma 14.3. *Under the hypothesis (A) made on the drift function α the stationary Fokker-Planck equation $L_0^* g \equiv \frac{v^2}{2} \frac{\partial^2}{\partial y^2}(yg) - \frac{\partial}{\partial y}(\alpha(y)g) = 0$ has a unique solution g such that $g(0) = 0$ for $y \leq 0$. It can be explicitly expressed as:*

$$g(y) = Cy^{-1} \exp\left(\frac{2}{v^2} \int_1^y \frac{\alpha(\xi)}{\xi} d\xi\right) = Cy^{\frac{2\alpha(0)}{v^2}-1} \exp\left(\frac{2}{v^2} \int_1^y \hat{\alpha}(\xi) d\xi\right)$$

for $y > 0$ and $g(y) = 0$ for $y \leq 0$. Here $\hat{\alpha}(y) = (\alpha(y) - \alpha(0))/y$ and $C > 0$ is a normalization constant such that $\int_0^\infty g(y) dy = 1$.

Proof. It follows by direct verification of the equation. The other linearly independent solution g_2 to the equation (14.80) has a nontrivial limit $g_2(0+) > 0$. $\diamond$

In what follows, we denote by $\sigma^2, D > 0$, and S the limiting mean value, dispersion and skewness of the stochastic process for the y-variable representing stochastic dispersion, i.e.

$$\sigma^2 = \int_0^\infty y g(y) dy, \quad D = \int_0^\infty (y - \sigma^2)^2 g(y) dy,$$
$$S = \frac{1}{D^{\frac{3}{2}}} \int_0^\infty (y - \sigma^2)^3 g(y) dy. \tag{14.84}$$

Notice that $D = -\int_0^\infty \int_0^y (\xi - \sigma^2) g(\xi) d\xi dy$. In the generalized CIR model with a stochastic volatility, the instantaneous interest rate (short rate) r will be modelled by the mean reverting process of the form (7.4) where the volatility of is replaced by a square root of a stochastic dispersion y, i.e.

$$dr = \kappa(\theta - r)dt + \sqrt{y}\sqrt{r} dw_r. \tag{14.85}$$

The differentials of the Wiener processes dw_y and dw_r are assumed to be independent throughout this section, i.e., $E(dw_y dw_r) = 0$. Then the corresponding partial differential equation for the bond price $P^\varepsilon = P^\varepsilon(t,r,y)$ has the following form:

$$\frac{\partial P^\varepsilon}{\partial t} + (\kappa(\theta - r) - \tilde{\lambda}_1(y,r)r^{\frac{1}{2}}\sqrt{y})\frac{\partial P^\varepsilon}{\partial r} + \frac{1}{2}ry\frac{\partial^2 P^\varepsilon}{\partial r^2} - rP^\varepsilon$$
$$+ \frac{1}{\sqrt{\varepsilon}}\left(-\tilde{\lambda}_2(y,r)v\sqrt{y}\frac{\partial P^\varepsilon}{\partial y}\right) + \frac{1}{\varepsilon}\left(\alpha(y)\frac{\partial P^\varepsilon}{\partial y} + \frac{v^2 y}{2}\frac{\partial^2 P^\varepsilon}{\partial y^2}\right) = 0, \quad (14.86)$$

$(t,r,y) \in Q_T \equiv (0,T) \times R^+ \times R^+$, where $\tilde{\lambda}_1, \tilde{\lambda}_2$ are the so-called market prices of risk (cf. [75, Chapter 7]). By a solution P^ε to (14.86) we mean a bounded function $P^\varepsilon \in C^{1,2}(Q_T) \cap C(\bar{Q}_T)$ satisfying equation (14.86) on Q_T. Concerning the structural form of market prices of risk functions $\tilde{\lambda}_1, \tilde{\lambda}_2$ we shall suppose that

$$\tilde{\lambda}_1(t,r,y) = \lambda_1 \sqrt{r}\sqrt{y}, \quad \tilde{\lambda}_2(t,r,y) = \lambda_2 \sqrt{y}$$

where $\lambda_1, \lambda_2 \in R$ are constants. It is worthwhile noting that the latter assumption is not restrictive as the original one-factor CIR model assumes such a form of the market price of risk (cf. Kwok [75]). We shall rewrite PDE (14.86) in the operator form:

$$(\varepsilon^{-1}\mathcal{L}_0 + \varepsilon^{-1/2}\mathcal{L}_1 + \mathcal{L}_2)P^\varepsilon = 0, \quad (14.87)$$

where the linear differential operators $\mathcal{L}_0, \mathcal{L}_1, \mathcal{L}_2$ are defined as follows:

$$\mathcal{L}_0 = \alpha(y)\frac{\partial}{\partial y} + \frac{v^2 y}{2}\frac{\partial^2}{\partial y^2}, \quad \mathcal{L}_1 = -\lambda_2 v y \frac{\partial}{\partial y},$$
$$\mathcal{L}_2 = \frac{\partial}{\partial t} + (\kappa(\theta - r) - \lambda_1 r y)\frac{\partial}{\partial r} + \frac{1}{2}ry\frac{\partial^2}{\partial r^2} - r. \quad (14.88)$$

Next we expand the solution P^ε into Taylor power series:

$$P^\varepsilon(t,r,y) = \sum_{j=0}^{\infty} \varepsilon^{\frac{j}{2}} P_j(t,r,y) \quad (14.89)$$

with the terminal conditions $P_0(T,r,y) = 1, P_j(T,r,y) = 0$ for $j \geq 1$ at expiry $t = T$. The main goal of this is to examine the singular limiting behavior of a solution P^ε as $\varepsilon \to 0^+$. More precisely, we shall determine the first three terms P_0, P_1, P_2 of the asymptotic expansion (14.89). We shall henceforth denote by $\langle \psi \rangle$ the averaged value of the function $\psi \in C([0,\infty))$ with respect to the density g, i.e., $\langle \psi \rangle = \int_0^\infty \psi(y) g(y) \, dy$. We shall also use the notation $\langle \mathcal{L}_2 \rangle$ standing for the averaged linear operator $\mathcal{L}_2$, i.e.

$$\langle \mathcal{L}_2 \rangle \equiv \frac{\partial}{\partial t} + (\kappa(\theta - r) - \lambda_1 r \sigma^2)\frac{\partial}{\partial r} + \frac{1}{2}\sigma^2 r \frac{\partial^2}{\partial r^2} - r \quad (14.90)$$

Lemma 14.4 ([116]). *Let $\psi \in C^1([0,\infty))$ be such that $\mathcal{L}_0 \psi$ is bounded. Then $\langle \mathcal{L}_0 \psi \rangle = 0$.*

Proof. Notice that the operator $\mathcal{L}_0^*$ is the adjoint operator to the linear operator $\mathcal{L}_0$ with respect to the L^2-inner product $(\psi, \phi) = \int_0^\infty \psi(y)\phi(y)\,dy$. It means that $\langle \mathcal{L}_0 \psi \rangle = (\mathcal{L}_0 \psi, g) = (\psi, \mathcal{L}_0^* g) = 0$ because the density g is a solution to equation (14.80). ◇

The following lemma will be useful when computing higher order term in series expansion (14.89).

Lemma 14.5 ([116]). *Let $F \in C([0,\infty))$ be such that $\langle F \rangle = 0$. Then, up to an additive constant, there exists a unique solution $\psi \in C^2((0,\infty)) \cap C([0,\infty))$ to the non-homogeneous equation $\mathcal{L}_0 \psi = \frac{v^2}{2} F$. Its derivative $\frac{\partial \psi}{\partial y}$ is given by*

$$\frac{\partial \psi}{\partial y}(y) = \frac{1}{yg(y)} \int_0^y F(\xi) g(\xi) d\xi.$$

Moreover, $\langle \mathcal{L}_1 \psi \rangle = \lambda_2 v \int_0^\infty F(y) yg(y) dy$. In particular, if ψ is a solution to the equation $\mathcal{L}_0 \psi = 0$ then ψ is a constant function with respect to the y-variable.

Proof. Using equation (14.80) for the limiting density g and inserting $\frac{\partial \psi}{\partial y}$ into the operator $\mathcal{L}_0$ we obtain that ψ is a solution to the equation $\mathcal{L}_0 \psi = \frac{v^2}{2} F$. Other independent solutions are not continuous at $y = 0$. The formula for $\langle \mathcal{L}_1 \psi \rangle$ follows from the definition of the operator $\mathcal{L}_1$ by applying integration by parts formula. □

Now we proceed with collecting the terms of the power series expansion of (14.87).
- In the order ε^{-1} we have $\mathcal{L}_0 P_0 = 0$. According to Lemma 14.5 we have $P_0 = P_0(t,r)$, i.e., P_0 is independent of the y-variable.
- In the order $\varepsilon^{-1/2}$ we have $\mathcal{L}_0 P_1 + \mathcal{L}_1 P_0 = 0$. Since $P_0 = P_0(t,r)$ we deduce $\mathcal{L}_1 P_0 = 0$ and so $\mathcal{L}_0 P_1 = 0$. By Lemma 14.5, $P_1 = P_1(t,r)$ is independent of y.
- In the order ε^0 we have $\mathcal{L}_0 P_2 + \mathcal{L}_1 P_1 + \mathcal{L}_2 P_0 = 0$. Since $P_1 = P_1(t,r)$ we have $\mathcal{L}_1 P_0 = 0$. Hence $\mathcal{L}_0 P_2 + \mathcal{L}_2 P_0 = 0$. Taking the average $\langle . \rangle$ of both sides of the latter equation we obtain $\langle \mathcal{L}_0 P_2 \rangle + \langle \mathcal{L}_2 P_0 \rangle = 0$. By Lemma 14.4 and the fact that P_0 is independent of y-variable we conclude $\langle \mathcal{L}_2 \rangle P_0 = \langle \mathcal{L}_2 P_0 \rangle = 0$. Therefore P_0 is a solution to the classical one-factor PDE equation for the CIR model satisfying the terminal condition $P_0(T,r) = 1$ for any $r \geq 0$. It is well known that the solution $P_0 = P_0(t,r)$ to the equation $\langle \mathcal{L}_2 \rangle P_0 = 0$ is given by the explicit formula:

$$P_0(t,r) = A_0(t) e^{-B(t)r}, \qquad (14.91)$$

where $A_0' = \kappa\theta A_0 B$ and $B' = (\kappa + \lambda_1 \sigma^2) B + \frac{\sigma^2}{2} B^2 - 1$, $A_0(T) = 1, B(T) = 0$, i.e.

$$A_0(t) = \left(\frac{2\phi e^{(\phi+\psi)(T-t)/2}}{(\phi+\psi)(e^{\phi(T-t)} - 1) + 2\phi} \right)^{\frac{2\kappa\theta}{\sigma^2}}, \quad B(t) = \frac{2(e^{\phi(T-t)} - 1)}{(\phi+\psi)(e^{\phi(T-t)} - 1) + 2\phi},$$

$\psi = \kappa + \lambda_1 \sigma^2$, $\phi = \sqrt{\psi^2 + 2\sigma^2}$ (cf. Kwok [75, Chapter 7]). Since $\langle \mathcal{L}_2 \rangle P_0 = 0$ we have $-\mathcal{L}_2 P_0 = (\langle \mathcal{L}_2 \rangle - \mathcal{L}_2) P_0 = (\sigma^2 - y) f(t) r e^{-B(t)r}$ where

$$f(t) = (\lambda_1 B(t) + \frac{1}{2} B(t)^2) A_0(t).$$

Hence $\mathcal{L}_0 P_2 = -\mathcal{L}_2 P_0 = (\sigma^2 - y) f(t) r e^{-B(t)r}$. According to Lemma 14.5 we have

$$\frac{\partial P_2}{\partial y} = -\frac{2}{v^2} f(t) r e^{-B(t)r} H(y), \quad H(y) = \frac{1}{yg(y)} \int_0^y (\xi - \sigma^2) g(\xi) d\xi. \qquad (14.92)$$

- In the order $\varepsilon^{1/2}$ we have $\mathcal{L}_0 P_3 + \mathcal{L}_1 P_2 + \mathcal{L}_2 P_1 = 0$. Since $\langle \mathcal{L}_0 P_3 \rangle = 0$ we have $\langle \mathcal{L}_1 P_2 \rangle + \langle \mathcal{L}_2 P_1 \rangle = 0$. The function $P_1 = P_1(t,r)$ is independent of the y-variable and therefore $\langle \mathcal{L}_2 \rangle P_1 = \langle \mathcal{L}_2 P_1 \rangle = -\langle \mathcal{L}_1 P_2 \rangle$. By Lemma 14.5 we have

$$\mathcal{L}_1 P_2 = \frac{2\lambda_2}{\nu} f(t) r e^{-B(t)r} y H(y), \quad -\langle \mathcal{L}_1 P_2 \rangle = K_1 f(t) r e^{-B(t)r},$$

where $K_1 = -\frac{2\lambda_2}{\nu} \int_0^\infty \int_0^y (\xi - \sigma^2) g(\xi) d\xi dy = \frac{2\lambda_2}{\nu} D$ is a constant (see (14.84)). Notice that the constant K_1 and the function $f(t)$ depend on the first two moments σ^2 and D of the stochastic dispersion only. Equation $\langle \mathcal{L}_2 \rangle P_1 = \langle \mathcal{L}_2 P_1 \rangle = -\langle \mathcal{L}_1 P_2 \rangle$ reads as:

$$\frac{\partial P_1}{\partial t} + \left(\kappa(\theta - r) - \lambda_1 r \sigma^2\right) \frac{\partial P_1}{\partial r} + \frac{1}{2} r \sigma^2 \frac{\partial^2 P_1}{\partial r^2} - r P_1 = K_1 f(t) r e^{-B(t)r}. \quad (14.93)$$

The solution P_1 satisfying the terminal condition $P_1(T,r) = 0$ for $r \geq 0$ can be found in the closed form:

$$P_1(t,r) = (A_{10}(t) + A_{11}(t)r) e^{-B(t)r} \quad (14.94)$$

where the functions $A_{10}(t), A_{11}(t)$ are solutions to the system of linear ODEs:

$$\begin{aligned} A'_{11}(t) &= \left(\kappa \theta B(t) + \kappa + \lambda_1 \sigma^2 + \sigma^2 B(t)\right) A_{11}(t) + K_1 f(t), \quad (14.95) \\ A'_{10}(t) &= \kappa \theta B(t) A_{10}(t) - \kappa \theta A_{11}(t), \end{aligned}$$

with terminal conditions $A_{10}(T) = 0, A_{11}(T) = 0$. We can analytically and also numerically compute A_{10}, A_{11} in a fast and accurate manner. This way we have obtained the term $P_1(t,r)$. In Figure 14.9 examples of numerical approximation of the term structure $R^\varepsilon(T - \tau, r) = -\frac{1}{\tau} \log \langle P^\varepsilon(T - \tau, r, .) \rangle$ corresponding to the second order expansion of the averaged value of $\langle P^\varepsilon(t,r,.) \rangle$, $P^\varepsilon(t,r,y) \approx P_0(t,r) + \sqrt{\varepsilon} P_1(t,r)$. We plot term structures starting from the short rate $r = 0.03$ (left) and $r = 0.031$ (right) for parameters $\kappa = 5$, $\theta = 0.03$, $\kappa_y = 100$, $\nu = 1.1832$, $\theta_1 = 0.025$, $\theta_2 = 0.1$, $k = 1/3$, $\lambda_1 = -1$, $\lambda_2 = -100$ and $\varepsilon = 0, 0.001, 0.01$ (black, red and blue curves).

Having P_1 and $\frac{\partial P_2}{\partial y}$ we can compute the term $\mathcal{L}_1 P_2 + \mathcal{L}_2 P_1$. With regard to Lemma 14.5 equation $\mathcal{L}_0 P_3 = -\mathcal{L}_1 P_2 - \mathcal{L}_2 P_1$ then yields a formula for $\frac{\partial P_3}{\partial y}$ and

$$\langle \mathcal{L}_1 P_3 \rangle = -\frac{2\lambda_2}{\nu} \left((\frac{2\lambda_2}{\nu} K_3 + K_1 \sigma^2) f(t) r e^{-B(t)r} + D(-\lambda_1 r \frac{\partial P_1}{\partial r} + \frac{r}{2} \frac{\partial^2 P_1}{\partial r^2}) \right)$$

where the constant $K_3 = \int_0^\infty \xi^3 H(\xi) g(\xi) d\xi = -\frac{1}{2} S D^{\frac{3}{2}} - \sigma^2 D$ depends on the first three statistical moments of the stochastic dispersion.
- In the order ε^1 we have $\mathcal{L}_0 P_4 + \mathcal{L}_1 P_3 + \mathcal{L}_2 P_2 = 0$. Proceeding similarly as before we have $\langle \mathcal{L}_0 P_4 \rangle = 0$ and therefore

$$\langle \mathcal{L}_1 P_3 \rangle + \langle \mathcal{L}_2 P_2 \rangle = 0. \quad (14.96)$$

We decompose the function $P_2(t,r,y)$ in the form

$$P_2(t,r,y) = \bar{P}_2(t,r) + \tilde{P}_2(t,r,y), \quad (14.97)$$

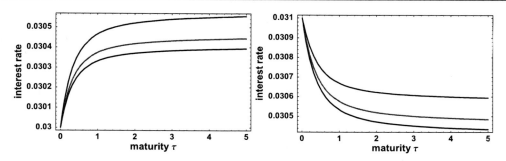

Figure 14.9. The approximate term structure $R^\varepsilon = R^\varepsilon(T-\tau, r)$ based on the first two leading terms of the bond price $P^\varepsilon \approx P_0(T-\tau,r) + \sqrt{\varepsilon}P_1(T-\tau,r)$ starting from the short rate $r = 0.03$ (left) and $r = 0.031$ (right) for several values of the singular parameter $\varepsilon = 0, 0.001, 0.01$ (lower, middle and upper curves), respectively. Source: Stehlíková and Ševčovič [116].

where $\bar{P}_2$ is the averaged value of P_2 and $\tilde{P}_2$ is a zero mean fluctuation, i.e., $\langle \tilde{P}_2 \rangle = 0$. As $\bar{P}_2$ does not depend on y, we have $\frac{\partial \bar{P}_2}{\partial y} = \frac{\partial P_2}{\partial y}$. Taking into account $\langle \mathcal{L}_2 \tilde{P}_2 \rangle = 0$ we obtain

$$\tilde{P}_2(t,r,y) = -\frac{2}{v^2} f(t) r e^{-B(t)r} \left(\int_0^y H(\xi)d\xi - K_2 \right)$$

where $K_2 = \int_0^\infty g(s) \int_0^s H(\xi)d\xi ds$ is a constant and the function H is given by (14.92). Now we can use decomposition (14.97) to evaluate $\langle \mathcal{L}_2 P_2 \rangle$. We have $\langle \mathcal{L}_2 P_2 \rangle = \langle \mathcal{L}_2(\bar{P}_2 + \tilde{P}_2) \rangle = \langle \mathcal{L}_2 \rangle \bar{P}_2 + \langle \mathcal{L}_2 \tilde{P}_2 \rangle$ because $\bar{P}_2$ is independent of y. Next we can determine $\langle \mathcal{L}_2 \tilde{P}_2 \rangle$ in the following form:

$$\langle \mathcal{L}_2 \tilde{P}_2 \rangle = -\frac{2}{v^2} K_4 f(t) r \left(-\lambda_1 \frac{\partial}{\partial r} + \frac{1}{2} \frac{\partial^2}{\partial r^2} \right)(re^{-B(t)r})$$

where $K_4 = \int_0^\infty \int_0^y H(\xi)d\xi (y-\sigma^2)g(y)dy$. It is worthwile noting that both constants K_2, K_4 depend on all nontrivial statistical moments of the stochastic dispersion. Equation (14.96) then becomes

$$\langle \mathcal{L}_2 \rangle \bar{P}_2 = -\langle \mathcal{L}_2 \tilde{P}_2 \rangle - \langle \mathcal{L}_1 P_3 \rangle = (a(t) + b(t)r + c(t)r^2)e^{-B(t)r}, \quad \bar{P}_2(T,r) = 0,$$

which is a partial differential equation for $\bar{P}_2 = \bar{P}_2(t,r,y)$ with a right hand side which can be explicitly computed from already obtained results in the closed form:

$$\bar{P}_2(t,r) = (A_{20}(t) + A_{21}(t)r + A_{22}(t)r^2)e^{-B(t)r} \qquad (14.98)$$

where the functions A_{20}, A_{21}, A_{22} are solutions to a linear system of ODEs. We omit details here.

In conclusion we have shown the following result which has been obtained by Stehlíková and Ševčovič in the paper [116].

Theorem 14.22 ([116, Theorem 2]). *The solution $P^\varepsilon = P^\varepsilon(t,r,y)$ of the generalized CIR bond pricing equation* (14.86) *with rapidly oscillating dispersion can be approximated, for small values of the singular parameter $0 < \varepsilon \ll 1$, by $P^\varepsilon(t,r,y) \approx P_0(t,r) + \sqrt{\varepsilon} P_1(t,r) + \varepsilon P_2(t,r,y) + O(\varepsilon^{\frac{3}{2}})$.*

The first two terms P_0, P_1 are independent of the y-variable representing unobserved stochastic volatility. They depend only on the first two statistical moments (mean value and dispersion) of the stochastic dispersion and other model parameters.

The next term in the expansion P_2 nontrivially depends on the y-variable. P_2 as well as its averaged value $\langle P_2 \rangle$ depends also on all nontrivial statistical moments of the stochastic dispersion.

The terms P_0, P_1, P_2 can be evaluated by closed-form formulae (14.91), (14.94), (14.98).

References

[1] Milton Abramowitz and Irene A. Stegun, Handbook of mathematical functions with formulas, graphs, and mathematical tables, ninth dover printing, tenth gpo printing ed., Dover, New York, 1964.

[2] Lee C. Adkins and Timothy Krehbiel, Mean reversion and volatility of short-term london interbank offer rates: An empirical comparison of competing models, *International Review of Economics & Finance* **8** (1999), no. 1, 45–54.

[3] Yacine Aït-Sahalia, Maximum likelihood estimation of discretely sampled diffusions: a closed-form approximation approach., *Econometrica* **70** (2002), no. 1, 223–262.

[4] E. L. Algower and K. Georg, Numerical continuation methods, Springer-Verlag, Berlin, 1990.

[5] Ghada Alobaidi, Roland Mallier, and A. Stanley Deakin, Laplace transforms and installment options, *Math. Models Methods Appl. Sci.* **14** (2004), no. 8, 1167–1189.

[6] J. Anděl, Základy matematické statistiky, MatFyzPress, Praha, 2005.

[7] Julia Ankudinova and Matthias Ehrhardt, Fixed domain transformations and highly accurate compact schemes for nonlinear Black-Scholes equations for American options, Nonlinear Models in Mathematical Finance: New Research Trends in Option Pricing (M. Ehrhardt, ed.), Nova Science Publishers, Inc., Hauppauge, 2008, pp. 243–273.

[8] _____, On the numerical solution of nonlinear Black-Scholes equations, *Comput. Math. Appl.* **56** (2008), no. 3, 799–812.

[9] M. Avellaneda, A. Levy, and A. Parás, Pricing and hedging derivative securities in markets with uncertain volatilities, *Applied Mathematical Finance* **2** (1995), 73–88.

[10] M. Avellaneda and A. Parás, Dynamic hedging portfolios for derivative securities in the presence of large transaction costs, *Applied Mathematical Finance* **1** (1994), 165–193.

[11] Guy Barles and Halil Mete Soner, Option pricing with transaction costs and a nonlinear Black-Scholes equation, *Finance Stoch.* **2** (1998), no. 4, 369–397.

[12] G. Barone-Adesi and R. E. Whaley, Efficient analytic approximations of American option values, *Journal of Finance* **42** (1987), 301–320.

[13] M. W. Baxter and A. J. O. Rennie, Financial calculus - an introduction to derivative pricing, Cambridge University Press, Cambridge, 1996.

[14] Fischer Black and Myron Scholes, The pricing of options and corporate liabilities, *The Journal of Political Economy* **81** (1973), no. 3, 637–654.

[15] Tomáš Bokes and Daniel Ševčovič, Early exercise boundary for American type of floating strike Asian option and its numerical approximation, *arXiv.org:0912.1321* (2009).

[16] L. A. Bordag and A. Y. Chmakova, Explicit solutions for a nonlinear model of financial derivatives, *Int. J. Theor. Appl. Finance* **10** (2007), no. 1, 1–21.

[17] Ljudmila A. Bordag and Ruediger Frey, Pricing options in illiquid markets: symmetry reductions and exact solutions, Nonlinear Models in Mathematical Finance: New Research Trends in Option Pricing (M. Ehrhardt, ed.), Nova Science Publishers, Inc., Hauppauge, 2008, pp. 103–130.

[18] Michael J. Brennan and Eduardo S. Schwartz, Savings bonds, retractable bonds and callable bonds, *Journal of Financial Economics* **5** (1977), no. 1, 67–88.

[19] _____, Analyzing convertible bonds, *Journal of Financial and Quantitative Analysis* **15** (1980), no. 04, 907–929.

[20] Damiano Brigo and Fabio Mercurio, Interest rate models—theory and practice, second ed., Springer Finance, Springer-Verlag, Berlin, 2006, With smile, inflation and credit.

[21] S. L. Byers and K. B. Nowman, Forecasting U.K. and U.S. interest rates using continuous time term structure models, *International Review of Financial Analysis* **7** (1998), 191–206.

[22] John Chadam, Free boundary problems in mathematical finance, Progress in industrial mathematics at ECMI 2006, *Math. Ind.*, vol. 12, Springer, Berlin, 2008, pp. 655–665.

[23] K. C. Chan, G. A. Karolyi, F. A. Longstaff, and A. B. Sanders, An empirical comparison of alternative models of the short-term interest rate, *Journal of Finance* **47** (1992), no. 3, 1209–27.

[24] Xinfu Chen, John Chadam, Lishang Jiang, and Weian Zheng, Convexity of the exercise boundary of the American put option on a zero dividend asset, *Math. Finance* **18** (2008), no. 1, 185–197.

[25] Youngsoo Choi and Tony S. Wirjanto, An analytic approximation formula for pricing zero-coupon bonds, *Finance Research Letters* **4** (2007), no. 2, 116–126.

[26] L. Collatz, Functional analysis and numerical mathematics, Academic Press Inc., New York, 1966.

[27] T.S. Corzo and E.S. Schwartz, Convergence within the eu: Evidence from interest rates, *Economic Notes* **29** (3003), 243–266.

[28] Georges Courtadon, The pricing of options on default-free bonds, *Journal of Financial and Quantitative Analysis* **17** (1982), no. 01, 75–100.

[29] J. C. Cox, Notes on option pricing i: constant elasticity of variance diffusions, Working paper, Stanford University, 1975.

[30] John C. Cox, Jonathan Ingersoll, and Stephen A. Ross, An analysis of variable rate loan contracts, *Journal of Finance* **35** (1980), no. 2, 389–403.

[31] John C. Cox, Jonathan E. Ingersoll, Jr., and Stephen Ross, A theory of the term structure of interest rates, *Econometrica* **53** (1985), no. 2, 385–407.

[32] John C. Cox and Stephen A. Ross, The valuation of options for alternative stochastic processes, *Journal of Financial Economics* **3** (1976), no. 1-2, 145–166.

[33] Min Dai and Yue Kuen Kwok, Characterization of optimal stopping regions of American Asian and lookback options, *Math. Finance* **16** (2006), no. 1, 63–82.

[34] Q. Dai and K.J. Singleton, Specification analysis of affine term structure models, *The Journal of Finance* **55** (2000), no. 5, 1943–1978.

[35] Rosario Dell'Aquila, Elvezio Ronchetti, and Fabio Trojani, Robust GMM analysis of models for the short rate process, *Journal of Empirical Finance* **10** (2003), no. 3, 373–397.

[36] J. N. Dewynne, S. D. Howison, I. Rupf, and Paul Wilmott, Some mathematical results in the pricing of American options, *European J. Appl. Math.* **4** (1993), no. 4, 381–398.

[37] L. Uri Dothan, On the term structure of interest rates, *Journal of Financial Economics* **6** (1978), no. 1, 59–69.

[38] G.R. Duffee, Term premia and interest rate forecasts in affine models, *Journal of Finance* (2002), 405–443.

[39] Bertram Düring, Michel Fournié, and Angsar Jüngel, High order compact finite difference schemes for a nonlinear Black-Scholes equation, *Int. J. Theor. Appl. Finance* **6** (2003), no. 7, 767–789.

[40] Erik Ekström, Convexity of the optimal stopping boundary for the American put option, *J. Math. Anal. Appl.* **299** (2004), no. 1, 147–156.

[41] Erik Ekström and Johan Tysk, The American put is log-concave in the log-price, *J. Math. Anal. Appl.* **314** (2006), no. 2, 710–723.

[42] C.M. Elliott and J.R. Ockendom, Weak and variational methods for free and moving boundary problems, Res. Notes Math., 59, Pitman, London, 1982.

[43] Athanasios Episcopos, Further evidence on alternative continuous time models of the short-term interest rate, *Journal of International Financial Markets, Institutions and Money* **10** (2000), no. 2, 199–212.

[44] J.D. Evans, R. Kuske, and J.B. Keller, American options on assets with dividends near expiry, *Mathematical Finance* **12** (2002), no. 3, 219–237.

[45] D.K. Faddeev and V.N. Faddeeva, Computational methods of linear algebra, Gosudarstv. Izdat. Fiz.-Mat. Lit., Moscow, 1963.

[46] Gaetano Fichera, Sulle equazioni differenziali lineari ellittico-paraboliche del secondo ordine., *Atti Accad. Naz. Lincei, Mem., Cl. Sci. Fis. Mat. Nat., Sez. I, VIII. Ser.* **5** (1956), 3–30.

[47] M. Fiedler, Special matrices and their applications in numerical mathematics, SNTL, Prague, 1981.

[48] H. G. Fong and O. A. Vasicek, Fixed-income volatility management, *Journal of Portfolio Management* **17** (1991), 41–46.

[49] R. Frey, Perfect option hedging for a large trader, *Finance and Stochastics* (1998), no. 2, 115–142.

[50] _____, Market illiquidity as a source of model risk in dynamic hedging in model risk, RISK Publications, R. Gibson Ed., London, 2000.

[51] Rüdiger Frey and Pierre Patie, Risk management for derivatives in illiquid markets: a simulation study, Advances in finance and stochastics, Springer, Berlin, 2002, pp. 137–159.

[52] Rüdiger Frey and Alexander Stremme, Market volatility and feedback effects from dynamic hedging, *Math. Finance* **7** (1997), no. 4, 351–374.

[53] R. Geske and H. E. Johnson, The American put option valued analytically, *Journal of Finance* **39** (1984), 1511–1124.

[54] R. Geske and R. Roll, On valuing American call options with the Black–Scholes european formula, *Journal of Finance* **39** (1984), 443–455.

[55] Robert Geske, The valuation of compound options, *Journal of Financial Economics* **7** (1979), no. 1, 63–81.

[56] Paul Glasserman, Monte Carlo methods in financial engineering, *Applications of Mathematics (New York)*, vol. 53, Springer-Verlag, New York, 2004, Stochastic Modelling and Applied Probability.

[57] Peter Grandits and Werner Schachinger, Leland's approach to option pricing: the evolution of a discontinuity, *Math. Finance* **11** (2001), no. 3, 347–355.

[58] G. Gripenberg, S.-O. Londen, and O. Staffans, Volterra integral and functional equations, *Encyclopedia of Mathematics and its Applications*, vol. 34, Cambridge University Press, Cambridge, 1990.

[59] A. T. Hansen and P. L. Jörgensen, Analytical valuation of American-style Asian options, *Management Science* **46** (2000), 1116–1136.

[60] Lars Peter Hansen, Large sample properties of generalized method of moments estimator, *Econometrica* **50** (1982), no. 4, 1029–1054.

[61] S.L. Heston, M. Loewenstein, and G.A. Willard, Options and bubbles, *Review of Financial Studies* **20** (2007), no. 2, 359.

[62] Thomas S Y Ho and Sang-bin Lee, Term structure movements and pricing interest rate contingent claims, *Journal of Finance* **41** (1986), no. 5, 1011–29.

[63] S.D. Hodges and A. Neuberger, Optimal replication of contingent claims under transaction costs, *Rev. Future Markets* **8** (1989), 222–239.

[64] T. Hoggard, A. E. Whalley, and P. Wilmott, Hedging option portfolios in the presence of transaction costs, *Advances in Futures and Options Research* **7** (1994), 21–35.

[65] John C. Hull, Options, futures, and other derivatives (5th edition), Prentice Hall, 2002.

[66] Martin Jandačka and Daniel Ševčovič, On the risk-adjusted pricing-methodology-based valuation of vanilla options and explanation of the volatility smile, *J. Appl. Math.* (2005), no. 3, 235–258.

[67] Michael Johannes and Nicholas Polson, Mcmc methods for financial econometrics, *Handbook of Financial Econometrics*, North-Holland. Forthcoming, 2002.

[68] H. E. Johnson, An analytic approximation for the American put price, *Journal of Financial and Quantitative Analysis* **18** (1983), no. 01, 141–148.

[69] M. Kabanov and M. M. Safarian, On Leland's strategy of option pricing with transaction costs, *Finance and Stochastic* **1** (1997), 239–250.

[70] I. Karatzas, On the pricing American options, *Appl. Math. Optim.* **17** (1988), 37–60.

[71] Ioannis Karatzas and Steven E. Shreve, Brownian motion and stochastic calculus, second ed., *Graduate Texts in Mathematics*, vol. 113, Springer-Verlag, New York, 1991.

[72] S. Kilianová and D. Ševčovič, Analytical and numerical methods for stock index derivate pricing, *Journal of Electrical Engineering* **55/s** (2004), 39–42.

[73] M. Kratka, No mystery behind the smile, *Risk* **9** (1998), 67–71.

[74] K. Kuske and J. Keller, Optimal exercise boundary for an American put option, *Appl. Math. Fin* **5** (1998), 107–116.

[75] Y. K. Kwok, Mathematical models of financial derivatives, Springer Finance, Springer-Verlag Singapore, Singapore, 1998.

[76] Yue-Kuen Kwok, Mathematical models of financial derivatives, second ed., Springer Finance, Springer, Berlin, 2008.

[77] Martin Lauko and Daniel Ševčovič, Comparison of numerical and analytical approximations of the early exercise boundary of the American put option, *The ANZIAM Journal* (2011).

[78] H. E. Leland, Option pricing and replication with transaction costs, *Journal of Finance* **40** (1985), 1283–1301.

[79] L. W. MacMillan, Analytic approximation for the American put option, *Advances in Futures and Options Research* **1** (1986), 119–134.

[80] R. Mallier, Evaluating approximations for the American put option, *Journal of Applied Mathematics* **2** (2002), 71–92.

[81] R. Mallier and G. Alobaidi, The American put option close to expiry, *Acta Math. Univ. Comenian. (N.S.)* **73** (2004), no. 2, 161–174.

[82] Terry A. Marsh and Eric R. Rosenfeld, Stochastic processes for interest rates and equilibrium bond prices, *Journal of Finance* **38** (1983), no. 2, 635–46.

[83] I. Melicherčík, L. Olšarová, and V. Úradníček, Kapitoly z finančnej matematiky, Epos, Bratislava, 2005.

[84] R.C. Merton, Theory of rational option pricing, *The Bell Journal of Economics and Management Science* (1973), 141–183.

[85] R. Mynemi, The pricing of the American option, *Annal. Appl. Probab.* **2** (1992), 1–23.

[86] Whitney K. Newey and Kenneth D. West, Hypothesis testing with efficient method of moments estimation, *International Economic Review* **28** (1987), no. 3, 777–87.

[87] K B Nowman, Gaussian estimation of single-factor continuous time models of the term structure of interest rates, *Journal of Finance* **52** (1997), no. 4, 1695–1706.

[88] K. Ben Nowman, The volatility of japanese interest rates: evidence for certificate of deposit and gensaki rates, *International Review of Financial Analysis* **11** (2002), no. 1, 29–38.

[89] Bernt Øksendal, Stochastic differential equations, sixth ed., Universitext, Springer-Verlag, Berlin, 2003, An introduction with applications.

[90] G.C. Papanicolaou, Stochastic equations and their aplications, *American Mathematical Monthly* **80** (1973), 526–545.

[91] Andrea Pascucci, Free boundary and optimal stopping problems for American Asian options, *Finance Stoch.* **12** (2008), no. 1, 21–41.

[92] Neil D Pearson and Tong-Sheng Sun, Exploiting the conditional density in estimating the term structure: An application to the cox, ingersoll, and ross model, *Journal of Finance* **49** (1994), no. 4, 1279–1304.

[93] A.R. Pedersen, A new approach to maximum likelihood estimation for stochastic differential equations based on discrete observations, *Scandinavian journal of Statistics* **22** (1995), no. 1, 55–71.

[94] Ingo Rechenberg, Evolutionsstrategie; optimierung technischer systeme nach prinzipien der biologischen evolution. mit einem nachwort von manfred eigen, Frommann-Holzboog [Stuttgart-Bad Cannstatt], 1973.

[95] Christoph Reisinger and Gabriel Wittum, Efficient hierarchical approximation of high-dimensional option pricing problems, *SIAM JOURNAL ON SCIENTIFIC COMPUTING* **29** (2007), no. 1, 440–458.

[96] Z. Riečanová, Numerical methods and mathematical statistics, SNTL, Prague, 1987.

[97] Richard Roll, An analytic valuation formula for unprotected American call options on stocks with known dividends, *Journal of Financial Economics* **5** (1977), no. 2, 251–258.

[98] P. Samuelson, Rational theory of warrant pricing, *ndustrial Management Review* **6** (1965), 13–31.

[99] Philipp J. Schönbucher and Paul Wilmott, The feedback effect of hedging in illiquid markets, *SIAM J. Appl. Math.* **61** (2000), no. 1, 232–272 (electronic).

[100] H. P. Schwefel, Contemporary evolution strategies, *Advances in Artificial Life, Third International Conference on Artificial Life, Lecture Notes in Artificial Intelligence Vol. 929* (V. F. Morán, A. Moreno, J. J. Merelo, and P. Chacón, eds.), Springer-Verlag, 1995, pp. 893–907.

[101] Hans-Paul Schwefel, Evolution and optimum seeking, Sixth-Generation Computer Technology Series, John Wiley & Sons Inc., New York, 1995.

[102] D. Ševčovič and A. Urbánová-Csajková, On a two-phase minmax method for parameter estimation of the Cox, Ingersoll, and Ross interest rate model, *CEJOR Cent. Eur. J. Oper. Res.* **13** (2005), no. 2, 169–188.

[103] Daniel Ševčovič, Analysis of the free boundary for the pricing of an American call option, *European J. Appl. Math.* **12** (2001), no. 1, 25–37.

[104] _____, An iterative algorithm for evaluating approximations to the optimal exercise boundary for a nonlinear Black-Scholes equation, *Can. Appl. Math. Q.* **15** (2007), no. 1, 77–97.

[105] _____, Partial differential equations and their applications, Iris, Bratislava, 2008.

[106] D. Ševčovič and A. Urbánová Csajková, Calibration of one factor interest rate models, *Journal of Electrical Engineering* **55/s** (2004), 46–50.

[107] Daniel Ševčovič, Transformation methods for evaluating approximations to the optimal exercise boundary for linear and nonlinear Black–Scholes equations, Nonlinear Models in Mathematical Finance: New Research Trends in Option Pricing (M. Ehrhardt, ed.), Nova Science Publishers, Inc., Hauppauge, 2008, pp. 153–198.

[108] K. R. Sircar and B. Papanicolaou, General Black-Scholes models accounting for increased market volatility from hedging strategies, *Applied Mathematical Finance* **5** (1998), 45–82.

[109] Robert Stamicar, Daniel Ševčovič, and John Chadam, The early exercise boundary for the American put near : numerical approximation, *Canad. Appl. Math. Quart.* **7** (1999), no. 4, 427–444.

[110] B. Stehlíková, Modeling volatility clusters with application to two-factor interest rate models, *Journal of Electrical Engineering* **56/s** (2005), 90–93.

[111] _____, Fast mean reverting volatility in Fong–Vasicek model of interest rates, *Journal of Electrical Engineering* **57** (2006), no. 7/S, 65–67.

[112] _____, Averaged bond prices for Fong-Vasicek and the generalized Vasicek interest rates models, *Proceeding of MMEI* (K. Cechlárová, M. Halická, V. Borbelová, and V. Lacko, eds.), TU Košice, 2007, pp. 166–175.

[113] _____, Averaged bond prices in generalized Cox-Ingersoll-Ross model of interest rates, *Proceedings of 5th Actuarial and Financial Mathematics Day* (M. Vanmaele, ed.), Koninklijke Vlaamse Acadamie van Belgie voor Wetenschappen en Kunsten, 2007, pp. 77–87.

[114] B. Stehlíková and D. Ševčovič, Approximate formulae for pricing zero-coupon bonds and their asymptotic analysis, *Int. J. Numer. Anal. Model.* **6** (2009), no. 2, 274–283.

[115] B. Stehlíková and D. Ševčovič, On a volatility averaging in a two-factor interest rate model, *Proceedings of Algoritmy 2005: 17th Conference on scientific computing* (Handlovičová, A. and Krivá, A. and Mikula, K. and Ševčovič, D., ed.), STU, Bratislava, 2005, pp. 325–333.

[116] Beáta Stehlíková and Daniel Ševčovič, On the singular limit of solutions to the Cox-Ingersoll-Ross interest rate model with stochastic volatility, *Kybernetika* **45** (2009), no. 4, 670–680.

[117] Elias M. Stein and Guido Weiss, Introduction to Fourier analysis on Euclidean spaces, Princeton University Press, Princeton, N.J., 1971, Princeton Mathematical Series, No. 32.

[118] R. G. Tompkins, Implied volatility surfaces: uncovering regularities for options on financial futures, *The European Journal of Finance* **7** (2001), 198–238.

[119] S. Treepongkaruna and S. Gray, On the robustness of short-term interest rate models, *Accounting and Finance* **43** (2003), no. 1, 87–121.

[120] Oldrich Vasicek, An equilibrium characterization of the term structure, *Journal of Financial Economics* **5** (1977), no. 2, 177–188.

[121] E. Vitásek, Numerical methods, SNTL, Prague, 1987.

[122] P. Wilmott, J. Dewynne, and S. D. Howison, Option pricing: Mathematical models and computation, Oxford Financial Press, United Kingdom, 1995.

[123] Lixin Wu, Yue Kuen Kwok, and Hong Yu, Asian options with the American early exercise feature, *Int. J. Theor. Appl. Finance* **2** (1999), no. 1, 101–111.

[124] Rongwen Wu and Michael C. Fu, Optimal exercise policies and simulation-based valuation for American-Asian options, *Oper. Res.* **51** (2003), no. 1, 52–66.

[125] Song-Ping Zhu, A new analytical approximation formula for the optimal exercise boundary of American put options, *Int. J. Theor. Appl. Finance* **9** (2006), no. 7, 1141–1177.

[126] Song-Ping Zhu and Zhi-Wei He, Calculating the early exercise boundary of American put options with an approximation formula, *Int. J. Theor. Appl. Finance* **10** (2007), no. 7, 1203–1227.

About the Authors

Daniel Ševčovič is Professor of Mathematics at Comenius University in Bratislava, Slovakia. His research interests are focused on applications of qualitative and quantitative methods for partial differential equations in various applied subjects including, in particular, the mathematical theory of financial derivative pricing. Beáta Stehlíková is Assistant professor at Comenius University in Bratislava. Her research interest is focused on mathematical analysis of interest rate models. Karol Mikula is Professor of Applied Mathematics at Slovak University of Technology in Bratislava. His research activities are focused on numerical analysis of nonlinear partial differential equations with applications in various fields of science and engineering.

List of Symbols

$_1F_1(.,.,.)$ - Kummer confluent hypergeometric function
$\langle . \rangle$ - expected value of a random variable
$\langle .|. \rangle$ - conditional expected value of a random variable
α, β - parameters of a drift function (chapter 14)
$(x)^+ = \max(0,x)$ - positive part of x
$S_f(t)$ - early exercise boundary
$rho(\tau)$ - early exercise boundary, $\tau = T - t$
$V^{ec}, V_{ec}, V^{ac}, V_{ac}, V^{ep}, V_{ep}, V^{ap}, V_{ap}$ - value of European and American call and put options
a, α, b, β - parameters of gamma distribution (chapter 8, 14)
a, a_1, a_2 - drift functions (chapter 14)
$A, B, A_1, A_2, B_1, B_2, C$ - shape functions of the interest rates
$Corr(.,.)$ - correlation of random variables
$E(.)$ - expected value of a random variable
EOC - experimental order of convergence
$f(.), g(.)$ - density functions
$F(.)$ - cumulative distribution function
$N(.,.)$ - normal distribution with specified parameters
$P, P(t,T,r), P(\tau,r), P(\tau,r_1,r_2), P(\tau,r,y)$ - price of bond
P^{ap}, P^{ap2} - approximations of bond prices
$\text{Prob}(.)$ - probability of an event
$\tilde{P}, \tilde{P}(\tau,r)$ - averaged bond price
r - short rate
r_1, r_2 - factors of the short rate
$R, R(t,R), R(t,T,r), R(\tau,r), R(\tau,r_1,r_2), R(\tau,r,y)$ - interest rate
R^{ap}, R^{ap2} - approximations of interest rates
$\tilde{R}, \tilde{R}(\tau,r)$ - averaged interest rate
T - maturity of a bond
$Var(.)$ - variance of a random variable
$Var(.|.)$ - conditional variance of a random variable
$w, w(t), w_t$ - Wiener process
w_{ij} - weights in the calibration (chapter 13)
$w(.)$ - weight function for drifts in volatility clustering model (chapter 14)
$\gamma > 0$ - parameter describing the dependence of volatility on the short rate
$\Gamma(.)$ - gamma function
$\Gamma(.,.)$ - gamma distribution with specified parameters

List of Symbols

$\theta > 0$ - long time limit of a mean reversion a process
$\kappa > 0$ - mean reversion parameter of a process
$\lambda, \lambda(t,r)\ \lambda(r), \tilde{\lambda}$ - market price of risk
$\mu, \mu(t,r), \mu(r)$ - drift function of a process
μ, σ^2 - parameters of normal distribution
$\sigma, \sigma(t,r), \sigma(r)$ - volatility function of a process
$\sigma > 0, v > 0$ - volatility parameter of a process
τ - time to maturity
$D(X)$ - dispersion of X
$P, P(t,T), P(t,T,r), P(\tau,r)$ - zero-coupon bond price
PRIBOR - Prague inter-bank offering rate
EONIA - Euro overnight index average
ES - Evolution strategies
EU - European Union
EURIBOR - Euro inter-bank offering rate
EURO-LIBOR - London inter-bank offering rate in EUR
BRIBOR - Bratislava inter-bank offering rate
CIR - Cox-Ingersoll-Ross
GMM - Generalized method of moments
HJM - Heath-Jarrow-Morton
β, ξ, ρ - transformed parameters (chapter 13)
$U, U(\beta, \xi, \rho)$ - loss functional
t - actual time

Index

American call option, 129
American options, 195
American put option, 129
analytic approximation, 203
arithmetic average, 75
Asian option, 75
ask price, 5
average rate option, 76
average strike option, 76
averaged term structure, 240

Banach fixed point theorem, 155, 157
Banach, S., 155
Barles and Soner model, 188
barrier option, 75, 80
bearish spread, 25
Bessel square root process, 95
Bessel, F.W., 95
bid price, 5
bid–ask spreads, 171
binary options, 27, 84
binomial tree method, 147
Black, F., 9
Black–Scholes equation, 22
Black–Scholes formula, 37
bond, 110
bought straddle, 25
Brennan–Schwartz model, 96
BRIBOR, 110
Brownian motion, 10
BUBOR, 110
bullish spread, 24
butterfly, 25

call option, 5
CEV model, 23
characteristic function, 100

CIR model, 95
compound option, 85
computed yield curves, 239
condor, 26
convergence model, 104, 120
Courant–Fridrichs–Lewy condition, 144
Cox, J.C., 95
Cox–Ingersoll–Ross model, 95, 96, 115, 231, 237
Cox–Ross model, 96
cumulative statistics, 240

Delta, 56
Delta of an option, 56
Dirac delta function, 97
discrete maximum principle, 145, 152
dispersion, 10, 240
Dothan model, 96
Dow–Jones index, 2
down-and-in barrier option, 80
down-and-out barrier option, 80
drift of the process, 93

early exercise boundary, 131, 132, 195
EONIA, 110
equilibrium model, 110
EURIBOR, 110
EUROLIBOR, 110
evolution strategy method, 241
exotic option, 75
experimental order of convergence, 215

feedback effects, 167, 178, 181
Fichera condition, 127
Fichera, G., 127
finite differences, 142, 143
fitness, 242

fixed domain transformation, 209
floating maximum rate lookback call option, 88
floating maximum strike lookback put option, 88
floating minimum rate lookback call option, 88
floating minimum strike lookback put option, 88
Fokker–Planck equation, 97, 98
Fong, H.G., 121
Fong–Vasicek model, 104, 121
forward, 4
free boundary, 132

Gamma, 58
Gamma equation, 188
gamma of the option, 58
Gauss, K.F., 153
Gauss–Seidel relaxation method, 153
generalized method of moments, 231, 232
geometric average, 75
geometric Brownian motion, 11, 77
goodness of fit, 243
greeks, 49

hedging strategy, 178
historical volatility, 49
Hoggard, T., 65

IBM, 2
illiquid markets, 167, 178
implied volatility, 52, 164
in-the-money, 5
Ingersoll, J.E., 95
interest rate model, 231
intrinsic value of the option, 7
Itō lemma, 77
Itō's lemma, 16
Itō, K., 16

Jensen inequality, 70, 74
jumping volatility, 189

Leland model, 65
Leland number, 68, 70

Leland, H., 65
LIBOR, 110
likelihood function, 242
limit interest rate, 94
linear complementarity, 136
long position, 38
lookback option, 75, 88
loss functional, 239, 240

market price of risk, 112, 244
Markov chain, 9
Markov process, 9
Markov property, 9
maximum likelihood function, 233
maximum likelihood ratio, 243, 244
maximum principle, 145, 152
mean reversion process, 94
mean value, 240
Merton, R.C., 9
Microsoft, 2
min-max calibration method, 231
minmax procedure, 242
multidimensional Black–Scholes equation, 45
mutation, 242

new CIR variables, 238
non-linear $\mathcal{R}^2$ ratio, 243, 244
non-linear regression, 243
nonarbitrage model, 122
nonlinear Black–Scholes equation, 167
Nowman's estimates, 233

offspring, 242
one-factor model, 110
option, 5
option premium, 5
Orstein–Uhlenbeck process, 94
out-of-money, 5

par value, 109
parameter reduction principle, 237
parent vector, 242
per annum, p.a., 50
perpetual option, 138, 207
plain vanilla options, 5

portfolio of bonds, 111
PRIBOR, 110
Projected successive over-relaxation method, 156
PSOR, 156
put option, 5
put–call parity, 38, 39, 139
put–call symmetry, 39, 140

RAPM model, 169
real market yield curve, 239
rebate, 81
restricted likelihood function, 242
return, 20
Rho, 60
risk adjusted pricing methodology, 169
risk aversion, 22
risk neutral probabilities, 149
risk neutral strategies, 21
risk premium, 7, 172
risk premium factor, 246
risk–neutral portfolio, 21
Ross, S.A., 95
Rubinstein, M., 149

Scholes, M.S., 9
selection, 242
sensitivity factor Delta, 56
sensitivity factor Gamma, 58
sensitivity factor Rho, 60
sensitivity factor Vega, 62
sensitivity factors, 49
sensitivity parameter Theta, 61
short position, 38
smooth mapping, 238
software Mathematica, 78, 243
sold straddle, 25
SOR method, 153
speed of mean reversion, 94, 244
spreads, 70
stochastic differential equation, 11
stochastic optimization, 241
stock–holding strategy, 169
swap, 124
swaption, 127

Taylor series expansion, 142
Theta, 61
time-weighted distance, 239
tracking error, 181
transaction costs, 65, 167
transformation, 238
trend of the process, 20, 93
trinomial tree method, 147
two-factor CIR model, 104
two-factor model, 117
two-factor Vasicek model, 104

unrestricted likelihood function, 242
up-and-in barrier option, 80
up-and-out barrier option, 80
utility maximization, 167

variance, 10
Vasicek model, 94, 96, 113, 231, 238
Vasicek, O.A., 113, 121, 231
vector of children, 242
Vega, 62
volatility smile, 55, 177

Whalley, A. F., 65
Wiener process, 10, 50, 77, 105
Wiener, N., 10
wild type population, 242
Wilmott, P., 65
WUBOR, 110

yield–curve, 109

zero-coupon bond, 109